Building control systems

CIBSE Guide H

Oxford Auckland Boston Johannesburg Melbourne New Delhi

Butterworth-Heinemann
Linacre House, Jordan Hill, Oxford OX2 8DP
225 Wildwood Avenue, Woburn, MA 01801-2041
A division of Reed Educational and Professional Publishing Ltd

A member of the Reed Elsevier plc group

British Library Cataloguing in Publication Data
A catalogue record for this book is available from the British Library

ISBN 07506 504 78

1 2 3 4 5 6 7 8 9 10

Typeset in Great Britain by Phoenix Photosetting, Chatham, Kent
Printed and bound in Great Britain by Bookcraft Ltd, Midsomer Norton

Contents

Foreword v
Preface vii
Acknowledgements viii

1 Introduction: the need for controls 1-1

1.1 Overview of the Guide 1-1
1.2 The modern control system 1-2
1.3 The global environment 1-2
1.4 The indoor environment 1-3
1.5 Energy conservation 1-3
1.6 Information technology and systems integration 1-4
1.7 Building operation 1-5
1.8 The benefits of a BMS 1-5
1.9 Summary 1-6
References 1-6

2 Control fundamentals 2-1

2.0 General 2-1
2.1 Control modes 2-1
2.2 Optimum start 2-7
2.3 Weather compensation 2-8
2.4 Stability and tuning 2-9
2.5 Artificial intelligence 2-13
2.6 Summary 2-14
References 2-14

3 Components and devices 3-1

3.1 Sensors 3-1
3.2 Actuators 3-7
3.3 Valves 3-8
3.4 Dampers 3-17
3.5 Motors 3-20
3.6 Pumps and fans 3-23
3.7 Control panels and motor control centres 3-23
3.8 The intelligent outstation 3-24
3.9 Summary 3-26
References 3-26

4 Systems, networks and integration 4-1

4.0 General 4-1
4.1 BMS architecture 4-1
4.2 Networks 4-5
4.3 Electromagnetic compatibility (EMC) 4-12
4.4 Systems integration 4-15
4.5 User interface 4-17
4.6 Summary 4-18
References 4-19

5 Control strategies for subsystems 5-1

5.1 Safety 5-1
5.2 Boilers 5-3
5.3 Chillers 5-7
5.4 Control of hydraulic circuits 5-14
5.5 Central air handling plant 5-28
5.6 Energy recovery 5-32
5.7 Mechanical ventilation 5-34
5.8 Variable air volume 5-41
5.9 Constant-volume room terminal units 5-46

5.10 Fan coil units 5-46
5.11 Dual duct systems 5-48
5.12 Chilled ceilings 5-49
5.13 Heat pumps 5-51
5.14 Natural ventilation 5-53
5.15 Lighting controls 5-56
5.16 Summary 5-61
References 5-62

6 Control strategies for buildings 6-1

6.0 General 6-1
6.1 Operating modes 6-1
6.2 Design techniques 6-2
6.3 Whole-building HVAC systems 6-4
6.4 Case studies 6-7
6.5 Summary 6-10
References 6-10

7 Information technology 7-1

7.1 Energy monitoring 7-1
7.2 Fault reports and maintenance scheduling 7-6
7.3 Summary 7-7
References 7-7

8 Management issues 8-1

8.1 Procurement options 8-1
8.2 Design and specification of a controls system 8-4
8.3 Tendering process 8-7
8.4 Commissioning 8-8
8.5 Operation 8-13
8.6 Occupant surveys 8-14
8.7 Cost issues 8-15
8.8 Summary 8-16
References 8-17

Appendix A1: Bibliography A1-1

Appendix A2: Tuning rules A2-1

A2.1 The PID control loop A2-1
A2.2 Digital control A2-1
A2.3 Tuning A2-2
A2.4 Step-by-step tuning procedure A2-3

Appendix A3: Glossary A3-1

Foreword

The CIBSE Applications Manual *Automatic Controls and their implication for systems design* was published in 1985 with the stated aim 'to provide building services engineers generally with the detailed guidance required to ensure that automatic controls are considered as an integral part of the design philosophy'. The technology, both hardware and software, of automatic control systems has advanced greatly since then and the need for a new edition is clear. The Department of the Environment, Transport and the Regions granted an award for the production of a new edition of the manual under the Partners in Technology scheme. The DETR provided 50% of the funding for the project; the Partners provided the balance, with the initial publishing costs met by CIBSE. A contact for the preparation of the draft was placed with the Department of Building Engineering at UMIST, and the work has been supervised by a Task Group, drawn initially from the CIBSE IT and Controls special interest group.

Partners	Ken Butcher	CIBSE
	Gerald McGilley	Landis and Staefa Division of Siemens Building Technologies Ltd
	Graham Nicholls	E Squared
	Martyn Harrold	Ove Arup
Task Group	Geoff Levermore	Dept of Building Engineering, UMIST, Chair
	Brian Flude	Consultant
	David Lush	Consultant
	Steve Loughney	Landis and Staefa Division of Siemens Building Technologies Ltd
	Andrew Martin	BSRIA
	Nick Skemp	Building Design Partnership
	Jeremy Warren	AspenTSC
	Roger Poole	British Airways
	Colin Ashford	BRECSU
	Tom Edge	Brian Clancy Partnership
	Peter Tunnell	Landis and Staefa Division of Siemens Building Technologies Ltd
	Alistair Cameron	R W Gregory
	Stuart Aynsley	Cylon Controls
	Ian Shuttleworth	Johnson Controls
	Peter McDermott	E Squared
	Peter Day	E Squared
	Malcolm Clapp	Consultant
	George Henderson	DETR Representative
Contract author	Donald McIntyre	Consultant
CIBSE	Ken Butcher	
	Robert Yarham	

Preface

Controls are poorly understood in our industry. The subject is also disliked in university courses as it can become highly mathematical. It is probably partly due to this lack of appreciation of controls and their commissioning that I have yet to find a building that works properly and satisfies the occupants after it has been commissioned. This Guide is an attempt to redress some of these problems. It explains the topic of controls with the minimum of mathematics. It builds on and updates the earlier 1985 Applications Manual *Automatic Building Controls*.

Some points did come up a number of times in the Task Group discussions that could not be quantified in the body of the Guide but are worth mentioning here;

- Adequate time should be allowed for the commissioning process. It can take between 1 day and 6 months to commission a control system, depending on the size and complexity of the system. Another general and difficult-to-define quantity is the time to commission a BMS point, but figures between 8 minutes and 60 minutes are often mentioned.

- The form of contracts and subcontracts can have major implications for controls and especially the time for commissioning; slippages and cost cutting can end with the design and commissioning being done poorly. Partnering may help this.

- Part-load performance details of the HVAC system to be controlled will aid the design and commissioning of the control system. Rarely do systems operate at, or get commissioned at, the near-extreme conditions of design.

- The controls should be considered from the earliest stages of the design process.

- Occupant feedback, via a self-assessed questionnaire, can be a useful design and commissioning tool.

- In many new buildings occupants need to be involved in the control process and also informed of the control process.

I would like to give my sincere thanks to the Task Group, which has provided many useful comments and much guidance, to Donald McIntyre for helping to get them into a coherent, easy to read document, and to Ken Butcher for his help throughout the preparation process.

Geoff Levermore July 1999

Acknowledgements

Preparation of the manual would not have been possible without the generous help of the many people who provided information and literature, and gave their time to discuss the project and read the many drafts. While it is not possible to mention them all by name, their contributions were highly valued.

1 Introduction: the need for controls

1.1 Overview of the Guide
1.2 The modern control system
1.3 The global environment
1.4 The indoor environment
1.5 Energy conservation
1.6 Information technology and systems integration
1.7 Building operation
1.8 The benefits of a BMS
1.9 Summary

This introductory section provides an overview of the Guide and will be of value when preparing the general case for a control system. It demonstrates the importance of controls in ensuring effective and efficient control of a building in order to:

- provide comfortable and productive working conditions
- provide the proper environment for industrial processes
- operate in an energy efficient manner
- be environment friendly.

1.1 Overview of the Guide

The first edition of this Guide was published in 1985 as an Applications Manual, under the title *Automatic Building Controls and Their Implications for Systems Design*. The many developments since then, particularly in the fields of microprocessor control and communications networks, have necessitated the production of an entirely new edition. The aim of the Guide has been restated since the first edition to reflect the growing importance of IT, and now reads: *to provide the building services engineer with a sufficient understanding of modern control systems and relevant information technology to ensure that the best form of control system for the building is specified and that proper provision is made for its installation, commissioning, operation and maintenance.*

The structure of the Guide is indicated in Table 1.1. This introductory section sets out the benefits to be gained from a modern building control system and will be of value in making the case that adequate provision be made at an early stage for a proper control system. The following section deals with the different types of control mode and their application in different situations; advice is given on the setting up and tuning of controllers to ensure stable operation.

Sections 3–6 deal with the practical design of control systems, starting with the hardware components, then their incorporation into control systems by linking them into networks, and then two sections on control strategies for HVAC systems and whole buildings. The Guide thus starts

Table 1.1 Organisation of the Guide

1 Introduction	The contribution that a modern building management system can make to the efficient and economical operation of a building
2 Control fundamentals	The basic types of control operation that are found in practice, ranging from the simple thermostat to microprocessor controlled self-learning algorithms. Guidance on the application of different control types and their tuning for optimum operation
3 Components and devices	The whole range of hardware components that constitute a control system, including sensors, valves, dampers, actuators, motors and basic controllers
4 Systems, networks and integration	The means by which components are brought together to form an operating control system. The various BMS architectures and the major standard protocols for bus systems. Characteristics of networks and the extension to full systems integration
5 Control strategies for subsystems	Control strategies for the fundamental parts of HVAC systems: safety interlocks, boilers, chillers, water and air systems, lighting
6 Control strategies for buildings	Control strategies for whole buildings. Avoiding conflict between subsystems. Illustrations of successful control installations
7 Information technology	The relation between BMS and IT. Energy monitoring and targeting, maintenance scheduling, facilities management
8 Management issues	The importance of the procurement method on the BMS design process. Commissioning, CDM requirements and cost issues

with the constituent components and moves up to complete systems. The user may prefer to consult the control strategies for systems of interest and then refer back in the Guide to obtain a fuller understanding of the component parts.

Section 7 deals with the relation between building management systems and information technology. The BMS and IT systems may share a communications network and the information gathered by the BMS can be used by the IT system for further purposes, enhancing the value of both systems. The final section shows the importance of the building procurement process in determining whether adequate resources are devoted to the design and installation of a suitable BMS and emphasises the necessity of taking control requirements into consideration at an early stage in the design process.

1.2 The modern control system

Good controls are essential for the safe and efficient operation of a modern building. The control system does more than keep the inside of a building comfortable for the occupants. It is required to keep the HVAC plant operating efficiently, to ensure that all plant operates safely in the event of any unforeseen circumstances, and it must be capable of two-way communication with the personnel charged with its operation. While it may be self-evident that modern highly serviced buildings require a sophisticated control system, it should be realised that simpler buildings relying on a boiler system and natural ventilation can still benefit from a modern BMS. The increasing emphasis on energy conservation and reduction of greenhouse gas emissions serves to increase the importance of efficient controls.

The late 1970s saw the introduction of digital data technology, in which information is transmitted not as an analogue electrical value, but as a number. Digital data transmission is less susceptible to error than analogue transmission and it is standard practice to construct the signal protocol in such a way that it is possible to detect whether an error has occurred during transmission. This was the beginning of direct digital control (DDC). It required the codification of rules by which values are converted to numerical messages for sending; such messages have to contain not only the value of the variable under consideration, but additional information such as the origin and destination of the message and error-checking information. Such conventions on the structure of the messages are the basis of data communication protocols. At the early stage of DDC, data handling was centralised and multiplexing circuits were used so that the central unit could contact each remote unit as required. As computing power rapidly increased, the functionality of the central control unit became more and more sophisticated, with the ability to handle increasing amounts of data and to perform additional functions such as the monitoring of energy consumption and the printing or reports.

The advent of the microprocessor allowed considerable computing power to be incorporated in a small device and meant that it was now no longer necessary for all control and monitoring functions to be carried out by a large centralised computer. Intelligent outstations placed round the building became capable of carrying out local control functions, while communicating with a central supervisor which could oversee their actions, receive any alarm signals and alter set points or operating times as required. There has been enormous progress in the field of data communication and the application of local area networks (LANs), which allow microprocessors and computers to communicate with each other over standardised networks. Communication may be extended to link together the operation of several buildings, which may be located miles apart, or even in different countries.

All these have contributed to the modern building management system. In this Guide the term 'control system' or 'building control system' is used to cover all control elements, including hardware, controllers, any linking network and central controllers. The term BMS refers to a system where components may communicate with each other and generally implies some form of central supervisor, which permits monitoring and control of the building from a single point. The period that saw the development of the BMS has also seen the rise in information technology (IT). A modern operation, whether it is office or factory, is likely to distribute and process large amounts of information dealing with the operation of the business. There may be advantages in linking IT and BMS, either for the economy of using shared networks or for the more efficient integration of management control over the many activities taking place in a building.

1.3 The global environment

The building industry is implicated in two major concerns about the possibility of global environmental change: global warming and damage to the ozone layer. Buildings are a major source of carbon dioxide emissions, whether directly by the consumption of fuel for space and water heating, or indirectly by the consumption of electricity for lighting, air conditioning and other uses. It is estimated that energy use in buildings in the UK accounts for about half of the total carbon dioxide emissions.

The destruction of the ozone layer is a distinct problem from climate warming, but the causes are linked. Chlorinated fluorocarbons (CFCs) are stable compounds which for years have been the most commonly used refrigerant in air conditioning applications. If released, CFCs interact with ozone in the upper atmosphere leading to a thinning of the ozone layer, which normally provides an effective barrier to excessive ultraviolet radiation from the sun. An international agreement has been reached to restrict the use of CFCs. The Montreal protocol allows the use of alternative refrigerants (HCFCs) with a much lower ozone depletion capacity than CFCs for an interim period, after which they would be banned as well[1]. Proposed EU legislation will further restrict the use of refrigerants. In summary:

— Any use of CFCs for new or existing systems is to be prohibited from 2001, including recovered and stockpiled CFCs.

— The use of HCFCs in new air conditioning plant will be phased out between 2001 and 2004, depending on the type of plant.

— The use of new HCFCs for maintaining existing systems is to be banned after 2010, though it will be possible to use recovered and recycled refrigerant.

— Recovery of CFCs and HCFCs on plant servicing or decommissioning will become mandatory and annual leak checks for larger systems will be compulsory.

There is therefore strong pressure to reduce energy consumption in buildings and to avoid the use of air conditioning wherever possible. This pressure may take several forms: further legislation can be expected, the price of fuel may be increased and responsible clients will incorporate environmental goals in the design brief. The *CIBSE Code of Professional Conduct* places a general duty on members to 'have due regard to environmental issues'[2] and the *CIBSE Policy Statement on Global Warming* recommends that members take positive steps to reduce global warming[3].

Any major reduction in energy use by buildings requires commitment by the client expressed in the design brief, followed by action at the design stage, where fundamental decisions on the form of the building and the use of air conditioning are made. Correct design and operation of building controls is essential to avoid waste of energy. Reduction in the use of air conditioning and the application of natural ventilation bring new challenges for effective building control and the maintenance of satisfactory internal conditions coupled with low energy use. The *CIBSE Guide to Energy Efficiency in Buildings*[4] has the objective of showing how to improve the energy performance of buildings. While primarily targeted at building services engineers, it is of use to all members of the building team.

1.4 The indoor environment

The function of the building services and their associated control system is to provide an environment within the building appropriate to the activities taking place therein. Several factors contribute to feelings of thermal comfort and their incorporation into the predicted mean vote (PMV) index is set out in European Standard *BS EN 7730*[5]. Recent research[6] strongly suggests that people adapt to their environment, allowing temperature settings to fall in winter and rise in summer. This has important implications for the design of naturally ventilated buildings. Guidance on the required conditions for a range of occupations and activities is given in CIBSE Guide A1[7]. Decisions on allowable excursions of the conditions outside the recommended comfort bands, e.g. during a summer heatwave, will have important repercussions on plant sizing. The specified control tolerances will affect the design and cost of the control system. The indoor environment affects not only comfort, but also productivity and health. Relevant legislation is referred to as appropriate in the text. *The Control of Fuel and Electricity*, Statutory Instrument[8], specifies a maximum heating level of 19°C in non-domestic buildings. The law has not been rigorously enforced. While it is difficult to substantiate precise claims of productivity gains, there is little doubt that comfortable conditions will have a beneficial effect. Surveys have consistently shown that the speed with which management attends to complaints is very important. Those companies where management attends promptly to problems are highly rated by the occupants[9]. The client's point of view for the specification of offices is represented in *Best Practice in the Specification for Offices*[10].

While the primary function of a building control system has been the control of temperature and humidity, the increased awareness of sick building syndrome (SBS) and other building related illnesses has emphasised the requirement to ensure good indoor air quality. The demands of energy conservation and healthy ventilation are sometimes in conflict, necessitating better attention to the control of ventilation to ensure a satisfactory compromise. More attention is being given to the quality as well as quantity of ventilation.

There are many buildings which house processes and operations which have their own special requirements for environmental control. Examples are low temperature for food preparation, high and controlled humidity for paper fabrication, clean rooms for electronic assembly. The pharmaceutical industry has its own special regulations for control of the environment, both for drug production and for animal housing. Companies producing goods for export may need to meet the requirements set down by the customer's country. It is outside the scope of this Guide to give the many regulations; it is the responsibility of the client or his representative to ensure that they are taken into account at an early stage in the design.

1.5 Energy conservation

Building controls have a vital role to play in preventing waste of energy. The amount of energy required to run a building is determined by:

- thermal efficiency of the building envelope
  - thermal insulation
  - airtightness
  - provision for passive solar gains
- requirements of the indoor environment
  - temperature schedule
  - ventilation needs
  - humidity control
  - indoor air quality
  - lighting requirement
  - hot water requirements
  - lifts and mechanical services
- processes within the building
  - IT equipment
  - industrial processes.

The above requirements taken together demand a level of base energy, which is the energy required to meet the business needs of the building operation. This provides a minimum level of energy expenditure. Any reduction in base energy requirement implies a change in building construction or use. The difference between actual energy expenditure and the base requirement represents avoidable waste. Examination of data from a number of UK buildings shows avoidable waste levels in the range 25 to 50%; in a well-managed building, avoidable waste levels of below 15% are achieved[11].

Avoidable waste has many causes, including:

- poor time and temperature control of the building interior
- ineffective utilisation of internal heat gains
- plant oversizing
- excessive ventilation
- low operating efficiency of the HVAC system
- poor system design and installation
- standing losses
- unnecessary use of artificial lighting and air conditioning.

The control system affects most of the above. Detailed applications will be found in the body of the Guide. Major contributions of the control system in reducing waste are:

- the limitation of heating and cooling to the minimum period necessary; this usually includes the use of optimum start controllers and some form of occupancy detection to avoid excessive out-of-hours use
- prevention of unnecessary plant operation and boiler idling
- monitoring to give early warning of malfunction or inefficient operation.

The establishment of a figure for base energy provides a clear figure to aim at and allows the performance of the building to be clearly and unambiguously stated in terms of avoidable waste. The base energy requirement must be estimated by a sound method, which is understood and respected by all parties involved. As far as possible, it should be broken down into components, to allow identification of areas which require action.

Fuel Efficiency Booklet 10[12] gives clear advice on the selection of an appropriate control system for energy saving purposes. It categorises control systems into four bands of ascending cost and complexity, shown in Table 1.2. The highest band gives the greatest potential energy savings, but may not be appropriate for all buildings or operating staff.

Band 0 is the minimum level of control required under the 1995 *Building Regulations*. Achieving this level of control in a building that has poor or non-existent controls will make considerable savings, and further savings of up to 20% may be made by the use of more sophisticated systems. Table 1.3, taken from the fuel efficiency booklet, gives rule-of-thumb assessments of cost savings that may be achieved. This table represents only a crude starting point. The decision on the type of control system to be installed in a building depends on more factors than simple energy payback savings and must be made on a deeper analysis of the costs and benefits to be expected.

1.6 Information technology and systems integration

A modern building contains several technical services in addition to heating and ventilation, such as lighting, lift control, security and access control, closed circuit television

Table 1.2 Banding classification of heating control systems, after Energy Efficiency Office[12]

Band	Time	Boiler	Distribution	Space heating	Hot water system
Output greater than 100 kW					
2	Optimiser plus time control of zones. Time control of HW storage	Boiler loading control. Off-line boilers isolated. Interaction with space control	Compensated with space temperature reset and separate zone circuits. Interaction with space control	All emitters with individual control, modulating where appropriate	Local gas fired water heaters or point of use electric units
1	Optimiser plus time control of zones. Time control of HW storage	Effective boiler sequencing and control strategies. High efficiency boilers	Compensated with space temperature reset and separate zone circuits	TRVs and room thermostats except in room with space reset sensor	Segregated HWS system, or top located or dual thermostats on calorifiers
0	Optimiser plus time control of HW storage	Effective boiler sequence control	Compensated with space temperature reset	None, or on emitters designed for separate control	Effective thermostats on calorifiers
–1	Timeswitch	None	No compensator	None or TRVs	Basic thermostat
Output less than 100 kW					
1	Optimiser plus time control of zones where appropriate. Time control of HW storage	Effective boiler sequencing and control strategies. High efficiency boilers	Compensated with space temperature reset and separate zone circuits	TRVs and room thermostats except in room with space reset sensor	Segregated HWS system, or top located or dual thermostats on calorifiers
0	Effective time control plus time control of HW storage	Effective boiler thermostat	Compensated with space temperature reset	None or on emitters designed for separate control	Effective thermostats on calorifiers
–1	Timeswitch	Boiler thermostat	Not compensated	None or TRVs	Thermostat

Table 1.3 Rule-of-thumb costs and benefits of control bands[12]

Control band	Capital cost increase over Band 0 (%)	Energy usage, compared with Band 0 (%)	Payback, over Band 0 base (years)	Comments
2	100–200	–20	2–4	Highly recommended for minimum energy usage
1	50–100	–10	1–2	Recommended for cost effective energy savings
0	0	0	N/A	Minimum
–1	N/A	≤ +50	N/A	Does not meet Building Regulations since 1985

(CCTV) systems, as well as the information technology network necessary for the user's business operation. All these services communicate within their own system using some form of network. There are major potential benefits if the various systems can communicate with each other, using the same communications network or a limited number of compatible networks:

— the reduction in cabling and infrastructure cost

— the ability of the systems to share information with each other.

This process is known as systems integration. At its most basic level, it means that devices from different manufacturers may use the same communications network, communicating with their peers and not interfering with other equipment. At the most advanced level, all systems within a building use the same communications network, exchange information with each other and are controlled from a single supervisor. For instance, the presence detectors of a lighting control system may feed information on out-of-hours occupancy to the security and access control systems. Full integration is also known as the intelligent building concept.

HVAC control systems operate in real time, ensuring proper operation of the environmental control system. The information generated may be fed into the information technology system where it can be used for the production of reports, energy monitoring and targeting and the preparation of maintenance schedules.

1.7 Building operation

A well-planned control system offers improved management of building services and can form the core of an integrated facilities management system, covering other building-related services such as access control, security, energy monitoring and targeting, information technology and maintenance. The amount of direct involvement by staff in the day-to-day running of an HVAC system has steadily reduced over recent years. However, it would be a mistake to assume that the control system can be left to take care of itself from the moment of handover. The client must choose from a range of options, from running the building services in-house with the client organisation's own staff, to outsourcing to a service bureau who may supervise efficient operation, deal with occupant requests and organise maintenance, all from a remote supervisor. Whichever form of organisation is chosen, there should be clear ownership of the control system with unambiguous responsibility for its successful operation. This requires a commitment by the client to ensure adequate resources for the operation and maintenance of the building control system; part of this commitment is the provision of proper training for staff. The organisation should ensure prompt and effective response to requests or complaints from the building occupants; several studies have confirmed the importance of rapid response in ensuring occupant satisfaction with their place of work.

The software which has been developed for BMS supervisors has greatly simplified the day-to-day management of even large BMSs and will show savings in operating staff costs compared with a simpler system which requires frequent attention. With the development of wide area networks, it is possible to have remote supervision. This enables skilled personnel to be located at a single site and able to monitor the performance of several BMSs in scattered buildings, leaving less qualified staff to carry out the daily operation on site. There will also be a saving in maintenance costs as the BMS is able to keep run-time records of all equipment, allowing maintenance to be planned effectively. Early warning of failure is available from monitoring. Plant life is extended by the reduction in hours of use that is obtained by scheduling, by reducing unnecessary device operation or unstable hunting and by reducing fan and pump speeds.

1.8 The benefits of a BMS

When deciding on the appropriate type of control system to specify for a building, it is necessary to remember that the

Table 1.4 Benefits of a BMS

Building owner	Higher rental value Flexibility on change of building use Individual tenant billing for services
Building tenant	Reduced energy consumption Effective monitoring and targeting of energy consumption Good control of internal comfort conditions Increased staff productivity Improved plant reliability and life
Occupants	Better comfort and lighting Possibility of individual room control Effective response to HVAC-related complaints
Facilities manager	Control from central supervisor Remote monitoring possible Rapid alarm indication and fault diagnosis Computerised maintenance scheduling Good plant schematics and documentation
Controls contractor	Bus systems simplify installation Supervisor aids setting up and commissioning Interoperability enlarges supplier choice

benefits of a modern control system are enjoyed variously by the different groups of users involved with the building. Table 1.4 lists some of the benefits to be achieved with an effective modern BMS. It goes without saying that these benefits will only be obtained if the system is properly specified, installed, commissioned, operated and maintained. It is the function of this guide to assist in achieving that goal.

1.9 Summary

Effective control of the heating, ventilating and air conditioning systems in a building is essential to provide a productive, healthy and safe working environment for the occupants. Without a properly functioning BMS the activities carried out in the building will be disadvantaged. Along with good building design and efficient HVAC plant, the BMS plays a vital role in the prevention of energy waste and reducing the environmental impact of the building.

Modern BMSs are based on intelligent controllers which may be programmed to carry out a wide range of control functions. Typically, a number of controllers are employed, each controlling an item of plant or an HVAC subsystem. The controllers communicate with each other and with a central supervisor over a local area network (LAN). The system manager is able to monitor and control the entire BMS from one point.

The scale and complexity of the control system should be appropriate to the building and its operation; highly effective and reliable control may be achieved with relatively simple control systems. However, when considering the cost effectiveness of a BMS, all the operational benefits that flow from a well-managed facility should be taken into account: not only energy saving but also the reductions in staffing cost, improved maintenance scheduling and the benefits of system integration with other building facilities. Such facilities as access control, security and lighting may be integrated with the BMS, giving total building management from one point. There is steady progress towards compatibility between products, so that devices from different manufacturers may share the same LAN and event interact directly with each other. The goal of freely interchangeable devices is termed interoperability.

References

1 Butler D J G *Phase-out of CFCs and HCFCs: options for owners and operators of air-conditioning systems* BRE IP 14/95 (Watford: Building Research Establishment) (1995)

2 *CIBSE code of professional conduct* (London: Chartered Institution of Building Services Engineers) (1995)

3 *CIBSE policy statement on global warming* (London: Chartered Institution of Building Services Engineers) (1990)

4 *Energy efficiency in buildings* CIBSE Guide F (London: Chartered Institution of Building Services Engineers) (1998)

5 European Standard *BS EN 7730: Moderate thermal environments — determination of the PMV and PPD indices and specification of the conditions for thermal comfort* (London: British Standards Institution) (1994)

6 Nicol F, Humphreys M A, Sykes O and Roaf S *Indoor air temperature standards for the 21st century* (London: E & F N Spon) (1995)

7 *Environmental design,* CIBSE Guide A (London: Chartered Institution of Building Services Engineers) (1999)

8 *Statutory Instrument 1980/1013: control of fuel and electricity (amended) Order 1984* (London: Stationery Office) (1984)

9 Leaman A PROBE 10 occupancy survey analysis *Building Services Journal* **19**(5) 37–41 (1997)

10 *Best practice in the specification for offices* (Reading: British Council for Offices) (1997)

11 Ashford C J. *Avoidable waste/base energy budgeting* BRESCU Workshop Document (Watford: Building Research Establishment (1997)

12 *Controls and energy savings* Fuel Efficiency Booklet 10 (Harwell: Building Research Energy Conservation Support Unit) (1995)

2 Control fundamentals

2.0 General
2.1 Control modes
2.2 Optimum start
2.3 Weather compensation
2.4 Stability and tuning
2.5 Artificial intelligence
2.6 Summary

This section introduces the concept of feedback and the control loop. It describes the basic control modes used in HVAC controls, including:

— simple on/off control
— proportional, integral and derivative control
— optimisers and compensators
— intelligent controls.

Issues of stability are dealt with and methods of tuning control loops for the best combination of response speed and stability are given. The section goes on to discuss advances in adaptive controls, which learn by experience how to optimise operation of a controller.

2.0 General

A control system consists of three basic elements: a sensor, a controller and a controlled device (see Figure 2.1). The sensor measures some variable such as temperature and transmits its value to the controller. The controller uses this value to compute an output signal, which is transmitted to the controlled device, which then acts to change the output of the load, which acts on the controlled system. In the majority of cases relevant to this Guide we are dealing with closed loop systems, where the controller is attempting to control the variable whose value is being measured by the sensor. The results of its actions are fed back to the controller input and the system is said to have feedback. In the example shown in Figure 2.1, the controller is attempting to maintain room temperature at a set point. A low room temperature results in increased output from the heater, which then raises the room temperature. This increase is detected by the sensor and transmitted to the controller, which alters its output accordingly to reduce the difference between set point and the measured value of the controlled variable. In the discussion of control modes that follows, it is implicitly assumed that the system is inherently controllable. Poor design may result in a system that is practically impossible to control; this will be discussed further below.

Open loop or feedforward systems operate without feedback. As before, the operation of the controlled device is a function of the value measured by the sensor, but this does not result in a change to the measured variable. A weather compensator is an example of open loop control, where an external air temperature sensor is used to control the flow temperature in a heating circuit used to heat a building. The control system has no way of knowing if the desired internal temperature has been achieved.

In practice, a control loop may have more than one input signal and more than one output signal. Groups of control loops can be chained together to create control sequences. The simple description above implies that the input and output of the controller are continuous variables and this is so for such variables as temperature. An important part of practical control systems is a set of complex interlocks, where the operation of one part of the system is contingent on the operating state of several other variables and systems. Many inputs and outputs are thus binary (on/off) in nature. When preparing a points list, it is conventional to refer to them as digital inputs and digital outputs (DI and DO); this does not imply DDC.

2.1 Control modes

Consider again the simple closed loop system of Figure 2.1. The way in which the control system responds to a change in the controlled variable is described by the control mode. Several control modes are in use and it is important to select the appropriate mode for the job in hand.

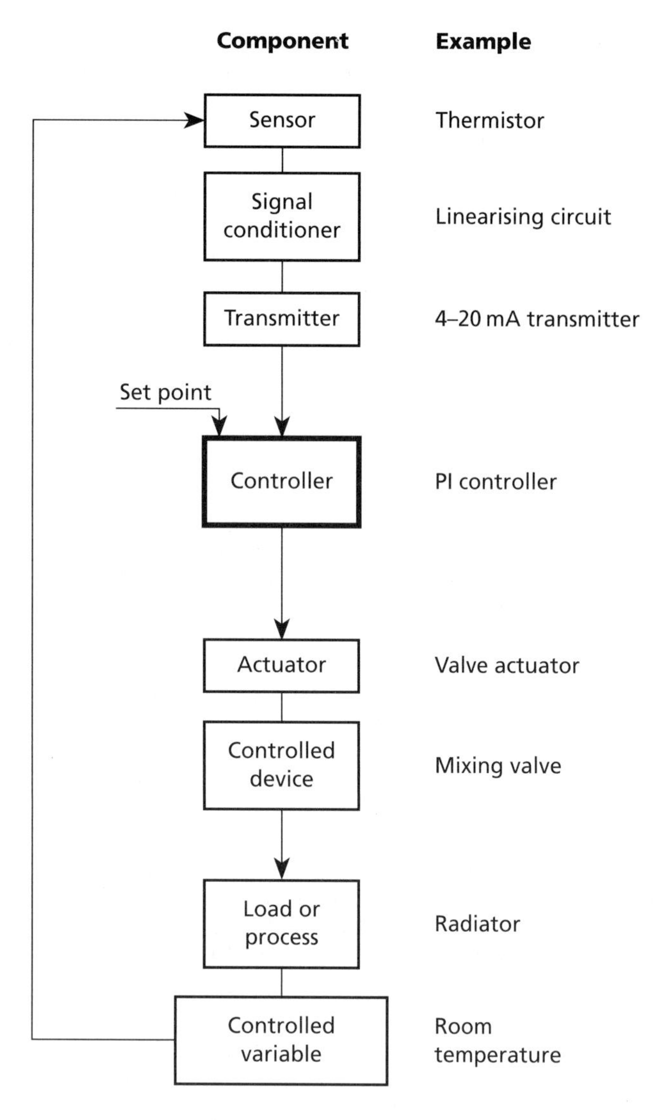

Figure 2.1 The components of a control system. In practice, some components may be combined

2.1.1 Two-position (on/off) control

In this mode, the controlled device gives either maximum or minimum output, typically on and off. Figure 2.2 illustrates two-position control for a simple heating system. It is desired to control temperature at the set point. For reasons that will become clear, it is necessary for there to be a temperature differential between switching on and switching off of the controlled device. With the heating on, the space temperature rises until the sensor output exceeds the set point. The heating then switches off and stays off until the temperature falls through the differential and reaches the lower limit, whereupon it comes back on and the cycle repeats. The temperature interval between the upper and lower limits is termed the differential gap or differential band; in American usage it may be referred to as the deadband. Within the differential gap the output may be either on or off, depending on the last switching operation. In accordance with present convention, the set point is taken to be the upper point of the differential gap; earlier conventions take it to be the centre point. The room temperature continues to increase for a time after the heating system has been switched off; this is caused by, for example, hot water present in the radiators. Two-position control results in a swing of temperature about the set point and a mean temperature that normally lies below the set point; some systems when operating under light loads may give a mean temperature above the set point. The swing may be reduced by reducing the differential, but at the cost of increased frequency of switching, with attendant wear on the system. The peak-to-peak variation in space temperature is termed the swing or operating differential and the differential of the controller itself, i.e. the differential that becomes apparent by turning the dial of the thermostat, is known as the mechanical or manual differential.

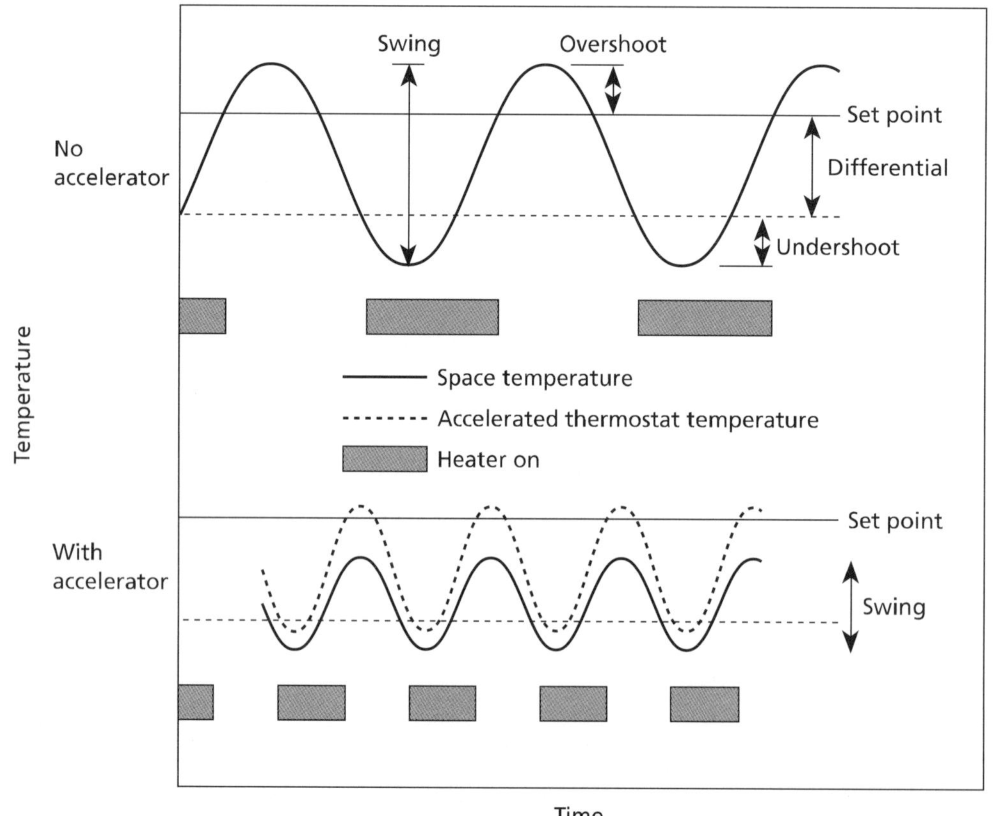

Figure 2.2 Two-position (on/off) control

The common domestic room thermostat is an example of a two-position controller. The inherent differential of the thermostat is of the order of 3 K, for mechanical reasons necessary to provide a snap action operation of the contacts to avoid arcing. The operating differential may be reduced by incorporating an accelerator heater in the thermostat. A low-powered heater within the body of the thermostat is wired in parallel with the load and comes on with the heating system. This has the result of increasing the temperature seen by the thermostat, resulting in earlier closure. The frequency of switching therefore increases, giving a lower operating differential and reduced temperature swings in the room. The effect of the accelerator is to reduce the mean room temperature achieved in practice below the set temperature and this control offset increases with load. This is equivalent to the load error found with proportional control and the action of the accelerated thermostat may be described as pseudo-proportional.

Floating control is a form of two-position control which requires that the controlled device can have its output increased or decreased by a slow-moving actuator. It is also known as three-position or tristate control. A typical example would be a motorised valve controlling flow of hot water. The valve moves slowly towards open or closed position during the application of a signal from the controller; with no signal, the valve stays where it is and holds its position. The output of the controller is now three rather than two position: increasing, decreasing and off (i.e. no change). Figure 2.3 illustrates this mode of control. When the room temperature exceeds the upper temperature limit, the controller signals the valve to start closing. The valve slowly moves towards the closed position, reducing the heat supply to the room. When the room temperature falls to the upper limit, the controller switches off and the valve stays where it is. The room temperature now floats within the neutral zone, until it crosses either the upper or lower temperature limit, whereupon the valve is driven in the appropriate direction. Such a system is designed to have a long operating time between fully open and closed positions of the controlled device; with a short operating time the action behaves like simple on/off control. Floating control is used for systems where the sensor is immediately downstream from the coil, damper or other device that it controls. It is not suitable for systems with a long dead time. A variant is proportional-speed floating control, where the further the value of the controlled variable moves outside the neutral zone, the faster the actuator moves to correct the disturbance. This is in fact very similar to integral action.

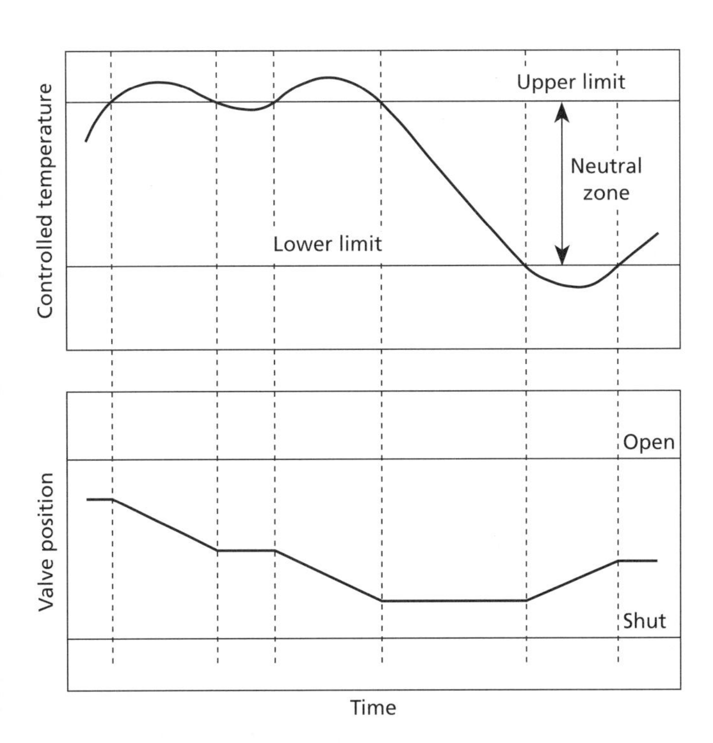

Figure 2.3 Floating control

2.1.2 Proportional control

Proportional control requires a continuously variable output of the controlled device. The control system produces an output which is proportional to the error signal, i.e. the difference between the value of the controlled variable and the set point. For the controller to produce an output to match the load on the system, it is necessary that there be an offset between the controlled variable and the set point. In steady-state conditions, a proportional controller produces an offset or load error, which increases with the load on the system. Figure 2.4 shows the operation of a proportional controller for a heating system. The control output increases from 0 to 100% as the input falls from the set point through the proportional band, also known as the throttling range. It can be seen that in steady-state conditions the equilibrium value of the control point will be below the set point and that this offset will increase with load, e.g. in colder weather when the heating load is greater. For cooling systems, the equilibrium value will be above the set point.

The proportional band may be expressed in units of the physical quantity being controlled, e.g. °C, %RH, pascal, or as a percentage of the controller scale range. If, for instance, the controller has a scale range of 0–80°C and a proportional band of width 20 K, the proportional band is 25%. The gain of a proportional controller is the reciprocal of the proportional band, expressed either in physical units, e.g. K^{-1} or non-dimensionally, e.g. a proportional band of 50% is equivalent to a gain of 2.

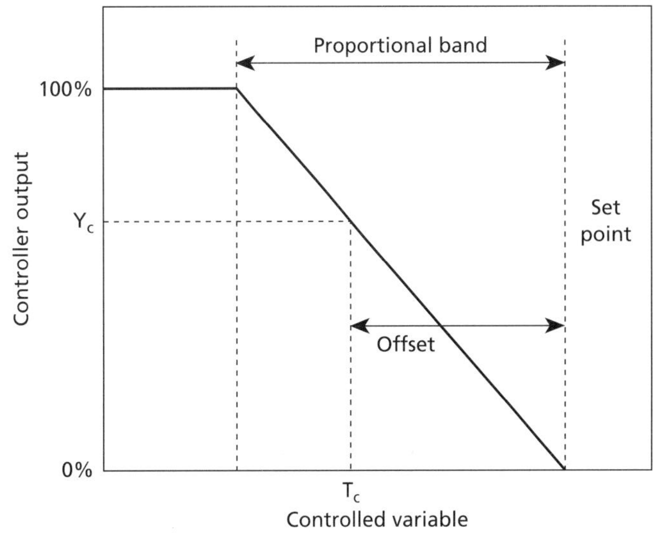

Figure 2.4 Proportional control. Diagram shows steady-state conditions with the controlled variable at T_c with a controller output Y_c. T_c is at an offset or load error below set point. (Sometimes the set point is in the middle of the proportional band)

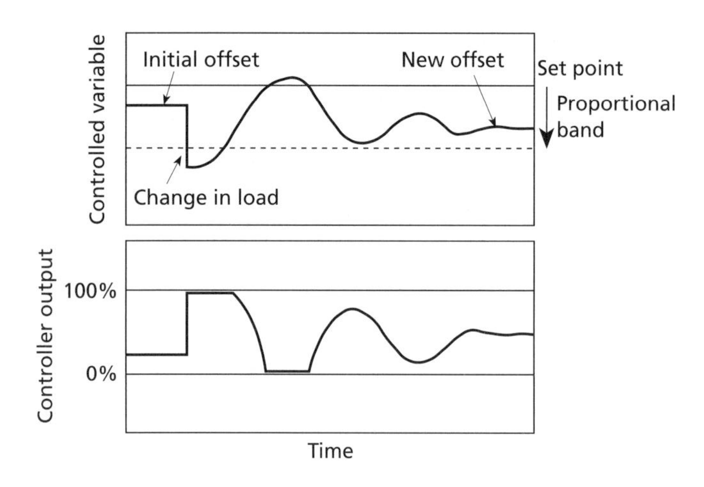

Figure 2.5 Response of a proportional controller to a sudden change in load

Figure 2.5 shows the response of a proportional control system to a change in demand. The value of the controlled variable follows a damped oscillation before settling down to the steady offset temperature. The amount of offset may be reduced by narrowing the proportional band, but at the risk of introducing instability; as the proportional band is reduced, the control action approaches on/off.

A form of proportional control known as time proportioning may be achieved even if the output device is only capable of a two-position output, e.g. high/low or on/off. The output from the controller varies the ratio of on/off times within a constant cycle period, e.g. if the cycle time is 10 minutes and the controller calls for 40% output, the output device will be switched on for 4 minutes and off for 6 minutes. The cycle time may be set independently; it should be sufficiently long to avoid any problems of wear caused by too frequent switching of the controlled device, but shorter than the response time of the overall system. The method is suitable for systems with long response times, where it will give much lower temperature swings than simple on/off control. The control behaviour is similar to a proportional system and will show a load error. Time proportioning control may be used for the control of electric resistive heaters where the switching frequency is limited by the requirement to avoid electrical disturbances on the supply(1).

2.1.3 Integral control

Integral control is not often found on its own, but is normally combined with proportional control in a PI controller. In its pure form it produces a rate of change of the output of the controller proportional to the deviation from the set point, or in other words, the output is a function of the integral over time of the deviation from the set point. When the controlled variable is at the set point, the rate of change of output is zero. The system should therefore settle to a steady-state condition, with steady output and zero offset. The control mode is similar to floating control, but with a zero width neutral zone and variable rate of change of output: compare with proportional-speed floating control. It is illustrated in Figure 2.6.

When integral control is used by itself, it must be used in systems with short time constants and fast reaction rates. It

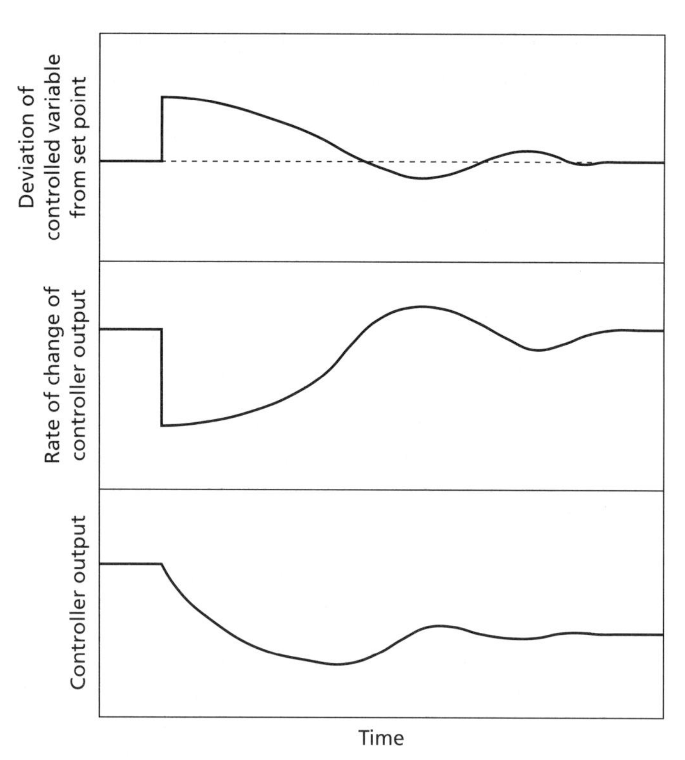

Figure 2.6 Pure integral control action. System is initially in steady state. Figure shows response to a step decrease in set point

is not suitable for a system with slow responses or long time lags, where it will over-correct. A typical controlled device is a valve driven by a variable-speed actuator, which gives the required variable rate of change of control response. A constant-speed actuator may be used where the controller provides a variable duration pulsed current to give effective variable speed. The speed of closure of the valve must be slow compared to the speed of response of the controlled system. The more common combination of proportional and integral control is discussed below.

2.1.4 PI — Proportional plus integral control

Adding integral control to a proportional controller compensates for the load error. This is probably the most widely used mode in HVAC control and when correctly set up is capable of providing stable control with zero offset. The controller integrates the deviation from set point over time and uses this value to adjust the control output to bring the controlled value back towards the set point. The proportional band may therefore be increased to give stable control; the load offset that would otherwise be introduced is eliminated over time by the integral action. The integral setting is characterised by the integral action time, which is the time it takes for the integral term of the control output equation to match the output change due to the proportional term on a step change in error. Alternatively, the integral setting may be characterised as the reset rate, which is the inverse of the integral time, and measured in resets per minute. Most PI controllers are interactive, where the integral gain is multiplied internally by the gain setting of the proportional action (see Appendix A2). The practical implication is that the proportional band may be adjusted without affecting the integral time. A non-interactive controller has independently adjustable gains for the proportional and integral actions, and so adjusting the proportional gain will alter the integral action time as defined above. A long integral time will increase the steady-state load error; in the limit of infinite integral action time the PI controller becomes a simple proportional controller. However, if the integral time is reduced to a value comparable to or less than the time constant of the controlled system, instability will result.

The output of the integral term depends on the past history of the controlled variable, and problems may result on start-up where the controller treats the preceding off period as a long-term error. This is known as wind-up. Wind-up will also occur if the controller output is 100% and the error remains positive; in this situation the integral action will continue to increase to a huge positive value. When the system becomes controllable again, a long period of negative error will be required to unwind the integral term and return to normal operation. Controllers incorporate anti-wind-up features to prevent this, either by locking the integrator at the pre-existing value whenever the controller output is at either extreme, or by limiting the integrator to some maximum value, typically 50% of full output. Similar problems can occur on starting up a system and some systems disable the integral action on start-up until the system is controlling within the proportional band.

2.1.5 PID — Proportional plus integral plus derivative

Derivative action provides a control signal proportional to the rate of change of the controlled variable. This has the effect of reducing control action if the controlled variable is rapidly approaching the set point, anticipating that the variable is about to reach the desired value and so reducing overshoot. It is therefore of value in systems with high inertia. Derivative action can cause problems in practice. If the measured variable is subject to rapidly varying random changes, the derivative action of the controller will produce an erratic output, even if the amplitude of the changes is small. See the note on derivative kick in Appendix A2. Derivative action is never used on its own, but is combined with proportional and integral action to produce PID control, also known as three-term control. A three-term controller is capable of maintaining a zero offset under steady conditions, while being able to respond to sudden load changes.

The gain setting of the derivative action is defined as the derivative action time, which is the time, usually measured in minutes, taken for the proportional term to match the derivative term when the error changes linearly with time. Derivative action is not normally required in HVAC applications and setting the derivative time of a PID controller to zero results in PI action. Three-term PID action is used mainly in process control applications.

Figure 2.7 shows the ideal characteristics of PID control on the behaviour of the controlled variable on start-up. With proportional control only, the output is a function of the deviation of the controlled variable from the set point. As the controlled variable stabilises, a residual load error results. With the addition of integral control, the controlled variable eventually returns to the set point, but there is still some overshoot before stable operation is achieved. Adding derivative control reduces the overshoot and the final set point is achieved in a shorter time.

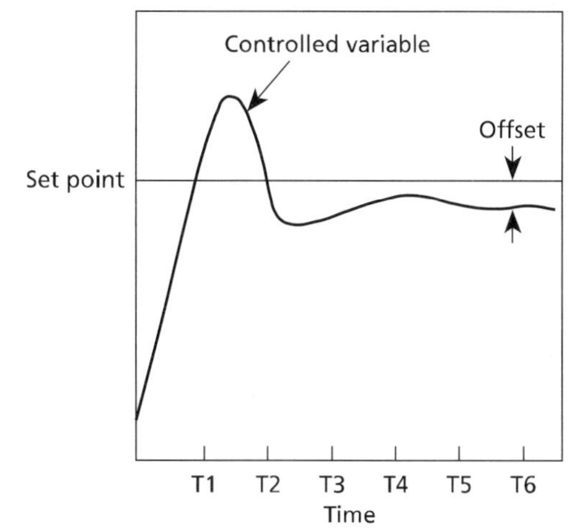

Proportional–integral control

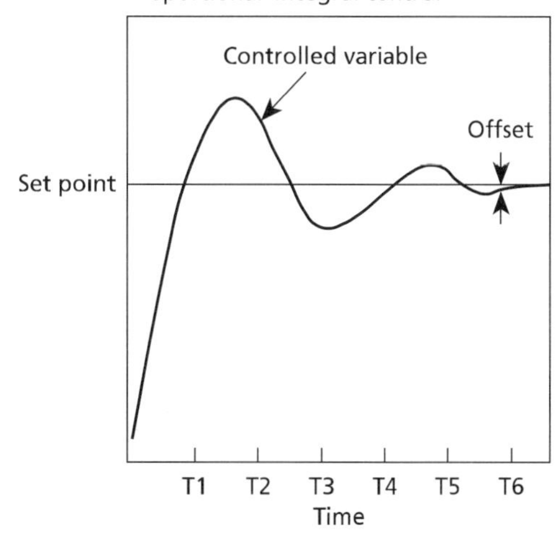

Proportional–integral–derivative control

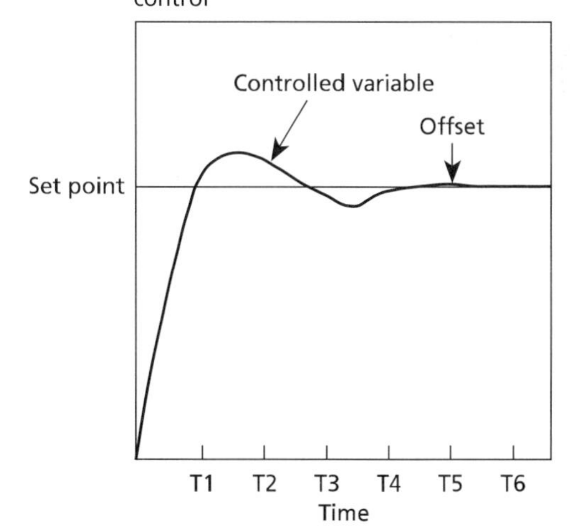

Figure 2.7 Illustration of PID control modes

2.1.6 Digital control

Microprocessor controllers operate by sampling values of the controlled variable at discrete intervals of time. The microprocessor then calculates the required controller output. For the most part, the processor mimics the analogue control modes described above. The controller is able to store past values of the controlled variable, which are needed to calculate derivative and integral terms. Programmable controllers are described in more detail in 3.8. One important difference between analogue and digital controllers is the effect of sampling rate. The frequency of sampling is limited by the speed of the processor and any multiplexing of the controller input, plus the ability of the network to transmit frequent messages. If the sampling rate is too low, instability may result if the controller is delayed in taking appropriate control action. Where there is a sufficiently fast sampling time, some controllers update the control output at intervals which are longer than the sampling interval; the interval between changes in output is known as the loop reschedule interval. Some controllers allow the loop reschedule interval to be adjusted independently from the sampling interval of the controller.

The control equations described in Appendix A2 are transformed in a digital controller into discrete time algorithms. The standard PID equation is called the position algorithm, since the position of the control element is related to the output signal. An alternative is the velocity or incremental algorithm. At each time step, the controller makes an adjustment to the position of the controlled device which is proportional to the change in the deviation of the controlled variable from the set point since the last sample, plus a term proportional to the deviation. Incremental control usually employs a slow moving actuator. At each time step, the controller calculates the required change in actuator position and sends a timed pulse to the actuator to move it the required amount. This control mode is similar to the floating control mode described in 2.1.1. It is found to work well and has the advantage of avoiding integral wind-up. The disadvantage is that the controller has no information on the actual position of the controlled device. If knowledge of the position is required, the controller can integrate the controller output and calculate the position of the controlled device. This integration calculation requires to be re-zeroed at intervals. This may be done by driving the actuator to a limit stop to provide a fix of its position. This may be done automatically each day during out-of-hours operation. The mathematical treatment of incremental control is given in Appendix A2.

2.1.7 Cascade control

For some applications it is an advantage to divide the controller into two subsystems: a submaster controller which controls an intermediate part of the controlled system, and a master controller which adjusts the set point of the submaster loop. A typical application is for temperature control of a large space, where the master controller controls the supply air temperature set point as a function of space temperature, and a submaster controller controls the supply air temperature by modulating the heating coil valve (Figure 2.8).

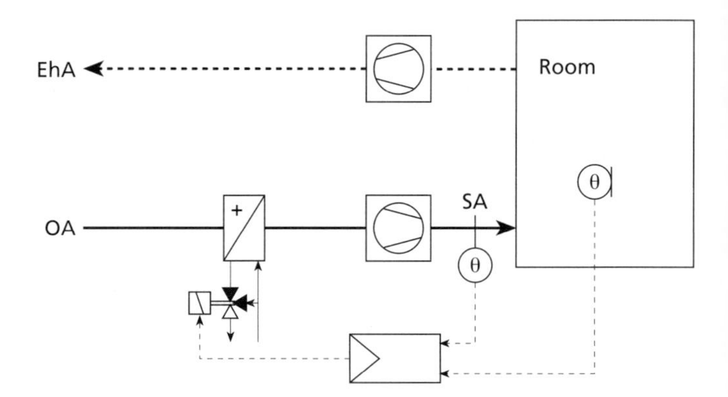

Figure 2.8 Cascade control. The room temperature is used to reset the set point of the controller controlling the supply air temperature

In its standard form, the submaster controller provides control of the supply air temperature against variations in incoming air temperature or fluctuating heating coil water temperature. The master controller resets the supply air temperature set point as a function of the space temperature using PI control. Care may be necessary to avoid instability if both loops use integral action. Some confusion in terminology may be found. An unambiguous term for this system of control is cascade control, and is used in this Guide. Cascade control is also commonly known in the UK as reset control. In the USA simple integral control may be referred to as reset, and a cascade controller is commonly referred to as master–submaster. The terms master–slave and primary–secondary are also used to refer to the two control loops. Cascade control is used when PI control alone is not suitable or will not provide stability, for instance where the space temperature responds slowly to variations in supply air temperature.

2.1.8 Time lags

In any feedback control loop, the response of the controlled system, as seen by a change in the sensor output, does not happen instantaneously upon a change in the controlled output. Two types of delay may be identified. A transport delay, also known as a distance–velocity lag, represents the time it takes for the heating or other medium to travel from its source to the point where its heat begins to be transferred to the controlled space. In large installations, distances can be very long and it can take some minutes for a change in water temperature at the boilerhouse to reach distant points in the building. The second type of delay, termed a transfer lag, depends on the time taken to increase the temperature of a component due to its thermal capacity. Consider the simple heating circuit of Figure 2.9. When the controller gives the signal to open the valve, hot water flows towards the heating coil, taking a time equal to the distance velocity lag to reach it. A series of first-order transfers then take place: primary water to heating coil, heating coil to calorifier water, water to cylinder material and finally cylinder metal to the sensor. The resultant response of the temperature sensor is shown. This is typical of a higher-order response, and may be approximated by a combination of dead time and first-order response as shown.

All lags contribute to poor control. Integral or floating control is unsuitable for systems with a significant dead time, since the controller will continue to change the output during the dead time, resulting in overshoot. PI control is then more suitable. The proportional control, with adequately wide proportional band, provides a stable control, and the integral action, with suitably long integral time, removes the load error.

2.1.9 Logic control

The use of microprocessor-based DDC controllers offers enormous freedom to the controls designer, since virtually any control strategy may be programmed into the controller software. In practice, digital controllers are based on the well-understood control modes described in this

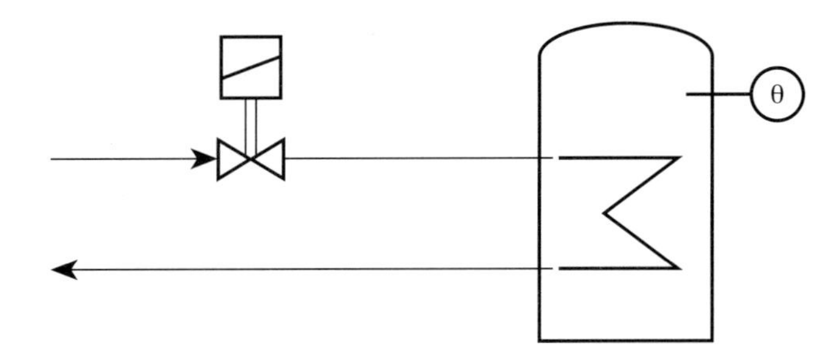

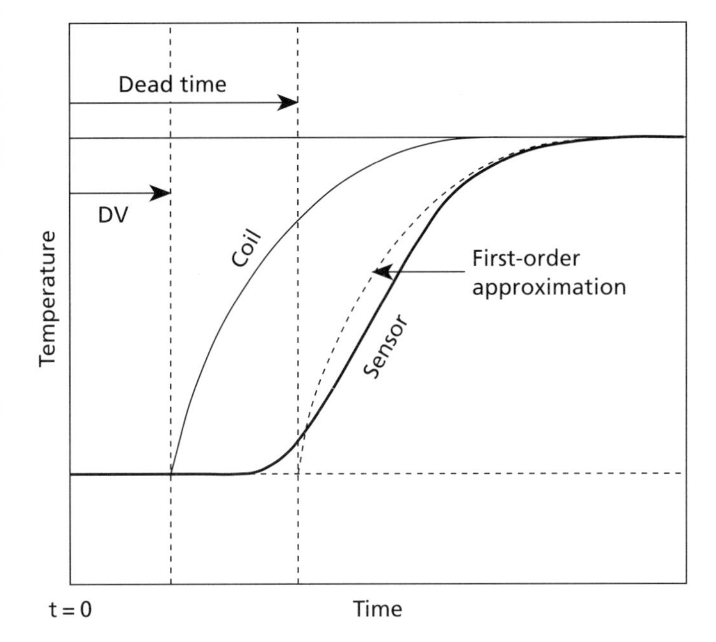

Figure 2.9 Response of system with dead time. On opening the valve at $t = 0$ there is a distance–velocity lag DV before the primary hot water reaches the heating coil. The coil then heats up with a first-order response. The sensor has a higher-order response, which may be approximated by a first-order response with dead time

section. The widely used universal controller, which incorporates pre-programmed control modules in its software is described in 3.8. The controller may be configured to meet the requirements of the actual control strategy to be implemented. Some examples of logic control are given in 5.8.2 and 5.14.3.2.

The controllers incorporate a number of logic control functions which may be used to improve control operation. Some examples are:

— *Hysteresis*. The hysteresis module only passes a change in input which is greater than a preset amount. It can be used to stop the control system responding to small fluctuations in the controlled variable, so reducing control action and wear.

— *Averaging*. The averaging module is used to produce a mean value of a number of inputs. For example, the system may be set up to control mean zone temperature, averaged over several temperature sensors. Sophisticated versions may be programmed to ignore extreme values.

— *Logic operators*. Logic modules provide the full range of Boolean AND, NOT, OR and XOR gates. They are used to provide software interlocks, e.g. preventing operation of a heating system when windows are open. Safety-critical interlocks should be hardwired.

— *Look-up tables*. Functional relationships can be provided in the form of look-up tables. Examples are the conversion of a thermistor resistance to a temperature or the software linearisation of a controlled element characteristic.

The full range of available modules is too great to list here. The range of modules is sufficient to cover most control requirements and the controller manufacturers provide examples of control strategies to assist in configuration. If required for special situations, it is possible to write control strategies using a high level programming language, such as BASIC or C.

2.1.10 Choice of control mode

When selecting the appropriate control mode, the following considerations should be taken into account:

— the degree of accuracy required and the amount of offset that is acceptable

— the type of load changes expected, including amplitude, frequency and duration

— the system characteristics, such as the number and duration of time lags and speed of response of subsystems

— the expected start-up situation.

In general, use the simplest mode that will meet the requirements. Using a complicated mode may result in difficulties in setting up and lead to poorer rather than better control. Derivative control is not normally required in HVAC systems. Its function is to avoid overshoot in a high inertia system by measuring the rate of approach to set point and reducing control action in advance. It is used in some boiler sequencers, where it will inhibit bringing an additional boiler on line if the rate of rise of water temperature shows that the operating boilers will achieve the required temperature on their own. Table 2.1 lists typical applications.

2.2 Optimum start

One of the most important functions of a building control system is time control, ensuring that plant is switched off when not needed. Substantial energy savings may be made by intermittent heating or cooling of a building compared with continuous operation. The savings achievable from

Table 2.1 Recommended control modes

Application	Recommended control mode
Space temperature	P
Mixed air temperature	PI
Coil discharge temperature	PI
Chiller discharge temperature	PI
Air flow	PI. Use wide proportional band and short integral time. PID may be required
Fan static pressure	PI. Some applications may require PID
Humidity	P, possibly PI for tight control
Dewpoint	P, possibly PI for tight control

intermittent heating compared with continuous heating depend on several factors. The savings will be greater in a lightweight building which cools and heats up quickly; any estimation of overall running costs must take into account all relevant factors. Heavyweight buildings are able to absorb peak gains and benefit from night cooling, which may outweigh any savings from intermittent heating. In general, intermittent heating and cooling will be more beneficial in the following situations:

- lightweight building (low thermal mass)
- short occupancy period
- generously sized plant.

Simple timeswitch control can be effective and is suitable for heating systems with a heat output of up to about 30 kW. Above this figure, an optimum start controller is recommended; above an output of 100 kW, optimum start control is required by the *Building Regulations*[2]. The time of switching on prior to occupancy is selected to ensure that the heating system has time to achieve a comfortable temperature at the start of the occupancy period. If this is correct in cold weather, the system will come on unnecessarily early in mild weather, giving higher energy consumption than necessary. Nor will a simple timeswitch be able to cope with the longer preheat period necessary after a weekend or holiday. An optimum start controller, or optimiser, is designed to calculate the latest switch on time under a range of operating conditions. Figure 2.10 illustrates the required control characteristic. During the unoccupied period, the plant normally operates to provide a minimum temperature to provide protection for the building fabric and contents. This is typically 10°C but may be lower. Separate frost protection must be provided for the heating system.

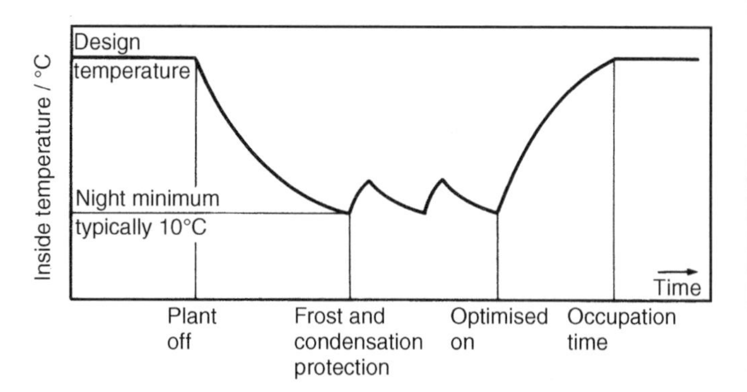

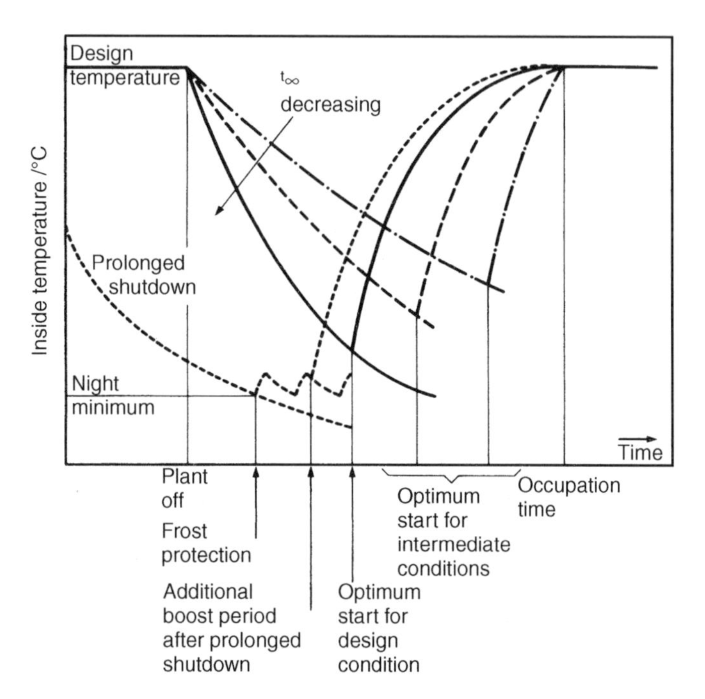

Figure 2.10 Optimum start control

The primary function of the optimiser is to calculate the latest switch on time. Several algorithms have been proposed. The most widely used is BRESTART[3]. This calculates the switch-on time as a function of both internal space temperature and external air temperature. Most controllers incorporate a self-learning or adaptive feature. By following the building performance over a few weeks, the controller sets its internal parameters to match the characteristics of the actual combination of building and heating system. The optimiser also selects maximum heat output from the heating system during the warm-up period by disabling any weather compensator that may be fitted. The boost is terminated and compensation restored when the building reaches the desired temperature.

An optimum stop function may be fitted, whereby heating or cooling is switched off before the end of the occupancy period, at a switch-off time calculated to ensure that the space temperature does not drift outside predetermined comfort limits by the end of the occupancy period. With air-handling systems, the zone air temperature will approach the building fabric temperature within about 15 minutes after switch-off. This may not provide comfortable conditions and so limits the usefulness of optimum stop strategies. Optimum stop is used less often than optimum start and the potential savings are less.

2.3 Weather compensation

A building heating system is designed to provide full heating on a design day; in practice an additional margin is allowed to provide extra power during the boost period of intermittent heating. The capacity of the heating system is therefore greater than required for operation in all but the coldest condition. For buildings heated by a conventional radiator system, operation during mild weather with the flow temperature at the full design value, typically 80°C, results in control problems, high temperature swings and consequent discomfort; it also results in wasteful heat loss from the hot water circuit. For buildings larger than domestic or the smallest commercial, it is required by the *Building Regulations* that weather compensation be provided to adjust the flow temperature in accordance with the outside temperature.

Compensation control allows the whole building to be controlled as one unit, or as a limited number of zones, thus eliminating the need to provide a large number of separate space temperature controls. It has the added advantage of limiting heat loss in the event of increased load, e.g. if windows are opened. If used as the only form of temperature control, it requires the radiator size to be carefully matched to the heat requirement of the building; since this is virtually impossible to do in advance, provision must be made for balancing the system and adjustment of the compensator control characteristic. A practical solution is to use weather compensation and trim local temperatures by the use of thermostatic radiator valves.

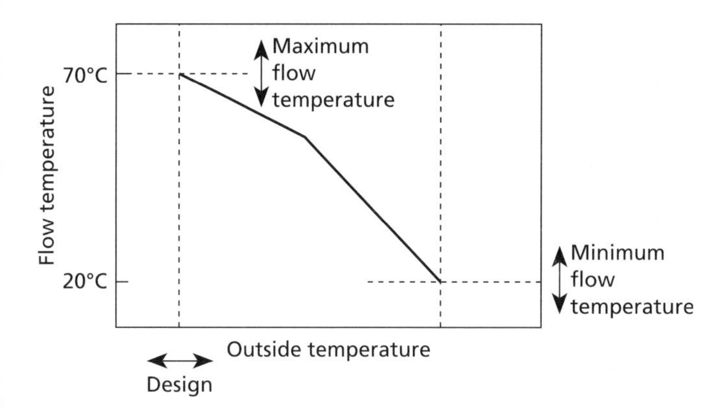

Figure 2.11 Weather compensation control characteristic with two adjustable slopes

The form of the weather compensator control characteristic is shown in Figure 2.11. The controller allows adjustment of the minimum and maximum flow temperatures and the slope of the characteristic curve. The heat output from a radiator is proportional to the 1.3 power of the difference between mean radiator temperature and room temperature and so the linear characteristic shown in the figure will tend to overheat in cold weather. Some controllers provide a two-slope or curved line to allow for this. This is discussed in detail in Levermore[(4)].

Figure 2.12 shows a common method used in smaller buildings of providing a compensated flow temperature. Water from the boiler is blended with cooler water from the secondary circuit return in a three-port mixing valve. The temperature sensor is downstream of the valve and responds quickly to temperature changes in the flow. For larger buildings or special circumstances, more complex control arrangements may be needed. Where different parts of the building respond differently to external climatic conditions, it will be necessary to subdivide operation into zones, and perhaps add additional external sensors, e.g. a solar detector to aid control of the southern side of the building.

The simple weather compensator described above is an example of open loop control; there is no feedback from zone temperature and the achievement of satisfactory temperature depends on the accuracy of setting up the characteristic relationship between outside temperature and flow temperature. The setting of the characteristic

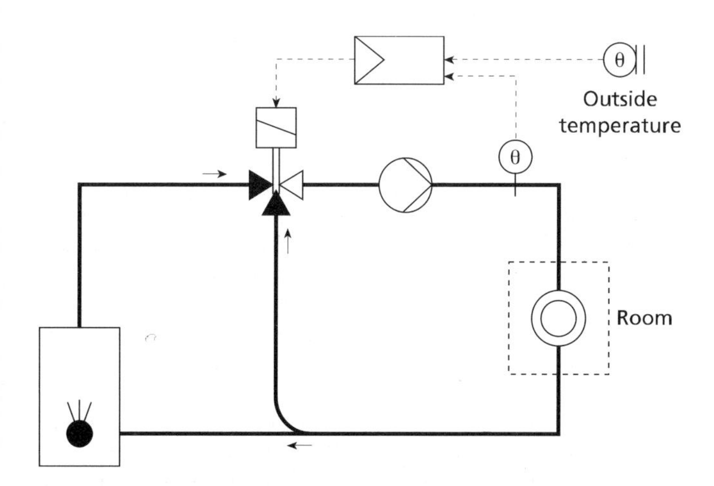

Figure 2.12 Weather compensator controlling flow temperature

curve should be reviewed whenever there is a change in building use, which may affect the level of internal gains, or modification such as fitting insulation. Variations in solar gain with season or changes in shading due to foliage may also affect the settings. There are two modifications to the simple weather-compensated circuit which give improved control and are becoming more commonly used.

— *Zone trim*. The standard weather-compensated circuit is used, with a zone trim applied to the compensated water flow temperature. The trim modifies the set point by a margin proportional to the difference between the measured zone temperature and the desired set point.

— *Cascade control of zone temperature*. The outside temperature sensor is not used, but replaced by an internal zone sensor. This provides the input to a PI controller which resets the water flow temperature. This gives stable effective control of indoor temperature with no offset.

2.4 Stability and tuning

The stability of a control system is concerned with its response to a disturbance. The disturbance may be a change in the external load, e.g. an increase in solar gain though the windows of a building. The HVAC system is required to react to bring the controlled variable (room temperature) back towards the desired value. For practical reasons, the stability of a system is usually considered in terms of its reaction to a sudden change in set point:

— *Stable*. After the change in set point, the controlled variable sooner or later settles down to a new steady value. On the way, there may be oscillations about the eventual steady-state value. All the systems shown in Figure 2.7 are characterised as stable; the presence of an offset from the set point (load error) is no disqualification.

— *Unstable*. The system does not achieve a steady state following a disturbance. There are two types of unstable response:

  — *Oscillatory*. The controlled variable continues to oscillate or hunt about the set point.

  — *Non-oscillatory*. The controlled variable continues to increase or decrease until it reaches a limiting value.

Non-oscillatory instability is unlikely to be produced in an HVAC system except by a design or installation error. For example, confusing the connection of room temperature sensors in different rooms will produce unstable control: an increase in heat load in Room A will reduce the heat input to Room B; the sensor in Room B will then demand more heat for Room A. Instability may be produced by the intervention of the occupants. If an occupant opens a window because the room is too warm and this results in cool outdoor air blowing over a poorly placed thermostat, heat input to the room may be increased.

On/off control inevitably produces an oscillating value of the controlled variable. If the proportional band of a proportional controller is reduced below a critical value, the control system goes into oscillation. When considering the effects of oscillation, it is necessary to distinguish variations in the controlled variable from changes in the position of the controlled device and the associated system output. The system output in an on/off system necessarily swings from 0 to 100%. However, the swing in the controlled variable may only be a fraction of a degree where an accelerated thermostat is used. A large variation in system output does not therefore necessarily imply unsatisfactory controlled conditions.

Most buildings and their HVAC systems have a high thermal mass and in consequence change temperature fairly slowly. As a result, building control systems may be relatively insensitive to poor setting up of the control system, and a control system that is hunting may produce conditions that are acceptable to the occupants. Hunting behaviour may only be evident on directly observing movement of the actuators or by detailed logging of system behaviour by the BMS. A control system which hunts is unsatisfactory for several reasons:

- It produces excessive wear on valves and actuators.
- Temperature cycling may produce undesirable effects on plant and equipment.
- It may produce instabilities in other parts of the system.

Some variation of control output and movement of the controlled device is inevitable as the system responds to small changes occurring within the building. The opportunity should be taken to observe the action of the controlled devices at a time when the HVAC system is controlling in a steady state. Any movement should be slow and of small amplitude. It is difficult to give figures which are of general application. However, as a rule of thumb, action should be taken to investigate the stability of the system if any controlled device is cycling with an amplitude greater than 20% of its full range, or is changing position by more than 20% in a period of 10 minutes.

The following factors must be taken into account to ensure a stable operating control system:

- correct tuning of controllers
- interactions between control loops
- stable operation must be ensured over all operating conditions, e.g. variation of system gain with load
- the influence of sampling period.

Interaction between control loops. An HVAC system may contain many control loops, the actions of which interact with each other. Consider, for instance a network of heat emitters controlled by two-port valves, being fed from a common secondary circuit. Each two-port valve is controlled by a local control loop which maintains a zone temperature. The movement of one valve in response to a change in load will alter the pressure seen by the other circuits, causing a change in flow, with a resultant alteration in the controlled variable and consequent movement of the vales. This in turn will change the pressure seen by the original two-port valve. It is possible for the circuits to interact in such a way that the valves do not achieve a steady equilibrium position, but continue to hunt.

A general principle that applies to any feedback control loop is that the sensor must measure the response of the controlled variable relevant to that loop and not any other. The effects of interchanging the connection of sensors in adjacent rooms is an obvious example. If one room is below its set point, additional heat will be supplied to the other room. This will result in a reduction in heat supply to the first room, giving maximum heating to one room and none in the other.

A more subtle variation of this problem occurs when the two controlled zones overlap. Consider the example of adjacent VAV boxes supplying a common space. Each box has its own temperature sensor, but the nature of the space and the air flow patterns is such that the two temperature sensors show the same value. If the set points of the two VAV boxes are not identical, one will show a small positive error and the other a small negative error. Since it is impossible for both to have zero error simultaneously, the effect of PI control will be to produce integral wind-up, which over time will result in one box being driven to maximum output and the other to minimum. This effect may be mitigated by using proportional-only control where there is a danger of adjacent boxes interacting. The use of intelligent VAV boxes or fan coil units allows groups to be programmed with a common set point, with one unit acting as a master, controlling the others as slaves.

Operating conditions. An HVAC system is designed to operate over a wide range of conditions. Most design work is done at design conditions, where the system is required to produce maximum output. For most of the year, the system operates at part load where control behaviour may be different from design conditions. Consider for instance a VAV system. The supply air is cooled by a cooling coil, which is modulated to control the air temperature. If the air flow is half the design maximum, then the change in valve position to produce a given change in air temperature will be about half that required at full flow, i.e. the gain of the system has doubled. There is therefore a risk of instability if the control loop is tuned at design conditions. To ensure stability, tuning should be done in conditions which produce a high system gain.

Sampling time. Digital control systems measure the controlled variables at intervals known as the sampling time. The sampling time is determined by:

- the time period required by the A/D converter
- any multiplexing between sensor inputs
- density of traffic on a bus system.

Too fast a sampling time will require an unnecessary amount of data handling by the BMS. However, too slow a sampling time may be inadequate to detect changes in the system in time to take appropriate control action. Where the sampled values are stored for evaluation, too slow a sampling rate can produce unsatisfactory or misleading information; a high frequency variation sampled at too low a frequency may be interpreted incorrectly as a low frequency variation; this is known as aliasing. The correct sampling frequency may be selected by reference to Shannon's sampling theorem, which states that provided a signal contains no component with a frequency higher than f_{max}, the signal may be reconstituted from a set of samples taken with frequency at least twice f_{max}. In practice, sampling frequencies of 10 times the theoretical value are used[4].

2.4.1 Tuning

The behaviour of a control loop under changing conditions is affected by the controller settings. Incorrect settings of the controller parameters may lead to unstable behaviour, resulting in large output swings, or else sluggish response and a deviation from the desired set point. Optimum performance from a controller depends on the correct settings for the control parameters. Two forms of response may be considered to represent the acceptable boundaries of sensitive and sluggish behaviour:

— *Quarter wave response*. Following a disturbance, the system response overshoots the new equilibrium value and approaches equilibrium in a series of damped oscillations. The amplitude of the first overshooting wave is four times that of the second.

— *Critically damped*. This response is the fastest approach to the new equilibrium value that can be achieved without any overshoot.

Values of the parameters may be found by carrying out in situ tests on the control loop. There are two common techniques used in estimation.

— closed loop ultimate cycling method (oscillation test)

— open loop reaction curve method (transient response).

Both require the behaviour of the control loop to be recorded; parameter settings may then be calculated from the measured response. Several calculation methods are available and may be selected to give the desired behaviour; they are summarised in Appendix A2. It must be emphasised that the parameters calculated in this way should be regarded as guide values. Since system gain often varies with operating conditions, control loop behaviour will also vary. To ensure stable operation under all conditions, tuning should ideally be done under conditions of high system gain. Where there is doubt, the parameters should be adjusted as follows to reduce risk of instability:

— proportional band increased

— integral action time increased

— derivative action time decreased.

2.4.2 Practical tuning

Before starting formal tuning procedures, it is advisable to check the system manually. If manual operation is unsatisfactory, the problems should be rectified before tuning the loop. Adjust the set point manually between several positions and observe:

— Is the process noisy: are there rapid fluctuations in the controlled variable?

— Is there appreciable hysteresis or backlash in the actuator?

— Is it easy to maintain and change a set point?

— In which operating region is the process most sensitive?

Some general principles should be borne in mind before starting to tune a control loop.

— In order for tuning procedures to work properly, the final output to the controlled device should have a proportional action. An on/off controller cannot be tuned satisfactorily.

— The controlled device should be within its operating range, i.e. not driven to fully open or closed, nor operating in a region where movement of the final actuator has little effect on the process output.

— Where cascade control is being tuned, tune the inner (submaster) loop first. Where a control loop depends on the stability of some other part of the system, tune the primary service first, e.g. tune a primary water circuit to ensure flow temperature stability before tuning any of the secondary loops.

— Some controllers operate more than one controlled device, e.g. a temperature controller may operate both heating and cooling coils. It may be necessary to disable operation of one of the outputs to prevent it being brought into operation during tuning if the controlled variable overshoots.

Determining the controller settings by the formal methods detailed in Appendix A2 may not always be practicable. Experience has shown the following adjustments will often give suitable settings.

— *Proportional band*. A conservative starting point for the proportional band is a value equal to the change in the controlled variable produced by a 50% change in plant output.

— *Integral time*. A conservative initial value for the integral time is to set it equal to the open loop time constant, estimated as the sum of the component time constants, e.g. actuator, sensor and HVAC process.

— *Derivative time*. Derivative control is applied when there are significant delays in the control loop. Try an initial value of derivative time equal to 50% of the loop dead time.

— *Sampling time*. Set to no more than 25% of the open loop time constant; greater sampling times may result in unstable control.

Guidance on typical initial controller settings will generally be provided by the manufacturer. Table 2.2 shows suitable settings for some common applications, which may be used if no better information is available.

2.4.3 Self-tuning controllers

The methods described above for tuning PID controllers are based on standard control theory, which makes assumptions about the linearity of the controlled system. In practice, the formal tuning methods may have disadvantages:

— *Effort*. Tuning control loops during commissioning is time consuming and it may be difficult to allocate the skilled effort required when under pressure of time. There is a temptation to leave the controllers at factory preset values, which are unlikely to be optimum.

— *Non-linearity*. The tuning methods assume that the control system is linear and that the system gain

Table 2.2 Typical settings for a PI controller

Controlled device	Controlled quantity	Proportional band	Integral time (min)
Heating coil	Zone temperature	2 K	0
Preheat coil	Duct temperature	3 K	4
ChW coil	Duct temperature	8 K	4
Humidifier	Zone RH	15% RH	15
Dehumidifier coil	Duct RH	15% RH	4
Thermal wheel	Duct air temperature	4 K	4
Run around coil	Supply air temperature	6 K	4
Recirculation damper	Mixed air temperature	4 K	4
Ventilation supply	Zone CO_2 concentration	100 ppm	10
Room terminal unit	Zone temperature	3 K	4
Supply fan	Static pressure	1000 Pa	1

remains constant. System gains vary with load and are likely to vary both over the day and over the year. Correct choice of valve or damper authority and characterisation helps to improve system linearity, but it is not always possible to achieve good linearity over the range of operations encountered.

Several means of automatic tuning of controllers have therefore been developed. There are different approaches:

— *Autotuning*. Autotuning software automates the tuning procedure by exciting a controlled response and calculating the optimum control settings. It is initiated by a controls engineer during commissioning or when required. Autotuning is applicable when the system characteristics are constant.

— *Adaptive techniques*. Where the operating characteristics of the control system vary substantially, it is desirable to use a form of self-tuning that acts continually to optimise the control settings. The different approaches are:

 — *Gain scheduling*. If the variation of optimum settings of controller parameters with plant operating condition can be established, then the parameters are automatically adjusted as a function of plant operation. Gain scheduling is effective when the operating dynamics of the system are well understood.

 — *Adaptive control*. Where the changes in plant dynamics are large and unpredictable, a controller which continuously tunes itself is desirable. Adaptive controllers use some form of plant model or pattern recognition. A commonly found form of adaptive control is to be found in optimum start controllers, which are described below.

2.4.3.1 Autotuning

Several manufacturers market controllers with an inbuilt autotuning facility. Software within the controller automates the tuning procedure by exciting a response from the plant and calculating the control parameters from the observed response. The controller is then set to the new parameters; the engineer may be required to accept them before they are applied. Before tuning, the process must be in a steady-state condition. The tuning action is initiated by the controls engineer, normally by simply pressing a button. The controller parameters remain fixed between tuning sessions. An autotuner is therefore unable to compensate for variations in system gain and the tuning process should therefore take place under representative operating conditions. In many HVAC systems there is interaction between different control loops. When using an autotuner it is important to select the order in which the loops are tuned to minimise any interactions. Start with fast-responding local control loops; when these have been tuned move to larger slower loops. The use of autotuning can give significant improvements in performance coupled with a substantial reduction in commissioning time compared with conventional PID controllers. Further details are to be found in Wallenborg[5] and Astrom *et al.*[6].

2.4.3.2 Adaptive techniques

The control characteristics of an HVAC system may vary with operating conditions. The characteristics and authority of valves and dampers affect linearity of response of a system and the achievement of a linear characteristic of a complex system over its whole operating range is difficult. The gain of a system will also change with operating conditions, e.g. the change in flow temperature produced by an optimiser or the change in inlet air temperature with season will produce changes in system gain. Since a high system gain tends to produce control instability, controllers must be tuned when the plant gain is high. This will ensure stability, but control performance may be sluggish in low gain conditions. There is therefore an advantage to using a controller which will retune itself automatically in response to changing conditions.

If the performance of the control system is well understood, it is possible to monitor the operating conditions of the plant and adjust the control parameters in accordance with a data table which relates control parameters to operating conditions. This process is known as gain scheduling, though it is not limited to adjusting the controller gain; derivative and integral times may be adjusted as required. Gain scheduling is not strictly a form of adaptive control, since no learning is involved. The relation between plant operating conditions and optimum controller settings must be known in advance. It is possible to build up a schedule of control parameters by initiating autotuning under a range of operating conditions.

If the changes in plant dynamics are large and unpredictable then a controller which continuously tunes itself has great advantages. Adaptive controllers learn the operating conditions of the plant and control system by observing the response to changes in set point or external disturbances. In order to protect the system against unreliable parameter estimates, additional software is used to supervise the system to prevent poor control performance in unpredictable situations; this is known as jacketing software. There are two basic approaches to adaptive control:

— Pattern recognition methods analyse the controlled response and recalculate new values of control parameters. Oscillations and offsets are measured and new control gains calculated when significant adjustments are required. A pattern recognition adaptive controller (PRAC) is described in Seem[7]. The method was developed for systems that can be modelled as a first-order plus dead-time response, which includes most HVAC systems. The controller is claimed to be easy to use and provide near optimal control. It has been used successfully in a number of HVAC situations.

— Indirect methods employ a model of the process which is continually updated by comparison with behaviour of the real plant. The model is then used to calculate updated values of the control parameters. This method may be combined with neural networks.

Adaptive controllers and self-tuning controllers have yet to find wide application in the HVAC industry, though they offer potential benefits in reduction in commissioning time, increased actuator life and improved environmental control. Their operation may be adversely affected if tuning is attempted in the presence of plant failure or periodic disturbances. An adaptive controller which continuously seeks to optimise its operation may act to compensate for plant degradation and so delay needed maintenance action.

2.5 Artificial intelligence

Conventional controllers, whether analogue or digital, rely on a precise mathematical relation between input and output. For a well-behaved system operating in a consistent environment, a well-tuned PID controller will give satisfactory control. However, many real systems involve uncertainties, non-linearities and variations in their operating environment. There may be multiple inputs which affect the desired control output. Such complications may be difficult to incorporate in a conventional controller. Advances in artificial intelligence and the availability of improved processing power are leading to the development of new types of controller. Fuzzy logic and neural networks show great promise in dealing with some aspects of HVAC control and their application is certain to grow.

Fuzzy logic is a means of decision making based on a set of rules of the type IF (A is true) THEN (take action B). The rules are written based on knowledge and experience of the system to be controlled and are expressed in near natural language[8]. Fuzzy logic differs from conventional logic in that a statement may have degrees of truth. To put it more formally, a point can have simultaneous degrees of membership of several fuzzy sets. This represents the way we deal with situations in real life; there is, for instance, no abrupt transition between 'comfortable' and 'cool'. Figure 2.13 illustrates the chain of operations in a fuzzy logic controller, for the simple example of a single-input, single-output controller; one of the advantages of fuzzy logic is its ability to deal with multiple inputs and outputs. The controller accepts a crisp input; crisp means an exact numerical value, such as the deviation of a measured temperature from the set point. The controller then fuzzifies the input, by establishing the degree of membership of the several fuzzy sets which have been defined in the controller. For instance, a measured room temperature may have 80% membership of the set 'comfortable' and 20% membership of the set 'cool'. The controller then applies the inference rules, which are of the form IF (the room is cool) THEN (set heating to half power). This results in an output which has membership of several fuzzy output sets; this output is then defuzzified to produce a crisp output value which is used to control the plant. The defuzzification process is analogous to the fuzzification process, where the degree of membership of several output sets is combined to produce a crisp output.

The main advantages of fuzzy control are that it does not require a model of the process to be controlled and that it is possible to incorporate the results of operational experience into the set of rules. These would be otherwise difficult to incorporate into a conventional control algorithm. Fuzzy controllers have been developed for HVAC systems and it has been found that they offer advantages of robustness, energy saving and fast response, compared with conventional PID control[9]. Simple fuzzy controllers are often found to exhibit undesirable oscillation when the system is controlling near the set point. It is possible to use the rate of change of the controlled variable as additional input to the controller. Extra fuzzy rules are added to modify the controller output according to the rate of change; this helps provide smooth control at conditions near the set point and is a common feature in fuzzy controller designs.

Neural networks attempt to reproduce the way the human brain learns by experience. In brief, a neural network device accepts data from a number of inputs, processes the data using a series of non-linear processing elements and produces a set of output data. What distinguishes a neural network from other types of processor is that it does not depend on a model or even an understanding of the process, but is capable of learning by experience. Learning algorithms are employed which adjust the internal parameters to optimise the performance of the network. Accordingly, a period of training is necessary before a neural network can achieve satisfactory performance. Neural networks are applicable where a high degree of non-linearity

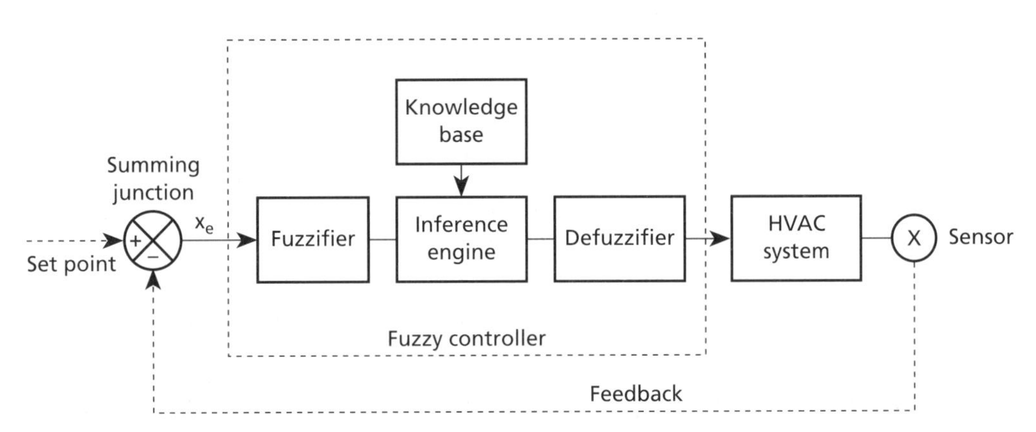

Figure 2.13 Components of a fuzzy controller

exists and there is a large amount of data available for training the network. Both these criteria apply to a BMS. A number of promising applications to HVAC systems have been identified[10]:

— *Condition monitoring*. Incipient plant failure may be detected by changes in performance. This requires knowledge of how the plant operates under a wide range of conditions and the ability to detect variations; a neural network can be trained to do this.

— *Optimisation*. The operation of a complete HVAC system is highly non-linear, making it difficult to optimise performance under a range of load conditions. A neural network can be applied to learn the system behaviour and then used off-line to investigate optimum control strategy.

— *Energy monitoring*. Neural networks have the potential to learn the consumption patterns against key variables such as occupancy, time of day, weather and process activity. Deviations from the expected pattern can then be detected, giving early warning of increased energy consumption.

2.6 Summary

The section sets out the basic concepts of open loop and feedback control loops. The fundamental control modes used in feedback control are:

— proportional control, which produces a control output proportional to the deviation of the controlled variable from the desired set point

— integral control, which produces a control signal proportional to the time integral of the deviation from the set point

— derivative control, which produces a control signal proportional to the rate of change of the controlled variable.

Proportional control can be used by itself to produce stable, effective control, but produces a load error, i.e. an offset from the set point. Integral action may be added to proportional control to give PI control, which eliminates the load error over time. Derivative control is used to anticipate overshoot in systems of high inertia. PID control can produce rapid response to changes in load with little offset, but is rarely used in building control systems.

Controllers need to be tuned to avoid the problems of instability on the one hand, or sluggish response on the other. Formal methods of estimating the optimum parameter settings are summarised. In addition, practical recommendations are given of values which are likely to be found satisfactory in practice.

Optimum start control is an algorithm that has been developed to turn on heating or cooling at the latest possible time to ensure that comfort conditions are achieved at the start of the occupancy period. The controllers are capable of learning the behaviour of the building and heating system and adjusting the settings accordingly; this is an example of adaptive control. The control of compensated heating circuits is described; flow temperature may be related to either outside weather in internal zone temperature. Developments in intelligent controls are described which are leading to the introduction of adaptive controllers which are capable of continuously retuning themselves to give optimum operation.

References

1 *BS 5839: Fire detection and alarm systems for buildings: Part 1 1988. Code of practice for system design, installation and servicing* (London: British Standards Institution) (1988)

2 *Conservation of fuel and power 1995* Approved Document L1 (London: Stationery Office) (1994)

3 Birtles A and John R A new optimum start algorithm *Building Services Engineering Research and Technology* **6**(3) 117–122 (1985)

4 Levermore G J *Building energy management systems* 2nd ed (London: E & F N Spon) (2000)

5 Wallenborg A O A new self-tuning controller for HVAC systems *ASHRAE Transactions* **97**(1) 19–25 (1991)

6 Astrom K J, Hagglund T, Hang C C and Ho W K Automatic tuning and adaptation for PID controllers – a survey *Control Engineering Practice* **1**(4) 699–714 (1993)

7 Seem J E Implementation of a new pattern recognition adaptive controller developed through optimisation *ASHRAE Transactions* **103**(1) 494–506 (1997)

8 Wang L *A course in fuzzy systems and control* (Englewood Cliffs, NJ: Prentice Hall) (1997)

9 Underwood C P *HVAC control systems: modelling, analysis and design* (Andover: E & F N Spon) (1998)

10 Barnard N I Neural networks: potential areas of application in building services *Building Services Engineering Research and Technology* **14**(3) B15–B18 (1993)

3 Components and devices

3.1 Sensors
3.2 Actuators
3.3 Valves
3.4 Dampers
3.5 Motors
3.6 Pumps and fans
3.7 Control panels and motor control centres
3.8 The intelligent outstation
3.9 Summary

This section deals with the hardware components that make up a control system. Guidance is given on the choice of suitable sensors and the importance of correct installation is stressed. Valve sizing and characterisation is covered together with choice of dampers. The use of variable-speed drives for pumps and fans is covered.

3.1 Sensors

Sensors form a vital component of any control system. Sophistication in the computing and software functions cannot compensate for inaccurate information provided by poor-quality or inappropriately mounted sensors. The types of sensor available for use in building control systems are reviewed and guidance on selection and installation is given.

3.1.1 General categories

The word sensor is generally used rather loosely to cover all processes between the measured variable and the input to the control module. The functions may be broken down into:

— *Sensing element*: a component that undergoes a measurable change in response to a change in the variable to be measured.

— *Transducer*: an active device that produces an electrical signal which is a function of the change in the sensing element.

— *Transmitter*: a device that produces an electrical signal that is a standardised function of the change in the physical variable and which can be used as an input to the control module.

In practice the function of the transducer and transmitter are often combined. Their function may be referred to as signal conditioning, which may include filtering to remove noise, averaging over time, or linearisation. In some systems the sensing element may be connected directly into the controller, e.g. thermocouple input, where the signal conditioning now takes place inside the controller module. The various combinations are categorised as follows.

Status sensors produce a binary (on/off) output depending on whether the measured variable is above or below a set point. The sensors are normally mechanical devices, where a physical movement of the sensing element causes switch contacts to open or close. Typical devices are thermostats, humidistats and pressure switches. The output may be connected to a digital input of a controller for status reporting or software interlock purposes. Safety-critical interlocks are usually hardwired directly to the item of plant which is to be switched off.

Where the condition being monitored is critical, it is important that it be measured directly. For instance, where it is important to shut down part of a system in the event of no air flow, the air flow in the duct should be monitored by a pressure sensitive switch; if the flow is non-critical it may be satisfactory to simply monitor the presence of an electrical input to the fan motor. The general form of output from a status sensor is the provision of voltage-free contacts.

Analogue sensors convert the value of the measured variable into an electrical signal which is input to other devices for measurement and control purposes. Analogue sensors may be subdivided into:

— Passive devices which consist of the sensing element only and do not contain a transducer. All signal conditioning is carried out in the controller to which it is connected. Examples include resistance type temperature sensors. No power supplies are required and the sensor is connected via field wiring directly to an analogue input on a controller.

— Active devices which incorporate signal conditioning within the sensor device and include a transmitter which converts the measured value to an industry standard electrical signal for connection via field wiring to an analogue input on the controller. The output of the transmitter commonly takes one of several standard forms shown in Table 3.1. Signal transmission employing

Table 3.1 Standard signals for transmission of sensor readings

Signal	Application
0–10 V DC	Standard for HVAC applications
4–20 mA	Common in process control
Voltage-free contact	For status indication
Pulse	Energy and flow measurement

a 4–20 mA signal requires only a two-wire connection and is suitable for use in hostile environments. It is commonly used in process control applications. A 0–10 V transducer requires both a power supply and a signal output, needing a three or four-wire connection. The lower cost of 0–10V systems leads to its widespread application in HVAC systems.

Intelligent sensors contain a microprocessor which converts the measured value or status of the measured variable into a digitally encoded signal for direct communication over a network for onward transmission to other intelligent devices for control and measurement purposes. In addition, an intelligent sensor may carry out additional data processing before transmitting the value, such as:

- checking upper and lower bounds
- calibration and compensation functions
- calculating derived values, e.g. enthalpy.

3.1.2 Selection requirements

The correct selection and location of sensors is vital to achieving the required performance from any control system. Sensor problems are the most frequent cause of control system malfunctions. A poor-quality sensor may suffer from drift or early failure, resulting in poor control and high maintenance costs.

3.1.2.1 General requirements

The requirements that must be borne in mind when selecting any type of sensor are set out in Table 3.2.

Sensor inaccuracy or failure is a major cause of problems in building management systems. Experience shows that the problems are usually caused by incorrect installation, rather than an intrinsic problem with the sensor itself.

3.1.2.2 Accuracy

The claimed accuracy for a sensor does not guarantee that the same accuracy will be achieved at the controller or BMS supervisor, nor that it will be maintained over the operating life of the sensor. The accuracy of the overall measurement system depends on several factors:

- *Accuracy of the sensing element*. The claimed accuracy of the element may not be available over the whole operating range or may be quoted under ideal conditions.
- *Sensitivity*. This is the smallest change in the measured variable that can be detected by the system.

Table 3.2 Sensor requirements

Sensor requirement	Checklist
Type	Status, analogue, intelligent
Sensed medium	Air, water, gas, oil
Sensed quantity	Temperature, pressure, velocity, humidity
Location	Space, duct, pipe
Housing	Accessibility, effect on accuracy and speed
Accuracy	Accuracy, resolution, hysteresis, repeatability
Operating range	The range over which the sensor performs accurately
Overload range	The range to which the sensor may be subjected without damage
Response time	Affected by sensor, housing and medium
Protection	Is protection required from a damaging environment?
Maintenance	Calibration requirements, ease of servicing and replacement
Interchangeability	Can sensor be replaced by another from the same or different manufacturer?
Cost	Initial cost and total ownership cost over life cycle

- *Interacting variables*. The condition of the sensor may be affected by other environmental variables, e.g. an air temperature sensor will be affected by thermal radiation or an RH sensor by local variations in air temperature.
- *Stability*. Sensors may drift with time and require checking. Stability is likely to be affected by operating conditions.
- *Hysteresis*. The sensor reading may be affected by its past history and speed and direction of change of the measured variable.
- *Mounting*. The mounting and location of the sensor will affect the reading.
- *Signal conditioning*. Associated transducers will introduce their own limitations to the accuracy achievable. Some systems 'filter' readings first and only transmit when the measured variable has changed by a specified 'filter factor'. This is used to minimise network traffic.
- *A/D conversion*. The discrimination of any analogue to digital conversion will set a limit to the achievable accuracy. Eight-bit conversion divides the range into 256 steps, 12-bit into 4096 steps. In the latter case, a measurement range of –50 to 150°C would have a step size of 0.05K.

3.1.2.3 Speed of response

Sensors need to respond sufficiently fast to changes in the measured variable so that stable and accurate control can be maintained. The speed of response is characterised by the time constant τ, which is the time taken for the signal output to change by 63% of the final change. The time constant of a sensor in practice includes the effects of its housing, the manner of mounting and the nature of the medium being measured. There may be additional delays introduced by the measurement system; the scan rate of the controller limits the speed of response of the system to a change in the measured variable. Increasing the relative speed of the fluid flowing past the sensor reduces the time constant, up to a speed of about 0.3 m/s for water and 2 m/s for air. Above these speeds there is little reduction in time constant. For accurate and rapid measurement it is possible to measure air temperature with an aspirated thermometer, where air is drawn past the sensing element by a small fan.

The time constant of the sensor should be considered in relation to the rest of the controlled system. Too low a time constant may give problems if short term fluctuations in the measured variable give rise to unwanted control action. This may be dealt with in the controller software, typically by incorporating an averaging function to extend the time constant. Too high a time constant may mean that the control system will respond too slowly to changes in the controlled variable; this cannot satisfactorily be compensated by the control software. The time constant of the sensor is only one part of the time lags involved in a heating and control system; the general problem will be discussed below.

3.1.3 Types of sensor

The following deals with the selection of sensors which are used in control systems. It does not cover more specialised sensors which are used in safety systems, such as gas or refrigerant leak detection. Nor does it include safety sensors which are normally included in plant, such as the fusible links used to detect over-temperature above boilers. Energy metering is covered in 7.1.3.1 and the use of occupancy detectors in lighting control systems is dealt with in 5.15.1.3.

3.1.3.1 Temperature sensors

Temperature is the most widely measured variable in HVAC applications and several types of sensor are available. Mechanical devices are in general cheap and reliable. The output is a physical displacement which is used to directly operate a switch; the most common example is the domestic room thermostat. The set point may be varied by physical adjustment. They are not used in automatic control systems, except for limit switches. The most widely used sensing elements are listed in Table 3.3; elements used primarily for specialist or technical purposes have not been included.

The specification of a temperature sensor reflects its application. An accuracy of 0.6 K achieved over a 15 to 25°C range is suitable for zone air temperature measurement, while an accuracy of 0.25 K is needed for control of chilled water temperature. There is little point in specifying greater accuracy than is realistically required; reliability is often more important in practice.

3.1.3.2 Humidity sensors

The measurement of humidity is important in air conditioning control, but presents practical problems of drift and contamination. Simple mechanical sensors are available based on the expansion of hair or a nylon film with increasing atmospheric relative humidity. These elements can be incorporated in humidistats, comparable in operation and application to mechanical thermostats. They are generally reliable and resistant to contamination, but have low accuracy, which worsens outside the range of 20 to 80% RH. While many types of humidity sensor exist, two types have come to dominate in HVAC applications (Table 3.4); the traditional wet and dry bulb psychrometer is not included as it is no longer used in automatic control applications. The capacitative polymer film sensor provides a direct measure of relative humidity and is in widespread use. The sensing element is normally protected by a membrane or netting filter; a sintered metal filter permits the use of an RH probe in air speeds of up to 40 m/s. Where an accurate measurement of dewpoint is required, particularly at low humidities or low temperature, the chilled mirror dewpoint sensor provides the most accurate measurement[1,2]. This detects the formation of condensation droplets on a small mirror surface which is cooled by the Peltier effect; the temperature at which condensation is detected provides a direct measure of dewpoint. A recent development detects the onset of condensation using electromagnetic means.

Table 3.3 Temperature sensors

Sensor	Element	Description	Types	Advantages	Disadvantages	Applications
Platinum resistance thermometer (PRT)	Thin metal film on substrate	Electrical resistance increases as a linear function of temperature. Three- or four-wire connection required	Pt100 and Pt1000, available as passive sensors or with inbuilt transmitter	Industry standard. Interchangeable elements, high accuracy, reliable, stable	Cost. Lead wire resistance and self-heating can introduce errors	Widespread
Thermistor	Semiconducting metallic oxides in bead, rod or disc format	Electrical resistance changes with temperature; highly non-linear	Various standards exist; some are supplier specific. Usually negative temperature coefficient (NTC)	Cheap, high sensitivity, small size, wide range available	Accuracy and drift. Short life unless protected against moisture. Signal conditioning required; may not be interchangeable	Widespread. Differential temperature
Nickel resistance thermometer	Wound coil of nickel or nickel alloy wire	Near linear change in resistance with temperature. Higher coefficient than platinum	Standards exist for Ni100 and Ni1000 elements. Alloy elements may be supplier specific. Available in special formats, e.g. duct averaging	Moderate to good accuracy. Field wiring has little effect on accuracy	May need signal conditioning. May not be interchangeable	Widespread

Table 3.4 Humidity sensors

Type	Element	Description	Types	Advantages	Disadvantages	Applications
Capacitive sensor	Thin polymer film acts as dielectric	Film expands with increasing RH, changing capacitance of sensor. Dedicated signal conditioning gives linear output with RH	Supplier specific, range of mountings available	Accurate, reliable and stable	High cost. Drift, particularly if exposed to high RH. Annual recalibration advised	Widespread
Dewpoint	Chilled mirror	Detects dew formation on mirror chilled by Peltier effect. Measurement head incorporates light source, detector and chilled mirror. Sophisticated control gear required	Supplier specific	Direct measurement of dewpoint, high accuracy	Size, expense. Maintenance requirements	High accuracy control

Where RH is measured and a measure of absolute humidity or enthalpy is required, it is necessary to take a simultaneous measurement of air temperature. The air temperature sensor should experience the same environment as the humidity sensor and ideally should be enclosed in the same housing. Sensors are available which incorporate both temperature and RH measurement within the same head and may have enthalpy and dewpoint output options, in addition to the direct measurement of temperature and RH. Measurement of humidity is subject to many uncertainties and it is unreasonable to expect an accuracy in practice of better than 2.5%. Further information on a wide range of sensor types, their errors and calibration methods are given in a guide published by the Institute of Measurement and Control[(3)].

In some applications, such as the control of chilled ceilings, it is necessary to anticipate the formation of condensation on a chilled surface. Two major types of sensor are used. One is a dedicated form of humidity meter. The sensing element is contained in a small metal housing which is held in good thermal contact with the cold surface. The attached control unit provides a digital signal output when the RH at the sensor reaches a preset level of 97% and so anticipates the formation of condensation. The other type detects the initial formation of condensation by monitoring the electrical resistance across an array of conducting tracks on a substrate held in good thermal contact with the cold surface.

3.1.3.3 Pressure sensors

Pressure is the most commonly measured quantity after temperature. Most pressure sensors consist of a diaphragm or bellows, which moves under the influence of the pressure applied across it. The movement is a function of the applied pressure and the measurement problem is fundamentally a measure of displacement.

Pressure may measured relative to atmospheric pressure, when it is termed gauge pressure, or else as a differential pressure across two points in a system. Building static pressure may vary appreciably over a building and the reference atmospheric pressure against which gauge pressure is measured must be carefully considered. Pressure within a building is affected by mechanical ventilation, door and window opening and the stack effect. An outside static pressure sensor should be mounted in free air at least 2 m above a surface and fitted with a wind shield. The bellows or diaphragm construction depends more on the working pressure than on the medium and most pressure sensors may be used with air, water or other media. The types of pressure sensor listed in Table 3-5 are classified by

Table 3.5 Pressure sensors, classified by transducer

Sensor	Element	Description	Advantages	Disadvantages	Applications
Capacitive	Diaphragm or bellows with capacitive transducer	Movement of bellows alters capacity	Low cost	Low output, signal conditioning required	Low pressure air. Duct static or filter differential pressure
Inductive	Diaphragm or bellows with inductive transducer	Movement of bellows operates a linear variable differential transformer	Rugged construction	Expensive; temperature compensation may be difficult	As for capacitative
Strain gauge	Diaphragm or bellows with strain gauge transducer	Strain gauge bonded to pressure sensing element	Rugged, linear output	Low output signal	High pressure, chilled water, steam
Potentiometer	Diaphragm or bellows connected directly to variable resistor	Movement of pressure sensing element operates a variable resistor	Inexpensive, high output	Low accuracy, large size, wear may shorten life	

transducer, rather than the pressure sensing element. Where fluctuating pressures or vibration are to be measured, use may be made of piezoelectric transducers, which can respond to sonic frequencies. A wide range of mechanical pressure switches is also available, which operate as status sensors and are used where safety interlocks are required, e.g. to detect fan operation.

3.1.3.4 Velocity and flow

Strictly, velocity implies both speed and direction. In practice, the direction of flow is determined by the duct or pipe and a speed measurement is all that is required. The flow in a duct may be measured directly or derived from a velocity measurement and knowledge of the cross-sectional area of the duct; corrections for temperature and pressure may be required. For well-developed flow in a straight pipe, it is possible to assume the velocity distribution over the cross-section of the duct. This requirement is generally satisfied by positioning the sensor so that there is a distance between the sensor and any disturbance of 10 pipe diameters before the sensor and five after. If this is not possible, multiple measurements will be required to provide some form of averaging. Several methods of velocity and flow measurement are available (Table 3.6). For control purposes it may be possible to monitor differential pressure across a fixed restriction, such as a heating coil or louvre; devices which have a variable resistance, such as filters or cooling coils, must not be used for this purpose.

3.1.3.5 Air quality

The use of demand-controlled ventilation (DCV) requires a measure of the condition of the indoor air. Carbon dioxide concentration is becoming the most generally accepted measure (Table 3.7). While it does not give a complete indication of the various contaminants, it gives an overall measure of the relation between occupancy and ventilation. Multi-gas sensors, based on a tin oxide layer, respond to a number of possible contaminants, but the integrated response does not necessarily reflect the air quality as perceived by occupants. They are therefore not generally recommended for DCV.[4,5]. Multi-gas sensors have applications in areas where there is a low concentration of people, but smells are a potential problem, for example restaurants and bars. Sensors sensitive to particular gases are required for specialised ventilation control applications, such as car parks. Indirect methods, such as counting the total building occupancy or predicting it from a work schedule, may be used instead of air quality measurement for ventilation control.

Table 3.6 Velocity and flow sensors

Sensor	Element	Description	Types	Advantages	Disadvantages	Applications
Pitot tube	Tube with opening at upstream end	The stagnation pressure of the air brought to rest in the tube is proportional to the square of the air speed. The difference between stagnation and static pressures is measured by a suitable pressure gauge	Can be obtained in the form of arrays which perform an averaging function over the area of the duct, e.g. Wilson grid	Inexpensive, but duct averaging sensor may be expensive	Requires unidirectional air flow above 3 m/s. Can become blocked with dirt	Air speed when direction is determined. Air flow rate for VAV units
Hot wire anemometer	Heated fine wire or thermistor bead	The sensor element is heated by an electric current to a constant temperature. The current is a function of the fluid speed	Available as grid for measuring flow rate in duct	Measures mass flow, resistant to contamination, short time constant. Can measure low flow rates	Fragile, expensive	Air flow. Available in arrays. Use thermistor type for non-directional flow
Orifice plate	A plate with a precisely defined hole is inserted into the duct	The pressure drop across the plate is proportional to the square of the velocity. Transducer may sense mass or volume flow rate	Calibrated plates with integral pressure tapping are available. Venturi devices operate on a similar principle; they have a lower pressure loss but are bulkier	Robust, no moving parts	Erosion of orifice will affect accuracy. Has to be fitted in straight section. Not suitable for low flows. High pressure loss	Can be used for air, water or steam
Turbine flow meter	Bladed turbine immersed in the fluid	The speed of rotation is proportional to flow. A magnetic detector produces a pulse train with frequency proportional to rotor speed	May be full bore, mounted between flanges, or a small insertion device mounted in the centre of the pipe	Moderate cost. Good accuracy. Wide range of velocities and low pressure loss	Susceptible to damage when used with water	Metering of water and natural gas. Heat metering

Table 3.7 Gas sensors

Sensor	Element	Description	Types	Advantages	Disadvantages	Applications
CO_2 sensor	NDIR optical cell (non-dispersive infra-red)	Measures the absorption of infra-red radiation by CO_2 in air	Wall-mounted housing with CO_2-permeable membrane for DCV applications	Small size, insensitive to humidity or other vapours	Cost	Demand-controlled ventilation for occupied spaces
Multi-gas sensor	Tin oxide film	Detects oxidation on a heated tin dioxide surface. Sensitive to heavy odours, smoke and solvent gases	Various wall and duct housings available	Sensitive to wide range of contaminants	Relative rather than absolute measure of contamination	Multi-contaminant situations. Carbon dioxide control preferred for occupied spaces
Specified pollutant	Non-dispersive infra-red	Uses IR absorption Sensor is specific to pollutant	Available for several gases	High accuracy and stability	High cost	Car park ventilation control (carbon monoxide and nitrogen dioxide); swimming pools (chlorine); air pollution monitoring (sulphur dioxide)
Obscuration	Optical cell	Detects absorption of light beam by suspended particles		High accuracy and stability	High cost	Roadway and tunnel ventilation control

3.1.4 Sensor mounting

A sensor can only measure the state of its own sensing element. Care must be taken to ensure that this is representative of the physical quantity being measured. The following general guidelines should be borne in mind:

- Ensure that the sensor is in a position representative of the variable to be measured.
- Take care that temperature and humidity sensors are screened from direct radiation, particularly sunlight.
- Take account of the active and inactive sections of the sensor probe.
- Ensure adequate immersion of the sensor in the medium. Heat conduction along a sheath can affect readings.
- Provide a tight sealing hole adjacent to every duct mounted sensor for the insertion of a test sensor.
- For pipe sensors use a separable pocket. Ensure that the sensor is a close fit in the pocket and use a thermal conductive paste.
- Use flexible cable. A rigid cable could disturb the position of the sensor.
- Sensor cables should be provided with a 'drip loop' to prevent water running into the sensor housing.
- Allow sufficient spare cable to allow for removal.
- Record the location of all concealed sensors.
- Fix a labelling plate next to every sensor, not on the device itself.
- Allow for stratification and use more than one sensor or an averaging sensor if necessary.
- Position duct temperature sensors downstream of supply fans.

A booklet on control sensor installation published by the Buildings Controls Group[6] gives excellent practical advice on the choice and installation of sensors.

3.1.5 Calibration

Sensors should be routinely checked and recalibrated. The frequency of checking depends on the nature of the sensor; it will be greatest for sensors in challenging environments and those subject to drift. Exposure to extreme conditions may affect calibration; this is a problem with RH sensors exposed to high humidities.

Sensor calibration may be performed by using the sensor to measure a standard condition, or to compare the sensor reading with that of a standard instrument. The former is more common in the laboratory that in the field. Standard relative humidities may be produced using saturated solutions of salts and it is possible to obtain a set of calibration vessels into which humidity probes can be inserted for checking. It is more common to check sensors by comparison against a reference standard. This should be a good quality instrument reserved for this purpose which has itself a certificate to prove that its calibration is traceable to a national standard[7]. The accreditation of calibration and testing laboratories in the UK is carried out by the United Kingdom Accreditation Service, which incorporates NAMAS (National Accreditation of Measurement and Sampling). Care must be taken to ensure that the sensor under test and the reference standard are in fact experiencing the same environment. Provision of test mounting points adjacent to pipe- and duct-mounted sensors is good practice and enables a rapid check to be carried out.

3.2 Actuators

An actuator responds to the output signal from a controller and provides the mechanical action to operate the final control device, which is typically a valve or damper. A wide range of actuators is available and the chosen actuator must:

— match the mechanical requirements of the controlled device

— match the characteristics of the control system, especially the output signal of the controller

— be suitable for its operating environment.

The major mechanical division is between linear and rotary actuators. Linear actuators are required for use with lift and lay valves. Rotary actuators tend to be used on shoe valves and butterfly valves. Actuator and valve are selected to be dimensionally compatible, so that the actuator spindle connects directly to the valve stem; many manufacturers provide matched combinations. Linear actuators may be used to operate dampers using a linkage to provide the required rotary movement. Movement is provided by an electric motor operating through a reduction gear train, producing relatively high torque at low speed. Where specified, a spring return mechanism is incorporated, which is wound as the motor operates. In the event of power failure, the actuator is returned to the specified position. It is now possible to obtain microprocessor-controlled devices for particular applications, e.g. a unit which combines differential pressure sensor, controller and damper actuator for the control of VAV boxes.

The actuator must be suitable for its operating environment. The manufacturer's data gives safe operating ranges of temperature and humidity and often the IP protection number (see Table 3.13). Where the operating environment is likely to be corrosive, as in a swimming pool or industrial setting, advice should be obtained from the manufacturer.

A range of modulating magnetic valves is obtainable, where a movable magnetic core moves in a solenoid and transfers the linear stroke directly to the integral valve. The valves are reliable and fast, capable of precise regulation, resulting in a high rangeability, coupled with a fast response time of only 1 s. The solenoid itself is operated by a 0–20 V phase cut signal, but in-built electronics allow operation from a standard 0–10 V DC signal input. The high rangeability and precise control makes these valves suitable for demanding situations and installations where the valve may be required to operate at low authority, requiring stable control at low valve openings.

Thermic actuators contain a solid expansion medium which is heated by an electrical resistance heater. When power is applied to the actuator, the heating effect causes the medium to expand, producing a linear motion of the spindle. On cooling, the spindle retracts by pressure of a built-in spring. Thermic actuators are robust and silent. Full stroke is about 3 mm and stroke time is a few minutes. Modulating control can be achieved with pulse width modulation; some models use a standard 0–10 V signal. Models are available designed to operate radiator valves.

Some gas valve applications use electro-hydraulic actuators, where a motor pumps hydraulic oil against a rolling diaphragm and piston. The piston then drives open the valve against the force of a spring. Once the valve is fully open, an internal relief valve is closed, the pump is switched off and the valve remains open without further application of power. To close the valve, the relief valve is opened and the spring forces the oil back into the reservoir.

3.2.1 Mechanical characteristics

A linear actuator requires the following quantities to be specified:

— *Thrust*. The thrust is measured in newtons and must be sufficient to close the valve against the differential pressure acting across the valve plug. In some hydraulic systems employing two-port valves, closing forces may be considerable, up to the full pump head. Manufacturers supply tables of maximum differential pressures for valve/actuator combinations.

— *Stroke*. The stroke of the actuator spindle must be sufficient to provide full operation of the valve. The stroke is normally adjustable to suit the valve; in many actuators this is self-set automatically.

— *Running time or stroke time*. This is the time taken for the actuator to travel over the full operating range. The low gearing of the electric motor used to drive the actuator spindle produces low operating speeds and the actuator operates at several seconds per mm travel, giving a stroke time of the order of minutes. This speed of movement may be an integral part of the control loop and is required to be input to the controller during configuration. Depending on the type of actuator motor and its control, the running time may be invariant or else dependent on the load.

Rotary actuators are used to operate dampers and butterfly or ball valves. The actuator produces a rotary movement which is coupled directly to the control element. The actuator has the following specification:

— *Torque*. The operating force is given as the torque in N m. Damper manufacturers supply tables of the required torque for their dampers as a function of damper area and air velocity in the duct.

— *Angle of rotation*. Rotary actuators typically have a maximum angle of rotation of just over a right angle. This angle may be adjusted to suit the required rotation of the valve or damper.

— *Running time*. As for linear actuators.

3.2.2 Electrical connections

An actuator requires some or all of the following connections:

— *Power to power the drive*. May be 230 or 240 V AC, 24 V AC, 24 V DC; 230 V is becoming an EU standard.

— *Control signal from the controller*. 0–10 V and 2–10 V DC are the most common. 4–20 mA is used in process control applications. 0–20 V phase cut may be used for magnetic actuators. A 24 V AC pulsed is also found.

— *135 Ω potentiometer*. Used by the Honeywell 90 series for both control input and positional feedback.

— *Position indication.* A 2–10 V DC signal to provide positional feedback or indication. Position feedback may also be provided by a potentiometer. End switches are used to provide positive open/close position indication.

The actuator may be fully modulating, where the position of the actuator is proportional to the control signal, or tristate, used for floating control, where the motor may be driven in either direction or stopped. Many actuators exhibit some degree of hysteresis and the relation between control signal and actuator position depends on the direction of travel. Most actuators have the facility to provide a positional feedback signal, indicating the actual position of the actuator. This may be used to drive a remote position indicator, or be incorporated in a control loop to provide positive positioning of the actuator; this overcomes any problems caused by hysteresis. In some cases where positive positioning is required, the position indicator may be mounted on the controlled device itself, e.g. damper. End switches may also be used in simple sequencing applications to divert a signal to another device when the first device is fully closed. The actuator may be configured so that either end of the travel is corresponds to a zero control signal. The spring return which operates on power failure may return the actuator to either the spindle extended or spindle retracted position. All actuators are fitted with a manual override, to allow for manual positioning of the actuator and control device during commissioning or maintenance.

3.2.3 Pneumatic actuators

Pneumatic actuators comprise a piston or diaphragm to which air pressure is applied to provide a linear displacement. A mechanical linkage is required where it is desired to produce a rotary movement, e.g. for damper control. The construction of the actuator and its method of connection to the valve or damper determines the direction of operation, i.e. whether increasing control air pressure results in opening or closing of the valve or damper. Most pneumatic actuators are of the single-action type where the force on the diaphragm is opposed by a spring and the net force applied to the valve or damper is the difference between them. When the air pressure is removed the spring will return the valve to the selected extreme position and this may be used for fail-safe requirements.

When a straightforward pneumatic actuator operates against large or variable forces the position of the actuator spindle may not be proportional to the signal pressure from the controller. To overcome this a pneumatic relay, known as a positioner, is fitted to the actuator; this uses main air line pressure to power the actuator and provide the operating force, but incorporates an all-pneumatic control to ensure that the actuator is correctly positioned for a given signal pressure. To do this a positive feedback of actuator position is used.

Pneumatic controllers provide reliable and fast operation and are still used extensively in the HVAC industry. For new installations they have been largely supplanted by DDC and pneumatic systems are now installed only in special situations. The necessity to provide a clean dry air supply can lead to maintenance problems. Where an existing pneumatic control system is being upgraded to electronic DDC control, it is possible to retain pneumatic operation of the actuators by using hybrid electro-pneumatic transducers which use pneumatic power to provide the operating force, but whose position is controlled by a standard electronic signal.

3.3 Valves

3.3.1 Hydraulic flow

Fluid temperatures used in commercial buildings are classified into five broad ranges (Table 3.8).

In SI units, pressure is measured in pascals. The bar is commonly used, where

1 bar = 100 kPa

It is sometimes helpful to visualise pressure in terms of head of water:

1 bar = 10.5 m head of water

Flow is properly measured in m^3/s in SI units. This results in small values which are awkward to handle and it is common to quote flow rates in litres per second (l/s) or m^3/h, where

$1\ l/s = 3.6\ m^3/h$

A control valve is subject to three types of pressure:

— *Static pressure.* This is the maximum internal pressure of the hydraulic system and for open systems will usually be governed by the height of the cold water tank in the roof above the lowest point of the system. The body of the valve and the gland through which the spindle moves must be able to withstand the maximum pressure of the system, which is equal to the static head of the system plus the pump pressure.

— *Differential pressure.* Valves should always be installed so that the plug closes against the flow; installation in the wrong direction could result in unstable operation with a danger of the valve snapping shut. The pressure produced by the pump in a circuit has to be capable of meeting the pressure drops at design flow rate produced by the resistance of all fittings, pipes, boilers valves etc. However, in some circuits using two-port valves, it is possible for a large differential pressure to be developed across a single valve under part load conditions; three-port valves nominally have the same pressure drop for all positions of the valve. The actuator must therefore be capable of driving the valve

Table 3.8 Working fluid temperature ranges

Classification	Temperature range (°C)
Chilled water (CW)	5–10
Low temperature hot water (LTHW)	20–90
Medium temperature hot water (MTHW)	95–120
High temperature hot water (HTHW)	120–200
Steam	< 16 bar

against a differential pressure equal to the maximum pump head. By the same token, the maximum differential pressure rating (MDP) of the valve must be greater than the maximum pump head.

— *Pressure drop*. There is a pressure drop across valve in normal operation. This is an important part of valve selection and is discussed below under valve authority.

3.3.2 Types of valve

Valves may be classified by their function into:

— *Regulating*: a valve which is adjusted during commissioning to provide a fixed resistance in a fluid circuit to give the design flow rate and ensure a correct balance. The adjustment is normally manual. A double regulating valve is provided with indicated positions of the valve opening and has an adjustable stop, so that the valve can be closed to isolate a circuit and reopened to the previously set position.

— *Commissioning*: a regulating valve which incorporates a calibrated orifice and pressure tappings for flow measurement during commissioning.

— *Flow limiting* (also known as automatic or dynamic balancing valves): valves which maintain a constant flow independently of the differential pressure across them. This constant flow is maintained over the regulation range of differential pressure, which is typically 10:1 or more. Valves are available with preset control cartridges for a predetermined flow rate, or can be externally adjustable for commissioning.

— *Differential pressure control*: a valve which automatically adjusts to maintain a constant differential pressure across part of a circuit.

— *Modulating*: a valve which is adjusted by the control system via an actuator to regulate hydraulic flow.

— *Safety shut-off*: a valve that is designed to close under spring pressure on power failure, critical condition (e.g. fire) or out-of-range process variable.

Modulating valves form the main subject of interest in this Guide. The complete valve assembly consists of:

— actuator, which converts the output signal from the controller into mechanical movement

— spindle, which transmits the actuator movement to the moving part of the valve

— valve body, containing the flow control device.

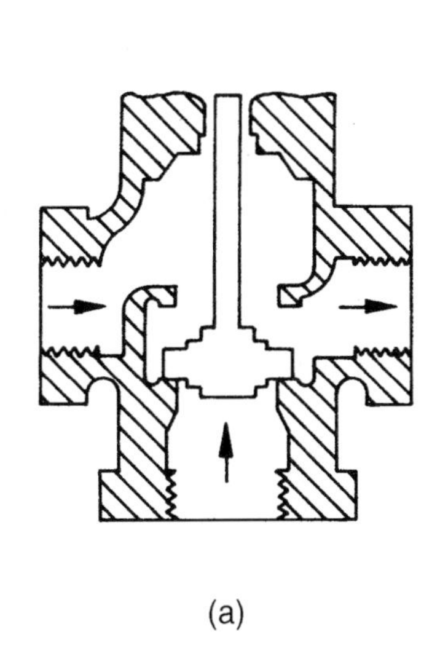

(a)

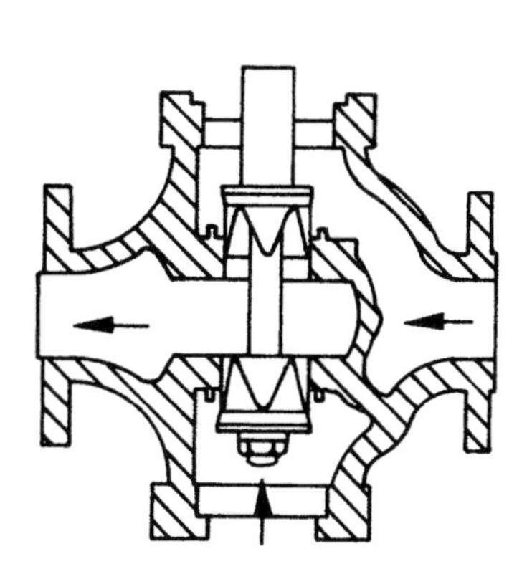

(b)

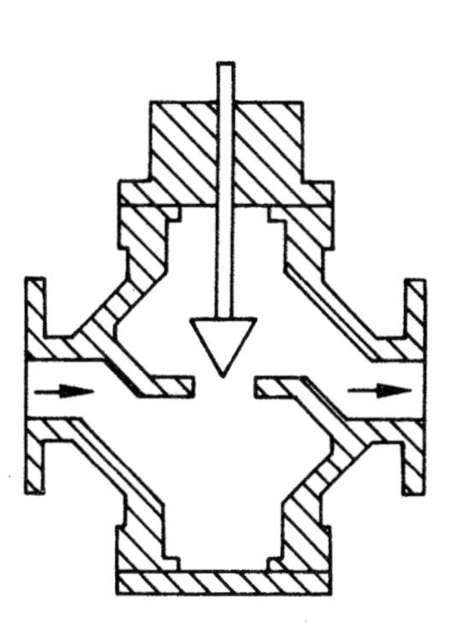

(c)

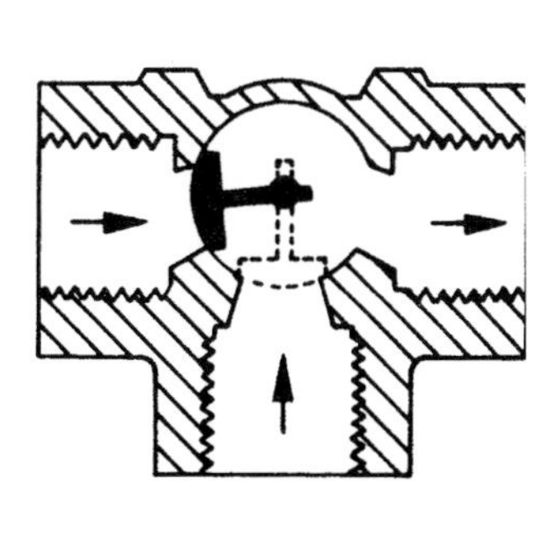

(d)

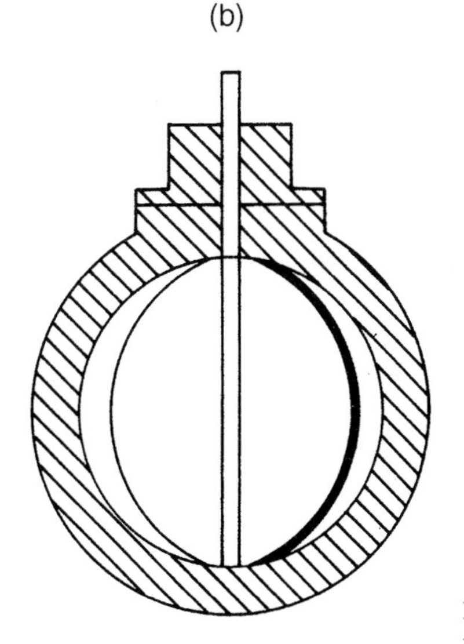

(e)

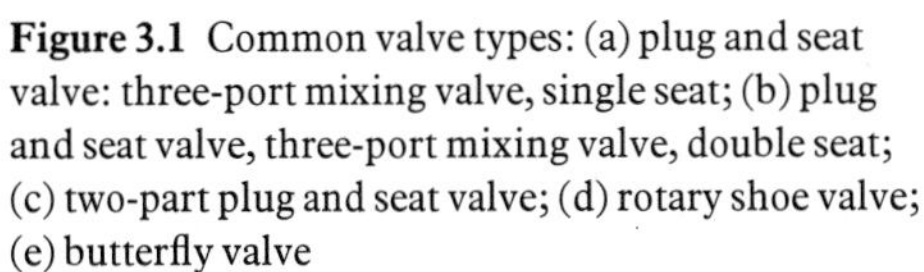

Figure 3.1 Common valve types: (a) plug and seat valve: three-port mixing valve, single seat; (b) plug and seat valve, three-port mixing valve, double seat; (c) two-part plug and seat valve; (d) rotary shoe valve; (e) butterfly valve

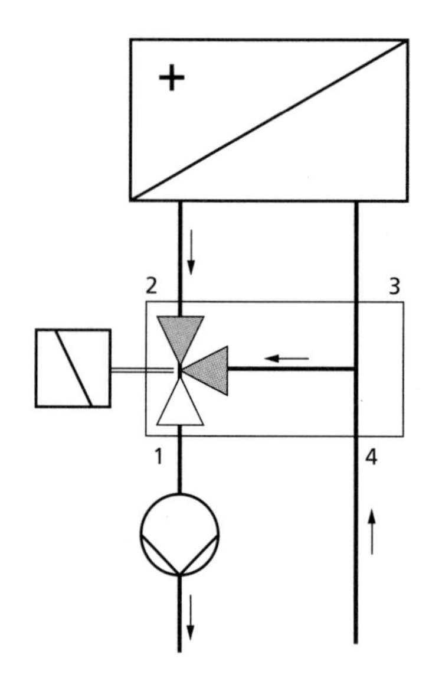

Figure 3.2 Four-port valve. A four-port valve incorporates the bypass and tee connections in the valve body giving compact installation

Several types of valve body are used and are illustrated in Figure 3.1:

— The butterfly valve is the simplest form of valve. The internal disc is aligned either in line or across the flow and is used for on/off control.

— Rotary shoe valves are used commonly on chilled and LPHW systems because of their compact size. In action, a rotating shoe is moved over the ports by a rotary actuator and provides modulating regulation, which may be characterised.

— Plug and seat valves are the most common and are available in a wide range of sizes and can be used for all heat transfer media. The valve plug is moved on and off the valve seat by linear movement of the spindle. The spindle moves through a gland, which must allow free movement and yet not leak under the static pressure of the system. The shape of the valve and seat determine the characteristic of the valve. They are also known as lift and lay or globe valves.

Valves are available in two-, three- or four-port configurations. Three-port valves have found widespread application in HVAC systems. Used with constant-speed circulation pumps they offer a range of well-established and trouble-free control solutions. A four-port valve performs the control function of a three-port valve but includes an internal bypass to produce compact connections; see Figure 3.2.

A three-port valve is provided with two inlet ports and one outlet port, when it is described as a mixing valve, or, less commonly, with one inlet and two outlet ports, when it is described as a diverting valve. Both mixing and diverting valves may be incorporated in a circuit for either mixing or diverting application. These terms may cause confusion and Figure 3.3 illustrates the meaning of the terms. In general, three-port valves available in the building services industry are designed to be used as mixing valves and are not suitable for use as diverting valves. Where a diverting valve is required, i.e. one inlet and two outlets, a double-seated or shoe type valve should be used to avoid the out-of-balance forces which would occur with a single-seated globe valve. Connecting a mixing valve as a diverting valve is likely to lead to unstable operation, since water pressure acts to slam the plug against the valve seat. Mixing applications are typically constant-flow, variable-temperature loads, such as a compensated heating circuit; diverting applications are typically variable-flow, constant-temperature circuits, such as heater batteries. Correct connection of a valve is important and may be crucial. There is not a consistent terminology for identifying valve ports. Figure 3.4 illustrates the more common conventions. This Guide refers to the ports as control, bypass and common and describes a three-way valve as open when the flow is between control and common ports, with the bypass shut. This applies to both mixing and diverting valves.

Two-port valves are used to throttle flow in a circuit. In a complex circuit the independent modulation of two-port valves can produce balancing problems and pressure variations throughout the system. However, when used with a variable-speed pump and with proper attention to system design, two-port valves offer reductions in both first cost and the cost of pumping. Suitable hydraulic control circuits are given in section 5.

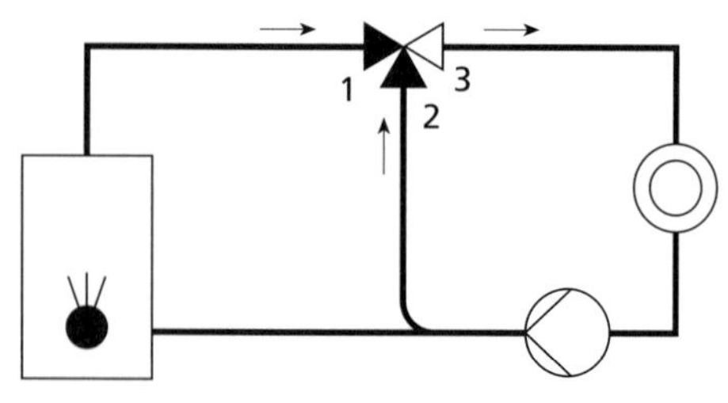

Mixing valve, mixing application

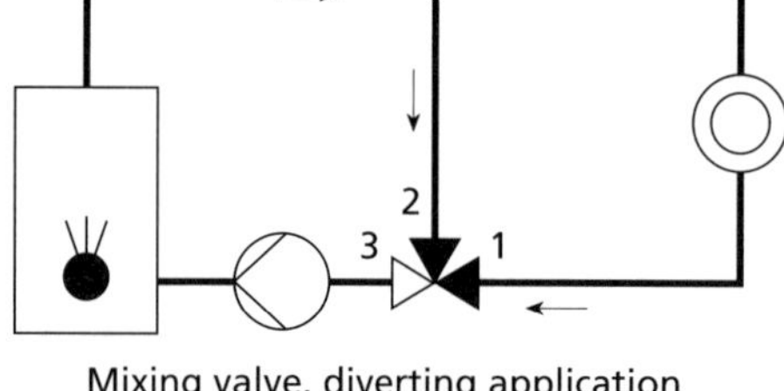

Mixing valve, diverting application

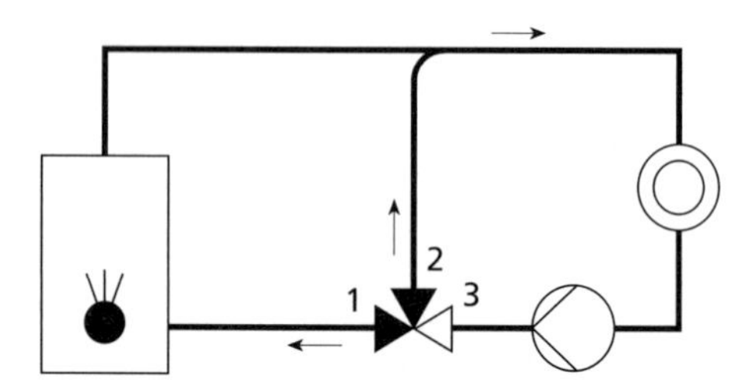

Diverting valve, mixing application (not recommended)

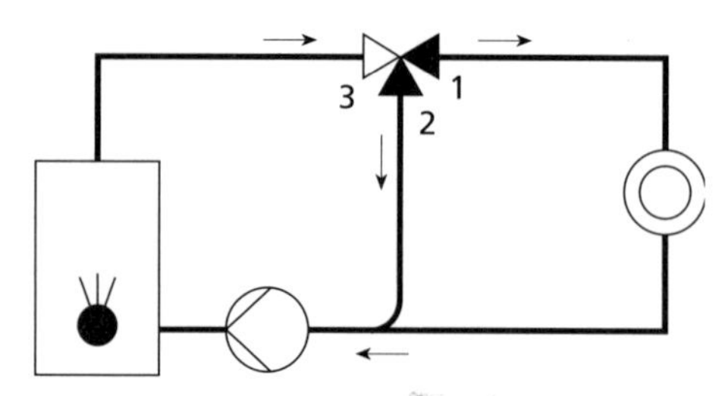

Diverting valve, diverting application (not recommended)

Figure 3.3 Three-port valves in mixing and diverting applications

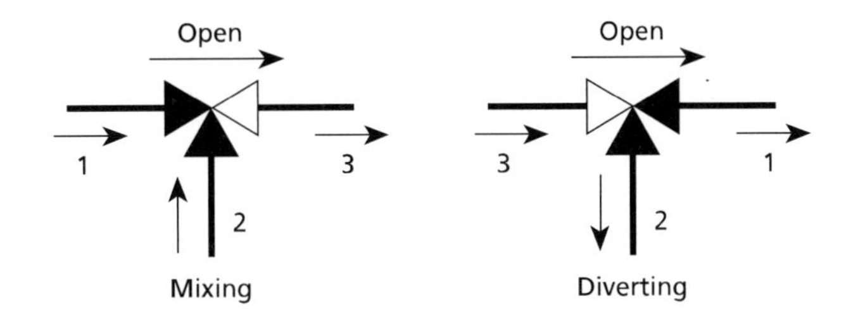

Port	Name	Alternative	USA	Europe
1	Control	Load	A	E
2	Bypass	Bypass	B	L
3	Common	Common	AB	C

Figure 3.4 Three-port valve terminology, showing designation of ports

3.3.3 Valve design

3.3.3.1 Characteristic

A control valve is one link in a chain of control, running from the error signal which is input to the controller to the change in output of the emitter; an example is shown in Figure 2.1. While not essential, a linear relation between the change in output from the controller and change in heat output from the emitter makes for a well-behaved and stable system. For most heat emitters, there is a strongly non-linear relation between the flow rate of the heat transfer fluid and heat output from the emitter, with output rising rapidly as the flow increases from zero, then flattening out at higher flow rates. For example, heat output from a radiator rises rapidly as hot water is admitted at low flow rates, but once the body of the radiator is hot, increasing the water flow rate will produce little increase in heat output. This is illustrated in Figure 3.5; the curves for a heating or cooling coil are similar. A valve which produced a flow rate proportional to spindle lift would therefore result in a non-linear system, operating over a narrow range of spindle movement and possibly resulting in unstable operation.

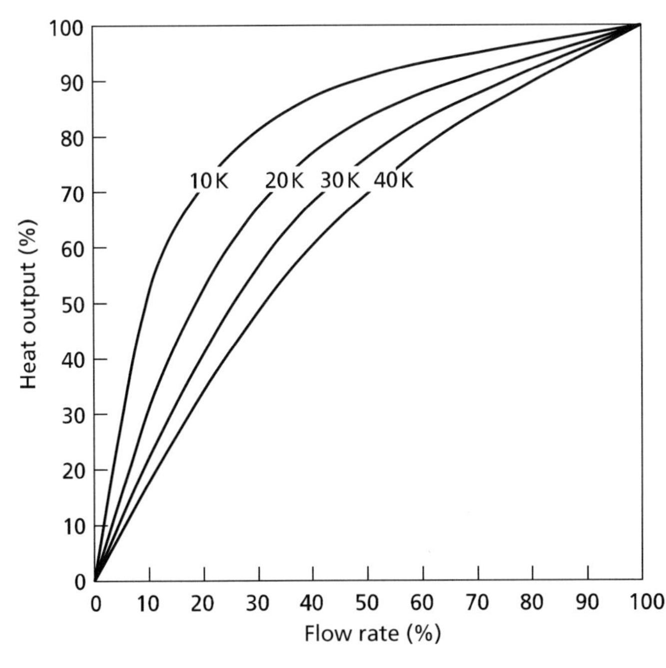

Figure 3.5 Radiator heat output as a function of flow rate, with temperature drop as parameter. Heating and cooling coils show a similar relationship

Valves are therefore often designed so that the resultant heat output from the emitter is approximately proportional to the spindle movement. This is done by shaping the plug so that the free area, and hence flow rate, is a defined function of spindle lift. This relationship is termed the valve characteristic. The most commonly encountered characteristics are:

— linear, where the orifice area is directly proportional to the valve spindle movement and the flow varies linearly with spindle lift

— characterised V-port, with a characteristic falling between linear and equal percentage

— equal percentage, or similar modified parabolic, where equal increments of valve spindle lift provide an equal percentage change of the area

— quick opening, where the flow increases very rapidly from zero for a small spindle movement, with a fairly linear relationship between flow and spindle movement. Such valves are used primarily for on/off service.

Figure 3.6 shows in stylised form the basic characteristic curves for the four types described above. The static characteristics are measured by measuring the flow as a function of valve position for a constant pressure drop across the valve. In a real circuit, the characteristic is modified by the authority of the valve; this is discussed further below.

Equal percentage valves have a theoretical characteristic which may be expressed as an equation relating flow through the valve to the degree of opening, measured at constant pressure drop. The degree of opening of a valve is variously referred to as the spindle lift, stem position, or percentage of full stroke:

$$Q = Q_0 \exp(Sn)$$

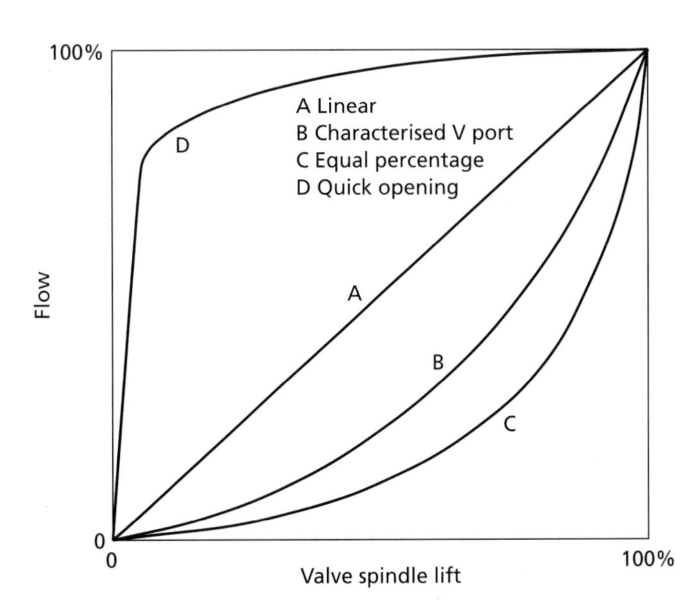

Figure 3.6 Stylised static valve characteristics

where

Q	flow through valve (nominal units)
Q_0	theoretical flow though valve at $S = 0$
n	valve sensitivity
S	spindle lift (1 = fully open).

The sensitivity n is the percentage change in flow through the valve produced by a 1% change in stem position. A typical value is about 4. The theoretical flow Q_0 though the valve at zero opening is a mathematical convenience and does not represent the actual flow when the valve is closed. Real equal percentage valves can be made to follow the theoretical characteristic reasonably well, but depart from it at low openings. A practical equal percentage valve has a characteristic of the form shown in Figure 3.7. Q_{min} represents the minimum flow at which the valve still provides reasonable control. Below this flow, the flow falls off rapidly and cannot be controlled reliably. Valves are normally designed to shut down rapidly below the minimum controllable flow. If the valve does not provide a tight shutoff, the residual flow is termed let-by. The ratio of the maximum controllable flow to the maximum flow is known as the rangeability R, sometimes referred to as the practical rangeability to distinguish it from the theoretical rangeability. A rangeability of 25 indicates a valve which will control according to its defined characteristic down to 4% of its maximum flow. The practical rangeability definition holds for all types of valve. For equal percentage valves it is possible to define an ideal rangeability

$$R_{ideal} = Q_{max}/Q_0 = \exp(n)$$

The ideal rangeability is directly related to the sensitivity.

Good rangeability is required when valves are required to control at low flow rates. This occurs in some circuits using two-port valves. Situations where valve authority is low will require operation at low spindle lifts. It is conventional to regard valves as having 'good' rangeability as follows:

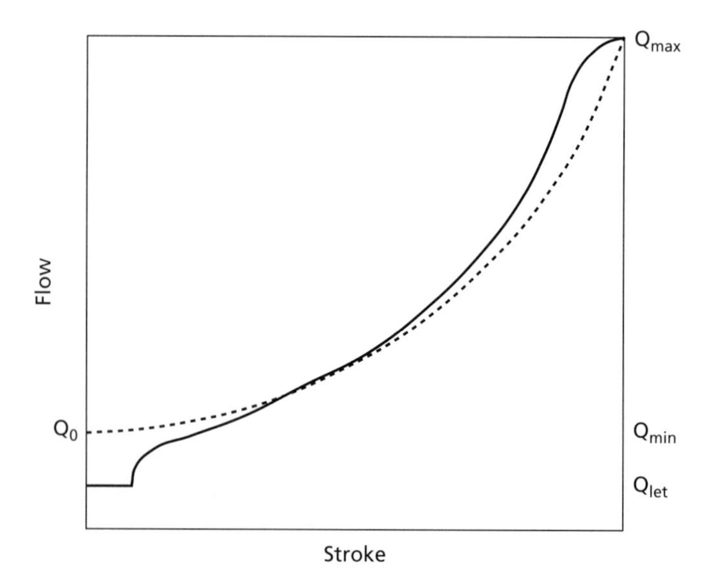

Figure 3.7 Valve rangeability. The dotted line shows a theoretical equal percentage valve characteristic. The solid line shows a practical valve, controlling over the range Q_{min} to Q_{max}. The practical rangeability $R = Q_{min}/Q_{max}$. The theoretical rangeability $R_{ideal} = Q_0/Q_{max}$. The residual flow Q_{let} is termed the let-by

— terminal unit valves: rangeability of 25:1
— screwed plant valves: rangeability of 50:1
— flanged plant valves: rangeability of 50:1
— magnetic solenoid valves: rangeability of 500:1.

Good control at low strokes requires an actuator that can position the valve accurately.

3.3.3.2 Flow coefficient

The size of a valve is described by the flow coefficient, which determines the rate of flow of fluid through the valve as a function of the pressure drop across it. The flow of fluid through a constriction is proportional to the square root of the pressure drop across the constriction. The flow coefficient of a valve is defined in SI units by the equation:

$$V_m = A_v \left(\frac{P_m}{\rho_{fm}}\right)^{0.5}$$

where:

V_m	volume flow rate (m³/s)
P_m	pressure drop across valve (Pa)
ρ_{fm}	fluid density (kg/m³)
A_v	flow coefficient of valve (m²).

Manufacturers often quote the capacity of a valve as the capacity index K_v. This is equivalent to the flow coefficient, but applied to water flow, with the flow rate measured in the so-called continental units of m³/h and pressure drop in bars. The flow equation becomes

$$V_m = K_v P_m^{0.5}$$

where:

V_m	volume flow rate of water (m³/h)
P_m	pressure drop across valve (bar)
K_v	flow coefficient of valve.

The older imperial unit C_v (equivalent to A_v) may also be met. The conversion factors and units are summarised in Table 3.9. It may be noted that 1 bar ≅ 1 kgf/cm². Manufacturers usually produce a range of valves with K_v values rising in geometrical progression, with each value of K_v 60% greater than the preceding one in the series.

3.3.3.3 Steam valves

Valves used for steam are a specialist application of two-port valves for compressible fluid flow. If a small constriction is introduced into a pipe-carrying steam, it has the effect of increasing the velocity of the steam passing through the constriction and reducing the steam pressure downstream of the constriction. The net result is that there is little effect on the overall flow of steam in the pipe. As the size of the opening is decreased, this effect is maintained until a maximum critical steam velocity is reached, when, for dry steam, the downstream absolute pressure is 58% of the inlet pressure. Further reduction of the area of the constriction results in reduced flow of steam. Thus, in sizing steam valves, the valve pressure drop may be assumed to be 40% of the absolute pressure at full load,

Table 3.9 Valve flow coefficients

	Notation	Flow	Pressure	Fluid density	A_v
SI (*BS 4740*)	A_v	m³/s	Pa	kg/m³	
Continental	K_v	m³/h	bar	Water	$28.0 \times 10^{-6} K_v$
Imperial	C_v	gal/min	psi	Specific gravity	$28.8 \times 10–^{6} C_v$
US	C_v	US gal/min	psi	Specific gravity	$24.0 \times 10^{-6} C_v$

immediately upstream of the valve. Using this figure and the given duty, the valve may be sized from manufacturers' tables. If the inlet pressure is below 100 kPa, applying the 40% rule gives unsatisfactory conditions downstream and, in such cases only, it is necessary to assume a smaller pressure drop, ignoring any possible loss in degree of control. Guidance on these reduced pressure drops is normally given in manufacturers' literature. Additional pressure drops due to pipe runs, isolating valves etc., mean that control valves must be sized on their inlet pressure at full load and not on the boiler pressure.

It is important to ensure that the heat exchange surface which is supplied by the valve is adequate to handle the full design load when the pressure at the inlet of the heat exchanger is 60% of the of the absolute pressure at the inlet of the control valve; any other pressure losses in pipework and fittings need to be taken into account. Care must be taken to ensure that suitable stream traps are incorporated. Failure to do so will result in unacceptably poor control under light conditions.

3.3.4 Valve selection

3.3.4.1 Authority

To provide good control, a control valve must be sized in relation to the circuit it controls so that the pressure drop across the valve is of the same order as that round the rest of the circuit. If the valve is too large when open, the resistance to flow in the circuit is dominated by the resistance of the rest of the circuit. Throttling the valve would initially have little effect on flow; control would then only be obtained over a small range of spindle lift when the valve is nearly closed. The control system is then operating under high gain conditions and instability may result. Conversely, too small a valve may have an unnecessarily high resistance and additional pump head would be required to maintain flow.

The relation of the valve size to the circuit is expressed quantitatively in terms of the valve authority. The concept of authority is first discussed with relation to two-port valves. In the simple circuit of Figure 3.8, a constant-head pump circulates water though a load and the flow rate is controlled by a two-port valve. At design conditions, the valve is fully open and the pump pressure is distributed between the differential pressure across the valve P_1 and the pressure drop round the rest of the circuit. The relation between valve lift and flow in the circuit depends on the resistance of the valve in relation to that of the rest of the circuit. This is expressed quantitatively by the authority N_{des}. The subscript 'des' emphasises that the authority is defined at design flow conditions; this is of significance in more complex circuits.

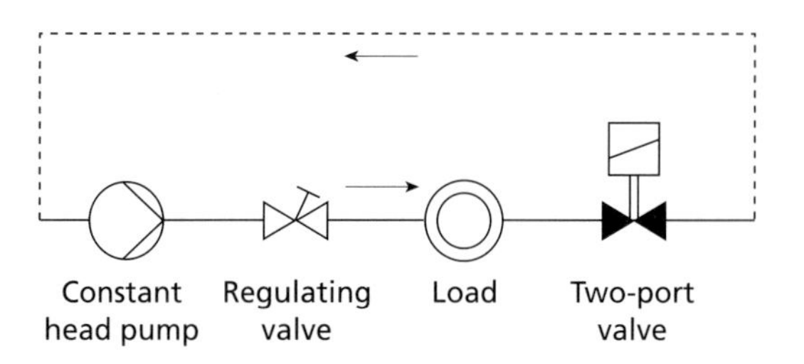

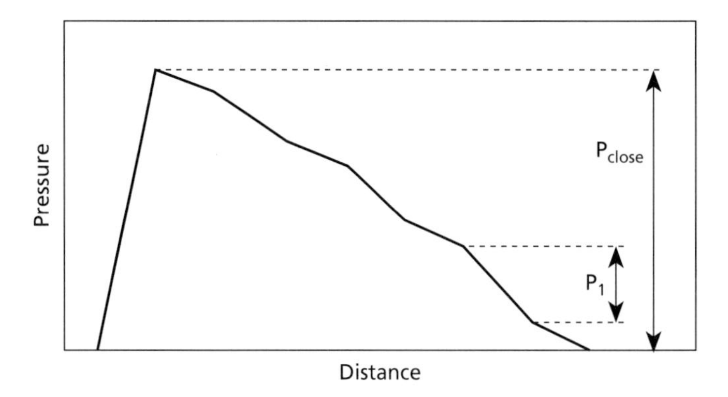

Figure 3.8 Valve authority in simple throttling circuit. Diagram shows pressure drop round a simple throttling circuit at design flow. The circuit has been drawn with a zero pressure drop in the return leg. The valve authority $N_{des} = P_1/P_{close}$

$$N_{des} = P_1/P_{close}$$

where:

- P_1 pressure drop across the valve in the fully open position at design flow
- P_{close} pressure drop across the valve in its fully closed position.

For the simple circuit shown, P_{close} is the pump pressure. Figure 3.9 shows the relation between flow in the circuit and valve lift for both linear and equal percentage valves. Reducing the valve authority distorts the valve characteristic, making control more difficult at low flow rates. A minimum authority of 0.5 is acceptable. Below an authority of 0.5 the characteristic is increasingly distorted away from the required shape. In other words, at design conditions the pressure drop across the fully open control valve should be about half the total locally available differential pressure.

Where the hydronic circuit has several branches, it is necessary to take into account the influence of control action taking place in other branches. Figure 3.10 shows several branches controlled by two-port valves, arranged in a ladder circuit. The valve authority of a single branch is calculated as described above, but with an extension to the definition of P_{close}:

> P_{close} is the pressure drop across the valve in its fully closed position, with all other branches remaining at design settings.

However, when the circuit is actively controlling, the valves in all branches act independently under the influence of their own control loops. In the worst case, all other branches may close down and the pressure across the branch under consideration rises towards the full pump pressure. In practice the pump head is a function of flow

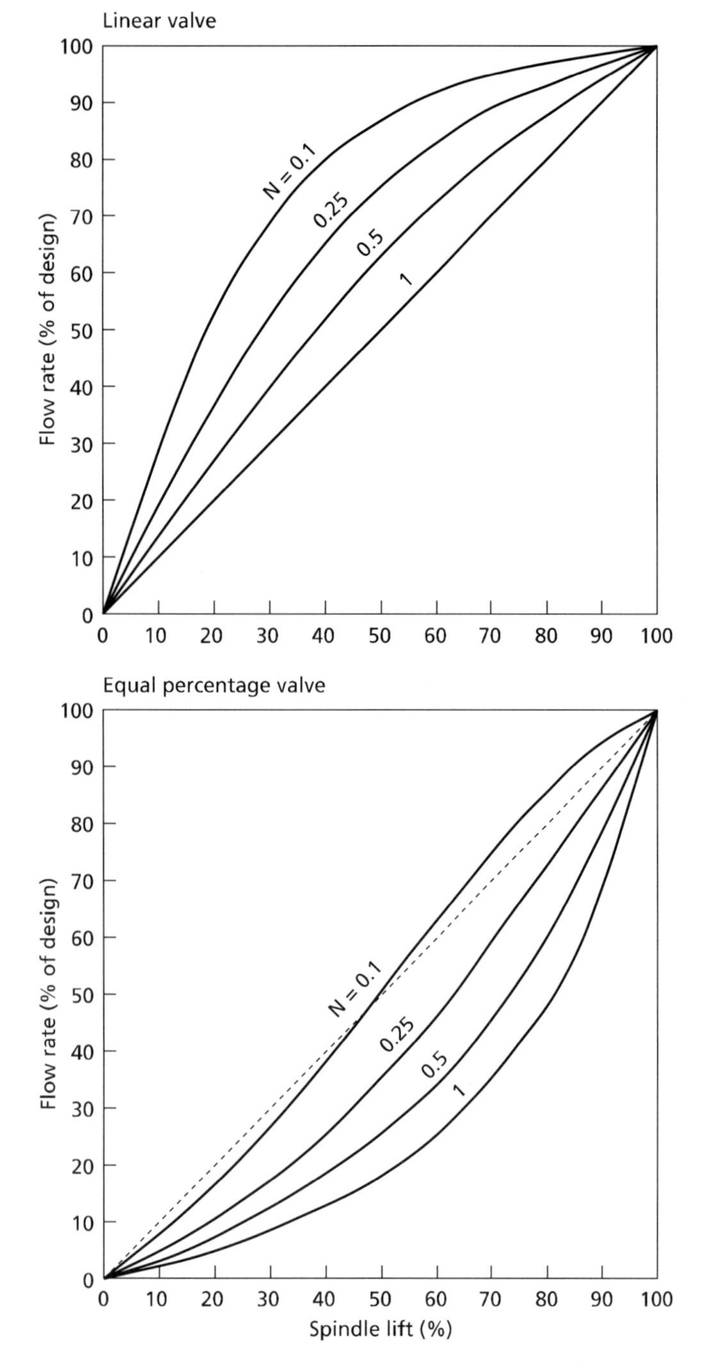

Figure 3.9 Flow through a control valve as a function of spindle lift, showing the effect of valve authority

rate; the consequences of this for control are discussed in section 5. This increase in pressure at the branch has the effect of reducing the valve authority. To reflect this, the minimum authority is defined as

$$N_{min} = P_1/P_{max}$$

where:

P_1 pressure drop across valve in the fully open position at design flow

P_{max} maximum pressure drop across the valve in its fully closed position, obtainable with any combination of valve positions in the rest of the circuit.

In general with all other valves shut, flow through the system drops towards zero and the pressure across the

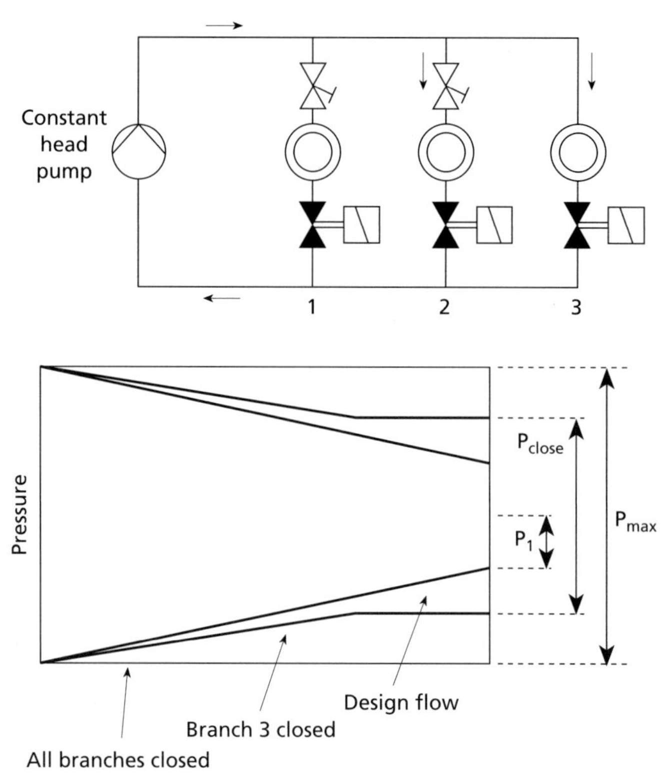

Figure 3.10 Authority in a multi-branch throttling circuit. Schematic three-branch throttling circuit, showing authority of valve in branch 3, which has a design pressure drop of P_1. Design authority $N_{des} = P_1/P_{close}$, minimum authority $N_{min} = P_1/P_{max}$

control valve under consideration rises towards the pump head P_{pump}. If automatic valves are fitted, P_{max} may be less than the pump head.

Valves should be selected so that

$$N_{des} \geq 0.5$$

$$N_{min} \geq 0.25$$

The treatment of three-port valves follows that of two-port valves. Figure 3.11 shows a three-port mixing valve in a circuit which is balanced to provide a constant flow though the heat exchanger as the valve modulates. The valve is said to be open when the flow is across the top of the tee, when the pump delivers water directly from the heat source to the

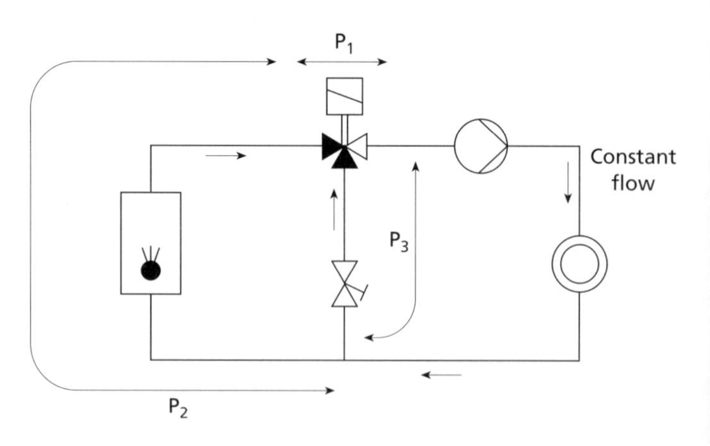

Figure 3.11 Authority of three-port valve in mixing circuit. $N = P_1/(P_1 + P_2) = P_1/P_3$, where P_3 is measured with the valve closed, i.e. in bypass position

heat emitter without recirculation through the bypass leg. The pressure drop across the valve in the closed position (i.e. full bypass flow) is denoted by P_3 and the authority is defined in the same way as for the two-port valve.

$$N_{des} = P_1 / P_3 = P_1 / (P_1 + P_2)$$

P_3 is constant because of the constant flow through the heat emitter. P_2 is the pressure drop round the variable flow part of the circuit, including the heat generator and associated pipework, but not the bypass or balancing valve.

The authority is equal to 0.5 if the pressure drop across the three-way valve when fully open is equal to the pressure drop in the variable flow part of the circuit, excluding the bypass.

The circuit of Figure 3.11 gives a variable flow through the heat generator which reduces to zero if there is no demand from the heat emitter. This has obvious practical problems which may be obviated by employing a separate pump in the primary circuit as shown in Figure 3.12. The bypass forms a virtual heat generator with almost no pressure drop. The authority of the three-way valve is now

$$N = P_1 /(P_1 + P_4)$$

where P_4 is the pressure drop round the variable part of the circuit, i.e. round the bypass pipework. P_4 is therefore small and the valve has an authority close to unity.

Where a three-port valve is used in a diverting application, as in Figure 3.13, the authority is defined in the same way. As before, it is the pressure drop round the controlled, i.e. variable-flow, section of the circuit that is used in the calculation of authority.

The above definitions of authority and the recommendation that the designer should aim at a design authority of 0.5 and a minimum authority of 0.25 will assist in the selection of an appropriately sized valve. It will not guarantee good control, especially where a number of variable flow branches interact. Where there is uncertainty a network analysis is recommended. Further details are given in the literature(8–10).

A three-port mixing valve is commonly used to provide a constant flow through the load in a circuit such as the one shown in Figure 3.11. The incorporation of a balancing valve in the bypass leg allows the resistance of the bypass leg to be set equal to the resistance of the heat generator circuit. In this way, the resistance seen by the pump is equal whether the three-port valve is fully open or fully shut and the flow is the same in both positions. However, to ensure constant flow at all positions of the valve, a further design parameter associated with three-port valves must be considered, the symmetry of the internal ports. Symmetrical design means that both control and bypass ports in a mixing valve have the same characteristic and the control and bypass connections may be reversed without affecting the control behaviour of the valve. However, when the ports both have an equal percentage characteristic, the total volume flow through the valve is not constant at all opening positions. This is illustrated in Figure 3.14, where it is seen that the common flow is reduced at intermediate opening positions. This is unsatisfactory for some situations and the solution is to use an asymmetrical valve to achieve constant flow. An asymmetrical valve has its internal ports designed so that the control port provides the desired operating characteristic, while the port connected to the bypass is designed to compensate to maintain a constant total flow through the valve independent of valve opening; see Figure 3.15. The ports of an asymmetrical valve are not interchangeable.

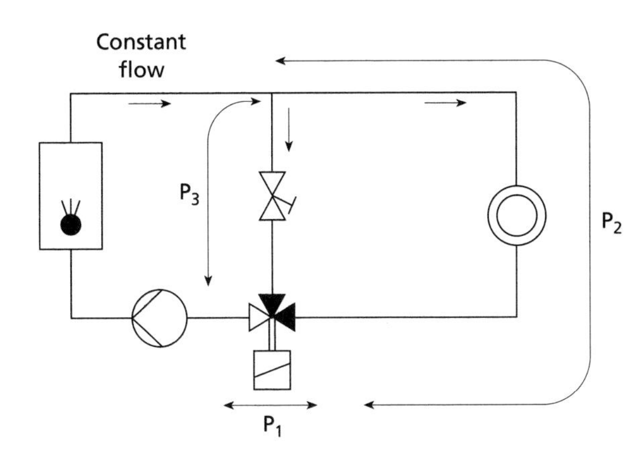

Figure 3.13 Authority of three-port valve in diverting circuit. $N = P_1/(P_1 + P_2) = P_1/P_3$, where P_3 is measured with the valve closed, i.e. in bypass position

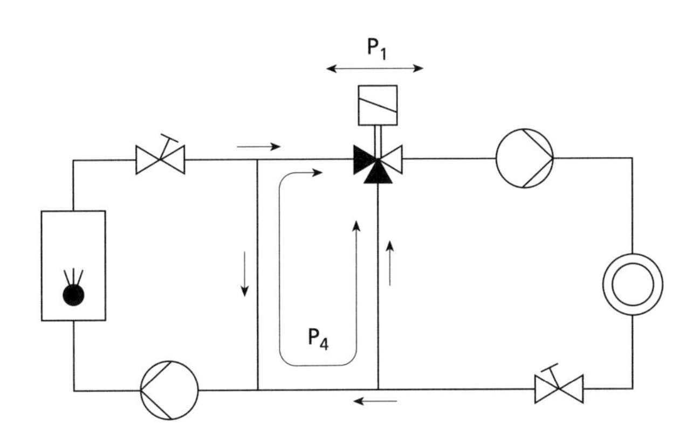

Figure 3.12 Mixing circuit with primary pump and bypass. The authority of the three-way mixing valve $N = P_1/(P_1 + P_4)$ and is close to unity

3.3.4.2 Sizing

The relationship between valve position and heat output may be illustrated using the graphical construction shown in Figure 3.16. The upper right quadrant shows the relation

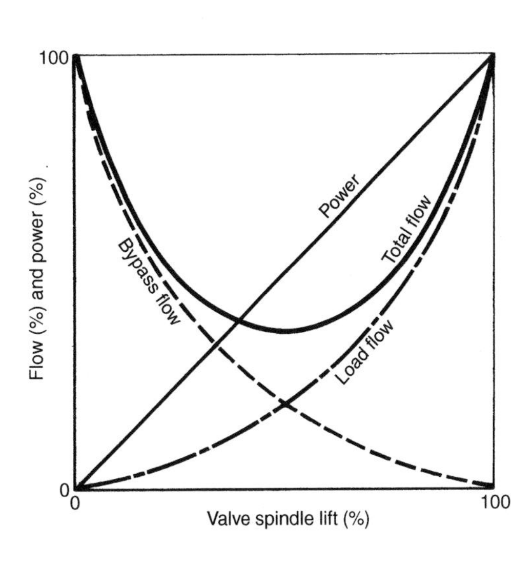

Figure 3.14 Curves for symmetrical three-port valve selected for power linear output

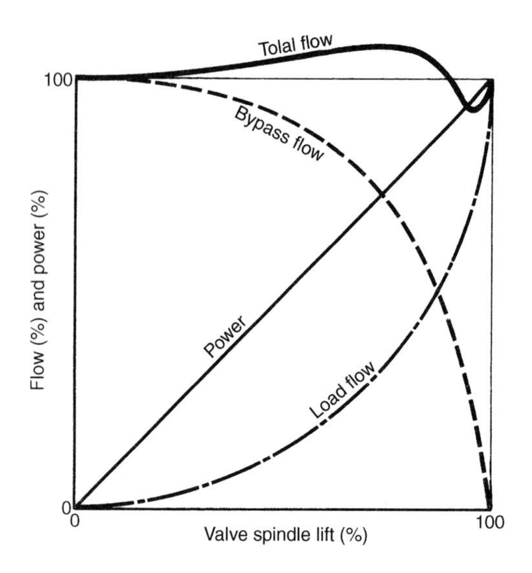

Figure 3.15 Curves for asymmetrical three-port valve selected for power linear output

between valve stroke position and flow for two types of valve characteristic for different authorities. The valve authority modifies the characteristic curves shown in Figure 3.6, which are drawn for a constant pressure drop across the valve, i.e. for an authority of unity. The bottom right quadrant shows the relation between flow through the heater emitter and heat output. Together, these allow the construction of the curve in the bottom left quadrant, which shows the relation between valve position and heat output. Two example curves are shown. Curve 2, representing an equal percentage characteristic with authority $N = 0.7$, gives an almost linear change of heat output with valve position. Curve 5, a linear characteristic valve with authority $N = 0.8$, is highly non-linear and shows why this type of valve is unsuitable in this application.

It is the task of the controls specialist to size and select control valves suitable for the HVAC system under

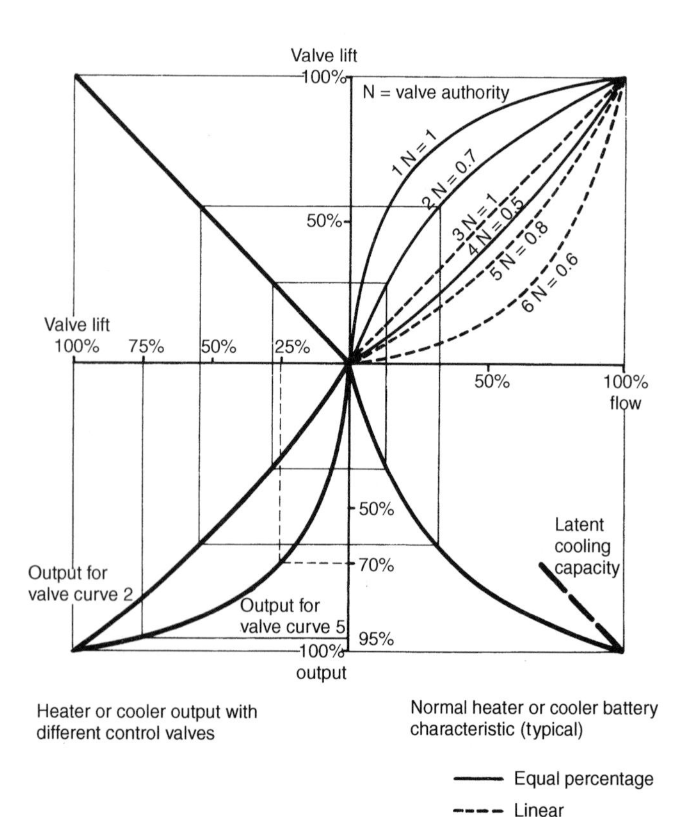

Figure 3.16 Stylised valve authority and power output characteristics

consideration. Once the circuit has been designed and the valve authority decided, the valve size can be selected from the pressure drop across the valve and the flow rate through it, both when the valve is fully open. Valve sizes are specified as the K_v value, see 3.3.3.2. The required valve size is given by the equation:

$$K_v = V_m / P_m^{0.5}$$

where:

- V_m flow through open valve (m^3/h)
- P_m pressure drop across open valve (bar).

This may be rewritten as

$$K_v = 36 V_{ml} / P_{mkP}^{0.5}$$

where:

- V_{ml} flow through open valve (l/s)
- P_{mkP} pressure drop across open valve (kPa).

The required valve with the nearest K_v value may be selected from a manufacturer's list. Choosing a valve with a lower K_v value will increase the authority. It must be emphasised yet again that the system design has a great influence on the possibility of designing effective controls. System design and controls should be considered together from the concept stage of the project. The design of hydraulic circuits is considered in more detail in section 5.

3.3.4.3 Cavitation

When a fluid flows through a restriction, its velocity increases and in consequence the static pressure decreases. The reduction in pressure may be sufficient to cause the formation of vapour bubbles. When these bubbles collapse, they create noise in the system. In addition, intense microjets of liquid are formed which can damage the surface of the valve and pipework. The condition for cavitation to occur is defined in terms of the pressure drop across the valve:

$$P_{max} = K_m (P_{in} - P_v)$$

where:

- P_{max} maximum allowable pressure drop across open valve (kPa)
- K_m valve recovery coefficient
- P_{in} absolute inlet pressure (kPa)
- P_v fluid vapour pressure at inlet temperature (kPa).

Values for K_m are a function of the valve construction, but are typically between 0.6 and 0.9. Values of water vapour pressure as a function of temperature are shown in Table 3.10. In general, cavitation problems are unlikely to be encountered if the pressure drop across the valve is less than 70 kPa, or if the fluid velocity is below 3 m/s.

3.3.4.4 Valve selection checklist

On water circuits, first decide whether two- or three-port valves are the most suitable and select a reliable control

Table 3.10 Vapour pressure of water

Water temperature (°C)	Vapour pressure (kPa)
10	1.2
20	2.3
30	4.2
40	7.4
50	12.3
60	19.9
70	31.1
80	47.3
90	70.1
100	101
110	143
120	199
130	270
140	361
150	476
160	618
170	792
180	1003
190	1255
200	1555

strategy. The following points should be taken into consideration when selecting a valve:

— Is the valve body suitable for the temperature and pressure of the fluid system? Remember the pressure is the sum of static and dynamic head.

— Ensure that the valve will pass the required flow at a pressure drop within the maximum differential pressure rating of the valve.

— Check for out-of-balance forces, particularly during closure of a two-port valve.

— Check whether tight shutoff is required; this is not usually possible with a double seated valve

— Check there is sufficient pump head to provide the pressure drop across the valve at the specific duty.

The above rules apply to all valves, including two-position on/off. For modulating valves, the following additional considerations apply:

— Select an equal percentage valve characteristic unless there is good reason to select an alternative.

— Ensure pressure drops though heat exchangers and associated pipework are known before control valves are selected.

— Select valves to provide an authority of at least 0.5 for diverting applications and 0.3 for mixing applications.

— Where possible, use heat transfer curves of flow against output to check possible anomalies and confirm the correct characteristic has been chosen.

— Ensure that the rangeability of the selected valves is large enough to provide stable control under low load conditions.

— Three-port valves with asymmetrical port characteristics should be used to maintain flow conditions.

Before finalising the selection:

— Check whether there have been changes to the specification of heat exchangers and pipework since the original design.

3.4 Dampers

3.4.1 Damper selection

Dampers are used to control air flow in ducts in a manner analogous to the use of valves in hydraulic circuits. The damper chosen for a particular situation must satisfy both the physical requirements of the application and also provide suitable control characteristics. The practice of simply selecting a damper to fit the available duct dimensions can lead to unsatisfactory control operation. The majority of dampers used for modulating control have a rectangular cross-section and provide control by rotating a set of blades. The blades may move in parallel or opposition to each other. Figure 3.17 shows different types of damper. Round dampers normally have a butterfly type blade and are used to control flow in ducts that have high static pressure and high velocity characteristics. Dampers

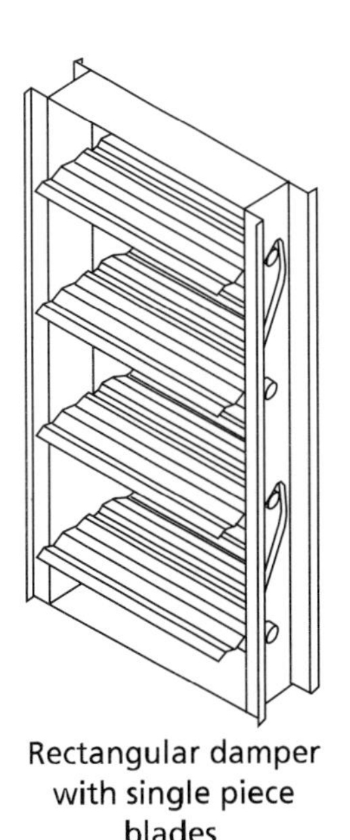

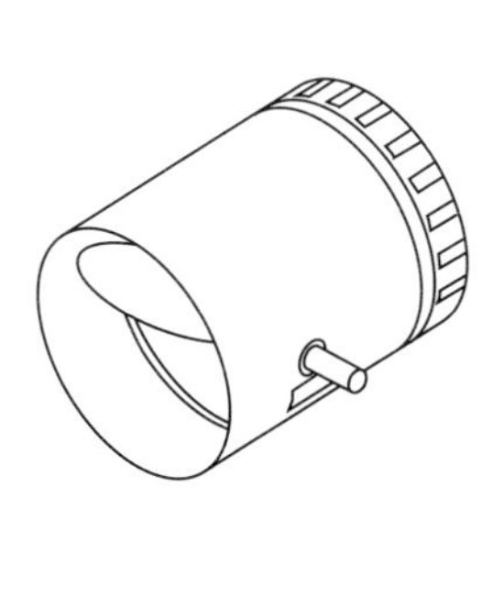

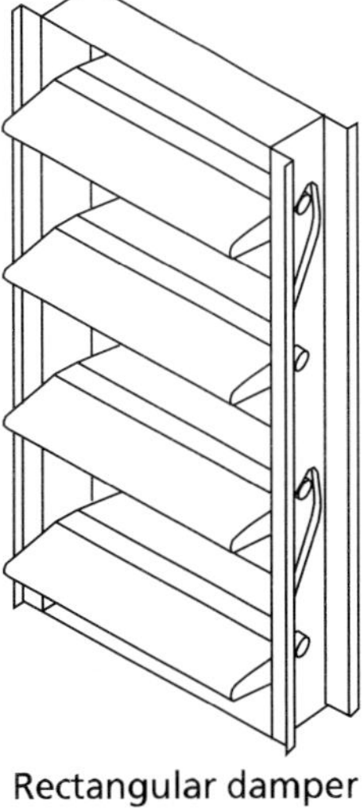

Figure 3.17 Types of damper

may be installed for the control of fire and smoke. Smoke dampers have to meet relevant criteria and are normally installed and operated independently of the HVAC control system; they are not considered further in this Guide. It is possible to obtain smoke and control dampers, which are capable of operating as modulating dampers, while retaining the necessary smoke control criteria after many operations. Damper sizing is covered in more detail below. Other factors to be taken into account when selecting a damper are:

— *Leakage rating*. Leakage through the damper in the closed position may be critical in such applications as fresh air intakes in cold climates and the design should specify the maximum acceptable leakage. It is difficult to achieve good shutoff with a damper and it is necessary to specify low leakage construction where required. Different classes of leakage are available and leakage is specified in terms of leakage volumetric flow per unit damper area at a specified pressure difference. Low leakage is obtained by the provision of seals and the use of stiffer blades; this may require a higher closing force from the actuator.

— *Velocity and turbulence*. As the air velocity in a duct increases, the damper blades experience higher forces, which may be sufficient to bend or twist the blade, or cause problems with the bearings and linkage. Dampers are given a velocity rating to indicate the maximum velocity in the duct; ratings may need to be reduced in turbulent conditions. Moderate turbulence may be found downstream of duct transitions or elbows. Severe turbulence can be found near the discharge of a fan and this can be sufficient to prevent satisfactory operation of a damper.

— *Pressure*. The maximum static pressure that can be developed across a damper occurs when the blades are closed. Dampers are given a maximum static pressure differential; operation above this value may give rise to excessive leakage and possible damage.

— *Torque requirements*. Two conditions must be considered when establishing minimum torque requirement of a damper. One is the closing torque which is needed to achieve the required maximum leakage rate. The other is the dynamic torque needed to overcome the effect of high velocity air flow over the blades. This will affect actuator selection.

— *Mixing*. Parallel blade dampers alter the flow direction of air passing through them. By directing the air flow to one side of the duct, they may cause stratification. In some cases, the change in direction of the air stream may be used to promote mixing where it joins another air stream.

3.4.2 Modulating dampers

The most common type of modulating damper has a rectangular cross-section, with a set of blades that may be rotated by a linkage connected to an actuator, to restrict the flow of air through the damper. The blades are linked to each other and may have parallel or opposed motion; see Figure 3.18.

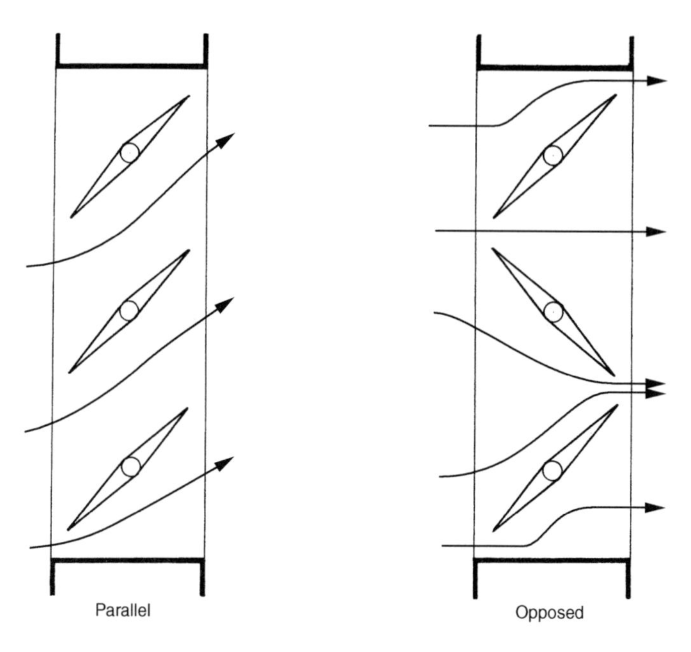

Figure 3.18 Parallel and opposed operation of damper blades

The relationship between damper blade rotation and flow through the damper for a constant pressure drop across the damper is shown in Figure 3.19. This is termed the inherent characteristic of the damper.

The 'ideal' damper characteristic would result in a linear relation between actuator movement and heat delivered to the conditioned space. The distribution of heat by air differs from that by water in that air systems are essentially open and all the energy in the air stream can be considered as delivered to the conditioned space. A linear damper characteristic is therefore desirable. Systems controlled by modulating dampers normally have a fan running at constant speed with a constant pressure drop across the system. In the simple ducted air system of Figure 3.20, where air is supplied by a fan running at constant pressure, the total pressure drop across damper and plant is constant

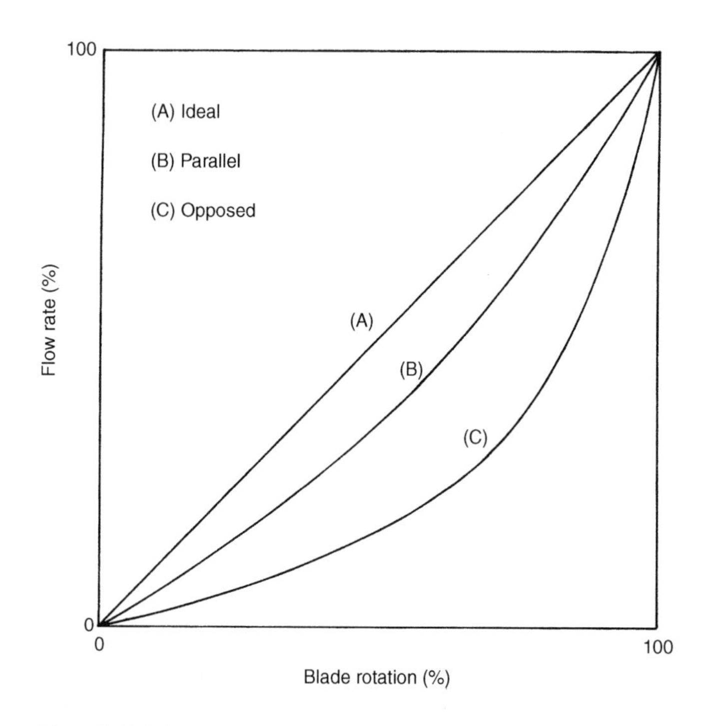

Figure 3.19 Inherent characteristics of damper types

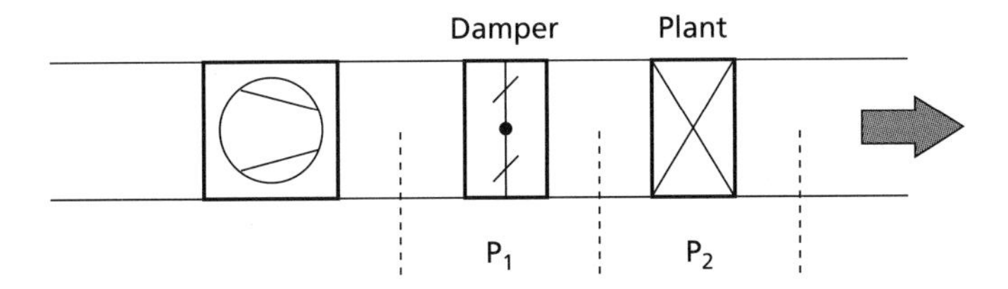

Figure 3.20 Damper authority $N = P_1/(P_1 + P_2)$

and so the pressure drop across the damper increases as the damper is closed; when fully closed it is equal to the static pressure of the fan.

The changing partition of pressure drop across damper and load as the flow changes results in dampers having an installed characteristic different from the inherent characteristic. The installed characteristic is affected by the size of the damper, described by the damper authority. This is defined in a similar way to that of valves, where the authority is given by

$$N = P_1/(P_1 + P_2)$$

where:

P_1 pressure drop across the damper in the fully open position

P_2 pressure drop across the rest of the circuit.

In many cases $(P_1 + P_2)$ is constant and equal to the pressure developed by the fan. American usage may refer to the damper characteristic ratio (α or sometimes δ) defined as

$$\alpha = P_2/P_1$$

The installed damper characteristic curves may be derived from a knowledge of the inherent curve and the authority. The pressure drop developed across a fitting is proportional to the square of the air flow through it and it is straightforward to show that for any intermediate opening position

$$N = \frac{(1/F_a)^2 - 1}{(1/F_i)^2 - 1}$$

where:

F_a ratio of actual flow in the intermediate position to actual flow at the fully open position

F_i ratio of flow in the intermediate position to flow through the fully open damper for the inherent characteristic.

Damper characteristic curves are supplied by damper manufacturers and typical examples of installed characteristics are shown in Figure 3.21 and Figure 3.22.

It can be seen that for an opposed blade damper the closest approach to linearity occurs with an authority $N = 5$ to 10%, while for parallel blade dampers it occurs when $N =$ 20%. The implication is that for a given pressure drop across the damper and system, the pressure drop across a parallel blade damper will be four times that across an opposed blade damper, with correspondingly higher energy consumption. Opposed blade dampers are therefore

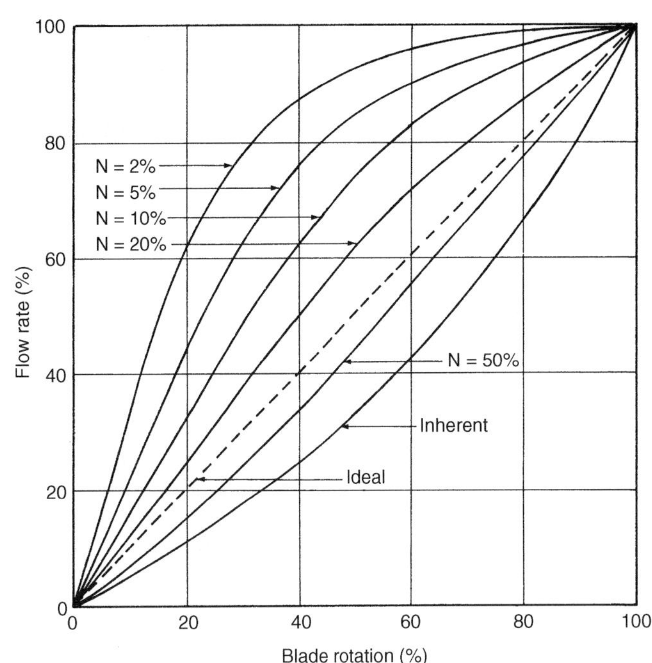

Figure 3.21 Installed characteristic of parallel blade damper

recommended in general for control purposes.

It is poor practice to choose a damper size based on duct size and convenience of location. This frequently results in oversized dampers with inadequate authority to provide proper control. The correct damper will often have smaller dimensions than the duct, giving additional benefits in reduced damper and actuator costs, together with lower leakage in the closed position. Where the damper is smaller than the cross-section of the ductwork, it may be installed with a baffle or blanking plate to take up the additional area. If the blanked off area exceeds about 30% of the total cross-section, it may be advisable to use a reducing section of

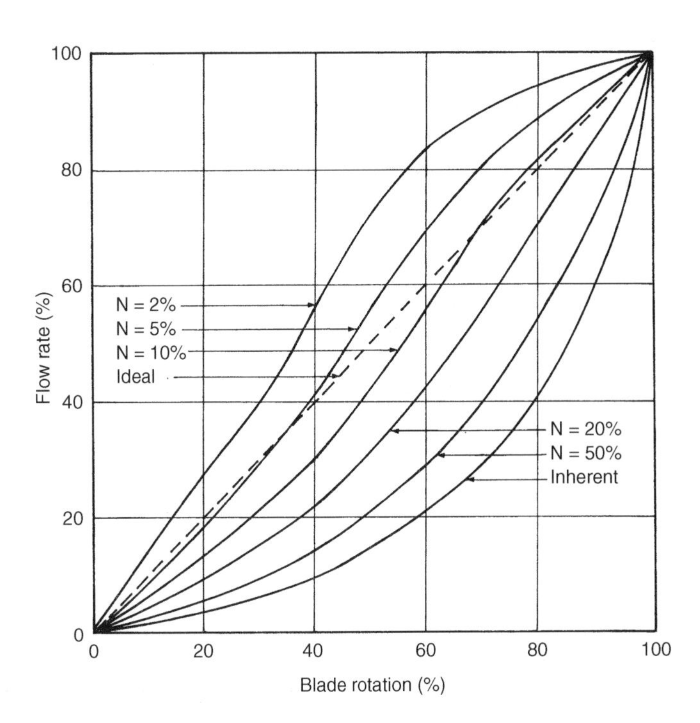

Figure 3.22 Installed characteristic of opposed blade damper

ductwork to avoid turbulence caused by the sudden restriction at the baffle.

3.4.3 Applications

In general, parallel blade dampers are used where there is little resistance to flow in series with the damper, and opposed blade dampers where the controlled section contains restrictions of any sort (Table 3.11). Useful treatments of damper applications are to be found in the literature(11,12).

3.4.3.1 Return air mixing

A common application is the control of return air shown in Figure 3.23. The system delivers a constant flow of air to the conditioned space; the dampers are used to vary the proportion of air that is recycled. The pressures at P_s and P_e are therefore constant and so for sizing purposes each damper may be considered separately in its own subsystem. The inlet and exhaust subsystems include the series resistance of weather louvres and bird screens and an opposed blade damper should therefore be chosen to give an authority of between 5 and 10%; Figure 3.22 shows that this will produce an approximately linear installed characteristic. Where louvres and screens are absent at inlet and exhaust, the lack of their resistance increases the authority of the damper and it may be appropriate to select a parallel blade damper to give a more linear characteristic. The return air path has little resistance other than the damper itself, and so the damper operates at a high authority. The correct choice here would be a parallel blade damper.

3.4.3.2 Face and bypass control

A common application of dampers controlling the flow of outside air uses two dampers in a face and bypass configuration, as shown in Figure 3.24. The system is designed to provide a constant pressure drop and hence constant combined flow rate, while the proportion of air flowing through the coil is varied by operating a linked pair of dampers. Since the pressure drop across the bypass damper is constant, a parallel blade damper is used to provide a linear characteristic with minimum pressure drop at full flow. Since the face damper has the resistance of the coil in series, an opposed blade damper is used of the appropriate authority to provide a linear characteristic. Starting with the known resistance of the coil, the face damper will be sized to give an authority of between 5 and 10% and the bypass damper then sized so that its resistance is equal to that of the combination of face damper and coil; this will give an approximately constant flow though the system at all operating positions. Consideration may be given to the use of parallel blade dampers for both sets to aid downstream mixing, as shown in the diagram.

Table 3.11 Damper applications

Control application	Damper type
Return air	Parallel
Outdoor air or exhaust air	
— with weather louvre or bird screen	Opposed
— without louvre or screen	Parallel
Coil face	Opposed
Bypass	
— with perforated baffle	Opposed
— without perforated baffle	Parallel
Two-position (all applications)	Parallel

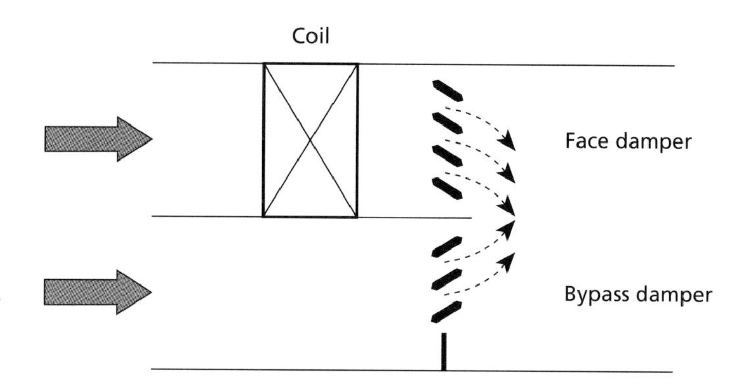

Figure 3.24 Face and bypass damper system, showing use of parallel blade dampers to assist mixing

3.5 Motors

3.5.1 Energy efficiency

Motors are major users of electricity and account for almost half the total electricity use in the UK. Recent developments in higher efficiency motors and variable-speed drives are described in the DETR *Good Practice Guide 2*(13), which offers good advice on the energy efficient use of motors, shown in Table 3.12. By far the most common type of motor used in HVAC applications is the three-phase squirrel cage induction motor. Three-phase power is supplied to the field windings in the stator, the fixed part of the motor which encloses the rotating rotor. The current in

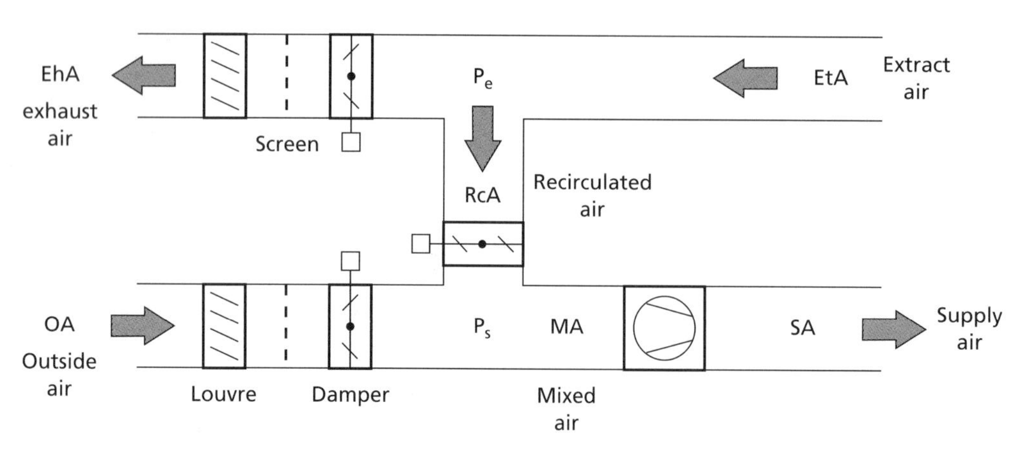

Figure 3.23 Damper selection for a recirculated air mixing circuit

Table 3.12 Energy saving checklist for motors

1	Is the equipment still needed? — Check that changing requirements have not eliminated the need
2	Switching the motor off — Time the switching according to a schedule — Monitor system conditions and switch off when motor is not needed — Sense the motor load and switch off when idling
3	Reducing the load on the motor — Is the transmission efficient? — Is the driven equipment efficient, e.g. gearboxes and belt drives? — Is the control system effective?
4	Minimising motor losses — Specify higher efficiency motors where feasible — Avoid oversized motors — Check power quality
5	Slowing down the load — Use variable speed drives where possible, for both control and regulation — Use multiple-speed motors where 2, 3 or 4 distinct duties exist — Check pulley ratios for belt drives

the field windings produces a rotating magnetic field which induces currents in the rotor conductors and pulls the rotor round with it. There are no electric connections to the rotor. The speed of rotation of the rotor is somewhat slower than the frequency of the applied field by an amount known as slip, expressed as a percentage of the synchronous speed, hence the description of the motor as asynchronous. Within the operating range of the motor, speed is approximately constant. The treatment here refers to three-phase induction motors, unless otherwise stated.

3.5.2 Starting

The low ohmic resistance of the field windings of an induction motor means that if a stationary motor is connected directly to the supply, it will draw a heavy starting current until the motor develops sufficient speed to produce a balancing back EMF. This starting current may be up to eight times the full load current and can produce problems for the electrical services within the building. Several methods have been developed to limit the starting current and provide controlled start-up of motors. The associated topic of variable-frequency drives is covered in 3.5.3.

3.5.2.1 Direct on-line (DOL)

In some situations it may be possible to simply connect the motor to the supply. This will produce a high peak starting current and in most cases a high starting torque. The control gear and contactors must accommodate the starting current. Switchgear designed to *BS EN 60947-4-1*[14] is based on a starting current of 7.2 times normal full load current. DOL starting is simple, but it use is limited to applications where:

- Low power motors are used and the supply can cope with the start current.
- The load driven by the motor can cope with mechanical shock produced by the high starting torque.
- A high starting torque is needed.

3.5.2.2 Star–delta starter

With this form of starter, the three-phase supply voltage to the stator windings can be switched between star and delta configurations. On starting, the supply is connected in star configuration, when the starting current is typically between 1.5 and 2.6 times the normal full load current and the peak starting torque is between 0.2 and 0.5 times the nominal operating torque. Under this condition the motor will accelerate to about 75% of full speed, when the connections are switched to delta configuration, allowing the motor to achieve full performance. The switchover is controlled by a timer, whose operating time is set during commissioning. It is normal to introduce a small delay before the delta connection, to avoid any arcing or transient currents. Its features are:

- low starting torque
- access to both ends of stator windings required.

3.5.2.3 Primary resistance starter

Starting current is limited by connecting a resistance bank in series with the motor windings. Once the motor has run up, the resistance is switched out and the motor is directly connected; the changeover is normally controlled by a timer. Starting current and torque are controlled by the choice of resistance; typically a peak starting current of 4.5 times full load and a peak starting torque of 0.75 operating torque. This form of starter is suitable for applications such as ventilator fans, where the load torque increases with speed.

3.5.2.4 Autotransformer

The motor is started at reduced voltage supplied from an autotransformer. This is a specialised solution, used for large motors above 100 kW.

3.5.2.5 Electronic soft start

Introduced relatively recently, this form of starter is rapidly growing in use. An electronic circuit is used to produce a gradually increasing voltage to the motor to produce a steady smooth acceleration. The units employ a thyristor bridge and by varying the firing angle of each set of thyristors, it is possible to control the output voltage. The supply frequency is unchanged and soft starters do not behave like inverter drives.

3.5.3 Speed control

The use of variable-speed drives (VSD) on motors for pumping fluids or moving air can lead to substantial energy savings[15]. The recent development of variable-frequency inverter drives has expanded the application of variable-flow systems, with attendant changes in control system design. This section summarises the main methods of motor speed control, concentrating on inverter drives.

3.5.3.1 Eddy current coupling

The eddy current coupling is an electromagnetic coupling which fits between a standard constant-speed AC induction motor and the load. The relative rotation of the motor and load generates the eddy currents to provide the coupling and so the maximum output speed is less than the motor speed. This form of speed control is well proven and reliable. It can produce high torque at low speed, though this is not often a requirement for HVAC applications.

3.5.3.2 Switched reluctance drive

This system requires a special motor with more stator than rotor poles. It is powered by a solid-state switched reluctance drive controller, which provides controlled DC pulses to the motor. While the components are similar to those used in inverter drives, the principle of operation is different. Sophisticated motor control is possible, with control of maximum and minimum speeds and acceleration rates. Control of starting conditions is available, making the use of a separate starter unnecessary. The number of suppliers is more limited than for inverter drives; the maximum motor size is about 75 kW.

3.5.3.3 Variable-voltage control

Speed control by varying the input voltage by a transformer is not normally recommended. The range of speed regulation is limited and efficiency is poor. Electronic variable-voltage drives are available which use phase angle control thyristors to provide a variable-voltage input to the motor. The cost of the drive is lower than an inverter drive, but the control and efficiency is poorer. A special motor matched to the controller is used and this form of control is only recommended for small pumps with built-in variable-speed drive.

3.5.3.4 Multi-speed motor

Multi-speed motors are available which are basically standard induction motors with additional windings which may be switched in or out, effectively changing the number of poles of the motor. The motor speed will therefore change in coarse steps, from 3000 rpm for a two-pole configuration to 750 rpm as an eight-pole configuration, less slippage. Two- or three-speed motors are common, with four-speed available. Multi-speed motors are simple and low cost and are found incorporated in small multi-speed pumps and fans. They are normally only available in fractional kilowatt sizes.

3.5.3.5 Inverter drive

Variable-speed inverter drives supply power of controlled frequency and voltage to the motor. Standard induction motors are used without modification and with recent developments and cost reductions this is now the most popular method for fan and pump speed control. The inverters produce a limited starting current for a soft start, making the use of a separate starting circuit unnecessary. Various types of speed control are available. There are three main types of inverter:

— pulse width modulation (PWM)

— pulse amplitude modulation (PAM)

— current source inverter (CSI).

Pulse width modulation inverters

The principle of the inverter is shown in Figure 3.25. Three-phase mains current is rectified and fed to a DC link circuit, which provides some smoothing. The rectifier may

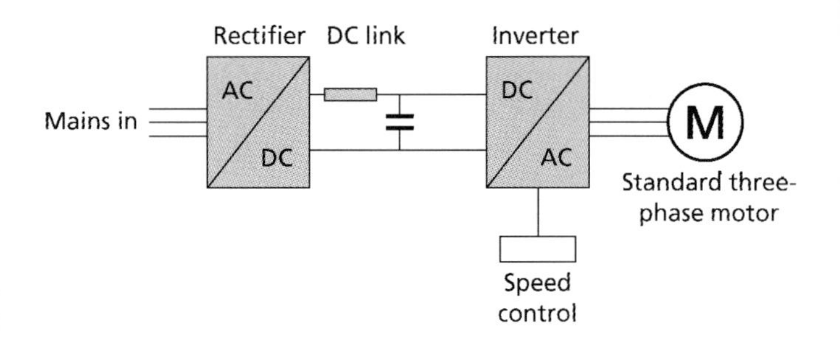

Figure 3.25 Principle of PWM inverter for motor speed control

be either uncontrolled, using diodes, or controlled, using thyristors. Uncontrolled rectifiers produce large disturbances and losses in the mains and the use of controlled rectifiers is generally recommended. The DC link contains induction coils and capacitors to provide a reservoir of smoothed current for the output stage. Recent developments have allowed the elimination of the DC link by the employment of a matrix inverter in the rectifier, which is a combination of power transistors controlled in such a way as to give a smooth output and eliminate the need for the DC link. This reduces the space requirement and also reduces the size of the EMC filter(16). The output bridge circuit uses transistors to supply a train of pulses. The output bridge operates at a switching frequency which must be substantially higher than mains frequency. By controlling the width and spacing of the pulses it is possible to produce a near sinusoidal supply to the motor of controlled frequency and amplitude; the frequency varies from a minimum of a few hertz to a maximum of mains frequency to give maximum rated motor speed. The higher the operating frequency of the output bridge, the more closely will the motor current approximate to a sine wave. With a low switching frequency (<5 kHz) the motor current waveform will not be smooth and the result may be undesirable motor noise. A higher switching frequency, now available up to 20 kHz, produces smoother running but requires more attention to EMC performance.

The inverter is controlled by driver and control circuits. The control circuit of the variable-speed drive ensures that the amplitude of the output is matched to the frequency; the output voltage is reduced at low frequencies to prevent excessive winding currents being drawn. Highest efficiencies are achieved when the voltage–frequency characteristic is matched to the type of load. The control gear also provides a soft start capability.

The processor may also incorporate localised control functions. The processor takes overall control of the inverter such as up/down ramping of the frequency, start/stop signal, motor protection and error handling. At the same time the processor may also take care of the control task, e.g. pumping application. The processor controls the inverter using a built-in PID regulator and a signal from a pressure sensor, to produce a constant pressure to the system. All adjustments to and communications with the surroundings are controlled by the processor, which also collects the data to be communicated to the user.

Open loop control. Open loop control operates without feedback. The inverter supplies the voltage and frequency which is expected to produce the desired speed. While adequate for many applications, this form of control is not suitable for sustained low speed operation, nor with fluctuating loads.

Voltage vector control (VVC). This is a sophisticated form of PWM inverter, in which the control circuit of the inverter monitors the output to the motor and digitally controls the timing of the output pulses to produce a high quality sine wave output matched to the motor type and load. It claims better speed control and efficiency than a simple PWM inverter.

Flux vector control. By obtaining information about the speed and position of the rotor, it is possible to calculate the torque and flux demand of the motor instant by instant and control the output bridge accordingly. This makes for a very accurate control of speed with rapid response to input signals. However, it is more expensive than the other varieties of PWM controller and requires some form of tachometer or position encoder on the motor, obviating the advantage of being usable with standard motors. It is not normally justified for building services applications. Sensorless vector control is a form of flux vector control which dispenses with the need for speed feedback from the motor by deducing the rotor behaviour from the winding current.

Pulse amplitude modulation inverters

The pulse amplitude inverter is similar to the PWM inverter in its basic principle, except that the output waveform is constructed from pulses of varied amplitude. With pulse amplitude modulation the output bridge of the inverter constructs an approximately sinusoidal voltage output of variable frequency. The output is made up from either six or 18 pulses per period; the six-pulse inverter gives a rather poor output waveform.

Current source inverter (CSI)

This is a simple low cost inverter, but has the disadvantages of producing harmonic disturbances on the mains and is of lower efficiency than other types of inverter. CSIs are not normally recommended for building services applications.

3.5.4 Variable-speed motors

The frequency inverter has become the standard form of variable-speed drive. Inverters are available as stand-alone units and extend up to the megawatt range. A stand-alone inverter drive offers the greatest choice of drive and speed control system, and when used with a feedback encoder gives great accuracy of torque and speed control. This is important for some process control, but not required in HVAC applications. The term 'variable-speed motor' is now generally used to imply a motor with integrated inverter drive. Such motors are available up to about 10 kW, though the upper limit is increasing. The choice between a variable-speed motor with built-in VSD and a separate inverter will be based on the following considerations:

- Variable-speed motor
  - Inverter and EMC filters built in, giving lower installation cost and reduced wiring runs.
  - Suitable for fan and pump speed control, less suitable for constant torque applications.
  - Simple product selection.
  - Inverter may be top or non-drive end mounted.
  - PID controller for pressure or temperature control can be incorporated.
- Stand-alone inverter[16]
  - More accurate control possible with use of feedback encoder.
  - Available in large sizes, up to megawatts.
  - Smaller bulk of motor may suit available installation space.
  - Suitable for all types of load.

3.6 Pumps and fans

The specification of pumps and fans is outside the responsibility of the controls engineer. A single pump may not be able to satisfy the full design flow and yet provide economical operation at part loads. The designer must also ensure that the pump motor is not overloaded under any possible operating condition. Several possible pump arrangements are available to the designer:

- multiple pumps in parallel or series
- provision of a standby pump
- pumps with two-speed motors
- variable-speed pumps
- distributed pumping, including primary, secondary and tertiary circuits.

This brief discussion is included to emphasise the importance of the chosen method of flow control to overall energy consumption. For a fan or pump operating in a fixed system with flow of constant density:

- Volume rate of flow is proportional to speed of rotation.
- Pressure varies as the square of the speed.
- Absorbed power varies as the cube of the speed.

Regulation of flow by the use of dampers or valves will therefore in general be less efficient than by the use of variable-speed drives. This topic is treated in detail in a BRECSU publication[15], from which Figure 3.26 is taken. The use of VSD in control applications will be considered in more detail in section 5.

3.7 Control panels and motor control centres

It is standard practice to group together low and medium voltage (e.g. 240 V single-phase and 415 V three-phase) control equipment and mount it in an enclosed cabinet. This gives protection for the equipment and operating staff and allows convenient connection of hardwired interlocks. Thus the cabinet will contain such equipment as relays, contactors, isolators, fuses, starters and motor speed controllers. Meters may be mounted in the surface of the cabinet as required. Control panels are classified as follows[17]:

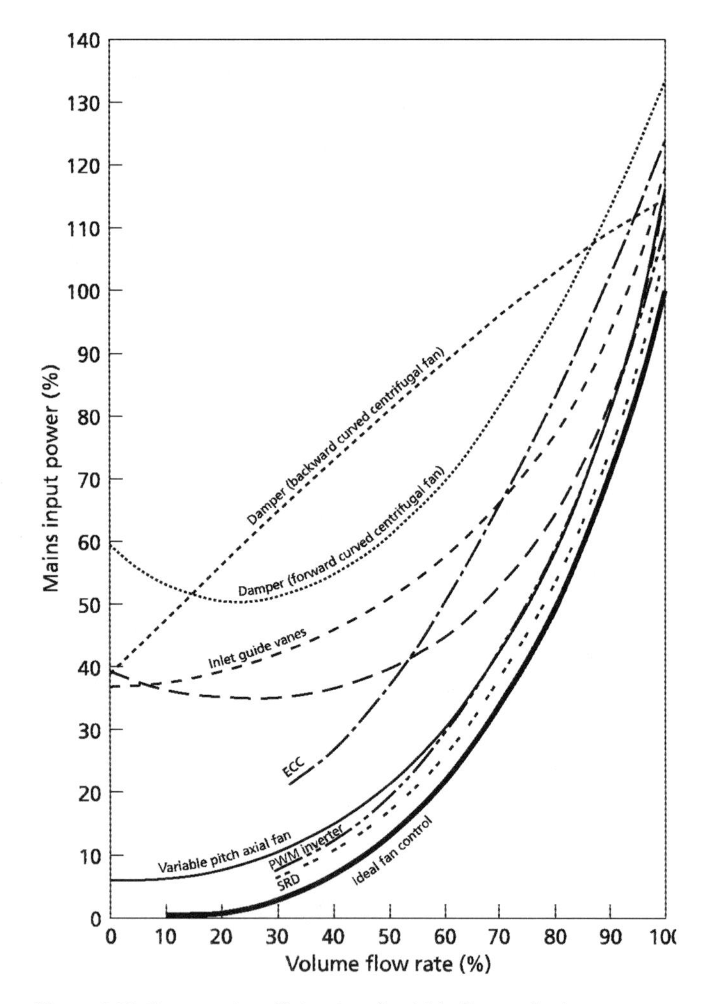

Figure 3.26 Comparative efficiencies of variable-flow methods

— *Form 2.* This is a conventional cupboard type box, with an isolator linked to the door. Opening the door of the cabinet isolates all components, which are commonly mounted on a DIN rail. There is no requirement to shroud connections.

— *Form 3.* The door of the cabinet is lockable but not interlocked to an isolator. All connections are shrouded and individual items of equipment, such as motor starters, have their own covers and isolators. This type of panel is used where several items of plant are served from the same control panel. One item may be isolated and serviced without switching off the rest.

— *Form 4.* The control centre consists of several separate cubicles, each serving a separate item of plant and with its own door and interlocked isolator. The construction serves to contain an explosion and allows work on individual plant items. The separation of electrical equipment also reduces EMC problems; see 4.3. The cubicle construction tends to restrict access for work. An alternative version mounts the equipment on withdrawable racks. This allows the equipment to be withdrawn for servicing or exchange.

With the increased use of inverter drives, there is a move to decentralise motor control, placing the inverter and associated control gear adjacent to the motor being controlled. Several variable-speed drives incorporate their own local controller, which may be connected to the BMS. Inverters require to be air cooled and placing several inverters in a control cabinet will require fan cooling.

Control panels usually have a separate section with its own door for the extra-low voltage (ELV) (e.g. 24 V) control system, which is not isolated. It is therefore possible to obtain access to the control wiring while the plant is in operation. There is normally a facility for plugging in a laptop PC or other interface for diagnosis and servicing. It is good practice to allow only ELV control cables to enter or leave the control centre. In particular, LV interlock cables should not enter the box, since they could remain live even when the box was nominally isolated. Some plant provides 240 V interlock signals, e.g. for boiler lockout. A relay at the plant should be fitted to provide a voltage-free contact for use by the control system.

Control panels and motor control centres are normally constructed by specialist subcontractors and delivered to site with internal wiring complete. Further information may be obtained from the regulations set out in *BS CEN 60439*[17].

3.8 The intelligent outstation

The intelligent outstation, or universal controller, is at the heart of the contemporary building management system (BMS). Known variously as a field processing unit, distributed processing unit, freely programmable controller or simply as a controller, it combines a stand-alone control capability with the ability to communicate over a network with the head end supervisor. Controllers are available from many manufacturers and share the characteristic of being highly versatile, able to accept a range of inputs, to drive different output devices and to be programmed to carry out most common control tasks. Where the controller is dedicated to suit a particular piece of plant, such as a VAV box or chiller unit, it is termed a unitary controller or dedicated function controller.

Controllers incorporate all or most of the following components and can usually accept additional modules as plug in units.

— *inputs*: a module accepting a range of different types of input

— *outputs*: voltage and current outputs, plus relay closures

— *communications*: the ability to communicate with the supervisor and other controllers

— *processor*: a microprocessor to perform the control operations

— *configuration modules*: a range of software modules which can be linked to carry out the required control functions

— *clock*

— *power supply*

— *local display*: either a built in display or provision for a plug in display unit

— *local supervisor*: provision to plug in a supervisor.

The components are housed in a case which is normally designed to be tamper proof. Protection against harsh environments is categorised by the ingress protection system. The degree of protection is categorised by the two-digit IP number[18] (Table 3.13). The first indicates the

Table 3.13 IP protection numbers IEC 144

IP	Foreign body protection against:	Water protection against:
0	None	None
1	Large foreign bodies	Vertical dripping water
2	Medium sized foreign bodies	Angled dripping water
3	Small foreign bodies	Fine water spray
4	Granular foreign bodies	Heavy water spray
5	Dust deposit	Water jet
6	Dust ingress	Water flow
7		Immersion
8		Submersion

degree of protection from dust, running from 0 (no protection) to 6 (dust tight). The second digit indicates water protection, running from 0 (no protection) to 8 (submersible). Controllers may operate in an electrically noisy environment and electromagnetic compatibility (EMC) is discussed in 4.3.

3.8.1 Configuration

The controller must be configured before it can be put into operation. This may be divided into:

- system configuration
- hardware configuration
- strategy configuration.

3.8.1.1 System configuration

The controller is part of a larger system, communicating with the supervisor and other controllers. Other devices may communicate with the controller, either to interrogate it for information or to download software or alter control parameters:

- local display panel
- remote display panel
- local or central supervisor
- another controller.

In addition, the controller can originate messages which are transmitted to other devices:

- alarms which can be received by the supervisor, printer or display
- communications transmitted to another controller.

A supervisor is a computer which allows an operator access to data held within controllers and also allows modification to the software held inside the controller. On a large BMS, there is a head end supervisor, which has access to the entire BMS network. It is normally possible to use a portable PC connected locally to a controller, which may be restricted to communicate with that controller alone. The supervisor may communicate in four modes, to restrict unauthorised access to the BMS configuration.

- *Read only*. The supervisor may be used to interrogate controllers and to display values and produce reports. No changes may be made to any system parameters.
- *Supervisor mode*. This mode enables the operator to change set points and operating times. It is normally password protected.
- *Configuration mode*. This mode enables the engineer to change parameters or set up and fundamentally modify the control strategy resident in the controller.
- *Upload/download*. This mode enables the operator to take a copy (i.e. upload) of the strategy defining file from the controller memory and then store on floppy disc. In the unlikely case of loss of memory in the controller, it is possible to download the file from the supervisor to the controller.

3.8.1.2 Hardware configuration

Typically, a controller is hardwired to a number of inputs and outputs, and communicates with the rest of the BMS over some form of network. These have to be set up correctly.

Inputs. Controllers can accept a range of analogue and digital inputs. Each input channel must be configured to match the connected input. Depending on the type of controller, this may be done by physically changing input modules, changing links or components. A universal input channel may accept any form of input, but still requires configuration by changing internal links. The range of analogue inputs includes:

- *Voltage*. This is usually 0–10 V DC to match the standard output of a sensor transducer.
- *Current*. This is usually 4–20 mA DC to match the standard.
- *Passive*. Some controllers can be configured to accept thermistors or resistance thermometers wired in directly to the input channel. Measurement circuitry is incorporated in the controller.
- *Digital*. This operates on a voltage-free contact closure and is used to monitor plant status. The controller provides a 24 V DC supply to energise the input signal.
- *Pulse*. This accepts pulses from pulsing energy meters, with defined minimum pulse width.

Outputs. Controllers provide a range of analogue and relay outputs, including:

- *Analogue voltage*. Usually 0–10 V. This signal may be used to control most actuators without further amplification. In some cases an intermediate driver module may be required.
- *Relay output*: changeover relays, used to energise controlled plant.
- *Time proportioning*: a relay output with variable mark/space ratio.
- *Pulsed*: 24 V AC pulsed operation, required by some actuators.
- *Resistive*: 0–135 Ω potentiometer output.
- *Network address*. The network node address of the controller must be identified by a by a unique number. This may be set either by a switch within the controller or during software configuration.
- *Communications*. Setting up the communications interface depends on the network.

3.8.1.3 Strategy configuration

Controllers are pre-programmed with a number of software modules. The modules exist as 'firmware' within the controller memory and are not re-programmable. In order for the controller to perform the desired function, the modules are brought into action by linking them together by so-called 'soft wiring'. In addition to the firmware, advanced controllers allow the writing of custom control routines in the controller's own programming language. The whole process is known as strategy configuration. Configuration is carried out using a proprietary program running on a PC. Each controller manufacturer has its own high level programming tool, which normally consists of a pictorial representation of the firmware modules contained in the controller, which may be linked on screen by 'soft wiring'. In addition to the interconnections of the modules, values of control parameters may be keyed in at this stage. The complete configuration may be downloaded into the controller. The program remains available for use in other controllers. A controller contains some or all of the software modules listed in Table 3.14.

3.9 Summary

The selection and correct use of appropriate components is essential to produce a reliable and satisfactory control system. The types of sensor available are summarised and advice given on selection. Sensors must be mounted in a position that is representative of the quantity being measured and which gives access for maintenance and calibration. Experience shows that incorrect installation is the major cause of problems with sensors.

Actuators convert the output from a controller into the mechanical movement of a valve or damper. The actuator chosen must have sufficient force to operate the controlled device; it is important to select an actuator with the required degree of positioning accuracy, particularly if the controlled device has to operate over its whole range. Modern actuators incorporate microprocessor control which gives accurate positioning and feedback of position to the control system.

Control valves for hydraulic circuits should be chosen to give adequate authority and a characterisation appropriate for the output device. The aim is to provide a final controlled output which is nearly linear with valve position; this will assist in providing satisfactory control.

Table 3.14 Typical controller software modules

Module	Example
Timer functions	Scheduling, time zones
Optimum start stop	Self-adaptive
PID control	Self-tuning control loop
Data logging	Stores input and output data
Metering	Utility energy metering from pulsed meter
Load management	Load cycling and maximum demand control
Actuator drivers	Time proportioning
Lighting	Presence or absence detection, dimming
Maths functions	Log, square root, enthalpy, sensor scaling
Knob	Input settings
Logic	Timers, counters, hours run
Pulse counting	Energy measurement
Sequencing	Controls module operation sequence

The same considerations apply to dampers. The section gives guidance on the choice between parallel and opposed dampers and emphasises the need to provide sufficient damper authority; this may require the damper area to be smaller than the duct. It is now possible to use standard induction motors with variable-speed drives. The use of VSD and two-port valves can lead to substantial savings in pump and fan energy consumption. The design of hydraulic systems incorporating two-port valves demands care to avoid problems at part load.

Lastly, the intelligent controller is described. This is the workhorse of most control systems. It provides local control of subsystems while communicating over a network with the building management system. Controllers contain a comprehensive range of pre-programmed control functions, which require to be configured for particular applications, using proprietary programming software provided by the controller manufacturer.

References

1 Carotenuto A, Dell'Isola M, Crovini L and Actis A Accuracy of commercial dew point hygrometer *ASHRAE Transactions* **101**(2) 252–259 (1995)

2 Carotenuto A, Dell'Isola A J and Vigo P A comparison between the metrological performances of humidity sensors *Transactions of Measurement and Control* **17**(1) 21–32 (1994)

3 *A guide to the measurement of humidity* (London: Institute of Measurement and Control) (1996)

4 Potter I N and Booth W B *CO_2 controlled mechanical ventilation systems* TN 12/94.1 (Bracknell: Building Services Research and Information Association) (1994)

5 Martin A J and Booth W B *Indoor air quality sensors* TN 1/96 (Bracknell: Building Services Research and Information Association) (1996)

6 *Control sensor installation* (London: Building Controls Group) (1996)

7 UKAS *Traceability of measurement: general policy M50* (Feltham: United Kingdom Accreditation Service) (1996)

8 Petitjean R *Total hydronic balancing* 2nd ed (Ljung, Sweden: Tour and Andersson) (1997)

9 Parsloe C J *The design of variable speed pumping systems* Draft 79000/2 (Bracknell: Building Services Research and Information Association) (1999)

10 Levermore G J *Building Energy Management Systems,* 2nd ed (London: E & F N Spon) (2000)

11 Honeywell *Engineering manual of automatic control* SI ed (Minneapolis: Honeywell Inc.) (1995)

12 Johnson *Damper application engineering*. Engineering Report H352 H352 (Milwaukee: Johnson Controls Inc.) (1991)

13 DETR *Good Practice Guide 2. Energy savings with electric motors and drives* Updated 1998 (Harwell: ETSU) (1998)

14 *BS EN 60947-4-1 Electromechanical contactors and motor starters* (London: British Standards Institution) (1995)

15 *Variable flow control* GIR 41 (Watford: Building Research Energy Conservation Support Unit) (1996)

16 ABB Getting it together *Electrical Review* **230**(12) 31–33 (1997)

17 *BS EN 60439-1 Specification for low-voltage switchgear and control gear assemblies* (London: British Standards Institution) (1994)

18 *BS EN 60 529 Specification for degrees of protection provided by enclosures* (IP code) (London: British Standards Institution) (1991)

4 Systems, networks and integration

4.0 General
4.1 BMS architecture
4.2 Networks
4.3 Electromagnetic compatibility (EMC)
4.4 Systems integration
4.5 User interface
4.6 Summary

Section 4 deals with the ways in which components are linked together to form a complete building management system. Large networks are divided into management, automation and field levels, each with their own communications protocols. The attributes of the major standard protocols are described. Systems integration allows communication between different building services systems, with consequent advantages of more efficient operation; the principles of integration are described, along with ways of avoiding possible contractual problems. Recommended installation practices to avoid problems of electromagnetic interference are described and good cabling practices set out. The principles of good user interface design are described.

4.0 General

The term building management system encompasses a wide range of control systems, ranging from dedicated controllers hardwired to the equipment they control, to large-scale distributed systems, extending over several sites or even countries. The development of larger integrated systems depends on the existence of communication protocols which allow devices from different sources to communicate with each other. The organisation of building management systems and the networks on which they depend is the subject of this section.

4.1 BMS architecture

The organisation of the various control elements into a comprehensive BMS is termed the system architecture.

4.1.1 Conventional controls

Electronic control systems were first developed using analogue signals to carry information between parts of the system and to provide the control action. Controllers were based on operational amplifiers, now replaced by microprocessors. The sensor could be connected directly to the controller, in which case the controller required the appropriate input circuitry to handle the signal input. Greater flexibility could be achieved by using a transducer to convert the sensor output to one of the several standard signal forms. This has the additional advantage of providing a stronger signal where transmission over a distance is required, where a small signal such as a thermocouple EMF might become corrupted. Outputs from the controller are standardised and are connected directly to the required actuator. Voltage-to-pressure and pressure-to-voltage transducers allow the use of pneumatic actuators with electronic controllers.

Controllers based on analogue inputs and outputs still form the basis of control systems for smaller buildings. Each controller is linked by direct connection to the controlled output device and performs a single function, i.e. a single control loop. Several independent control systems may coexist within a building, e.g. a boiler control incorporating time switch and flow temperature control plus independent zone temperature controllers. Conventional controllers share the following characteristics:

- single function controller, hardwired to sensor and actuator
- analogue inputs and outputs
- sensor signal conditioning required for long cable lengths
- standardised controller output
- no central supervision
- cost-effective for simple buildings.

4.1.2 Centralised intelligence

The advent of the microprocessor allowed the development of direct digital control (DDC) systems. The controller function is carried by a software program, which can be written to execute any desired control characteristic. Many control loops can be handled by the same processor. Changes to the control function can be made by changes to software alone, without any requirement to change hardware. The development of adaptive and self-learning control algorithms becomes possible.

The first systems using DDC employed a centralised controller (Figure 4.1) incorporating all the processing capability in one unit, known as the central processing unit (CPU). The CPU is connected to some form of user interface known as a supervisor, which allows the operator to view the status of the system and change control parameters. All sensors are brought back to the central unit. Where the path lengths are long this will normally require signal conditioning at the sensor. The signal conditioner combines a transducer, which converts the sensor output to a standard form, and the transmitter which amplifies the signal for transmission over long distances of wire. At the central controller, the incoming signal is conditioned by an input card, and converted to a digital form by an analogue-to-digital converter before being sent to the microprocessor for processing. The treatment of the output signal from the controller follows the same pattern in reverse. The output from the microprocessor is converted to analogue form, and converted to a standard form before being sent to the transducer at the controlled device.

In control systems with centralised intelligence:

- All sensor inputs are wired to central controller (CPU).
- Control functions are executed digitally.
- Software may be changed without hardware change.
- Reporting and parameter changes are made via the supervisor.
- The wiring cost is high.
- The system is vulnerable to controller malfunction.
- They have been superseded by the intelligent outstation, i.e. distributed intelligence.

4.1.2.1 Dumb outstation

The provision of separate wiring for each sensor and actuator is expensive and this has led to the introduction of local outstations, also known as data-gathering panels. Sensors are connected into the outstation which performs signal processing and communications functions. A so-called dumb outstation (Figure 4.2) does not carry out any control function. Connection to the central controller is by a data transmission cable. The CPU is required to address each point in turn; this is known as polling and in a large system results in slow operation. The features of a dumb outstation are:

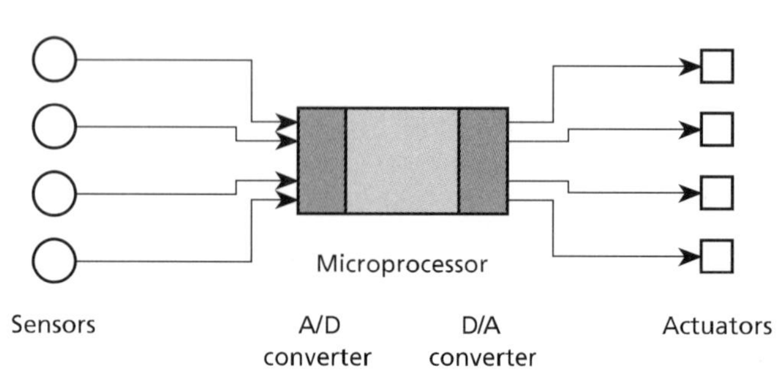

Figure 4.1 DDC with centralised controller and hardwiring

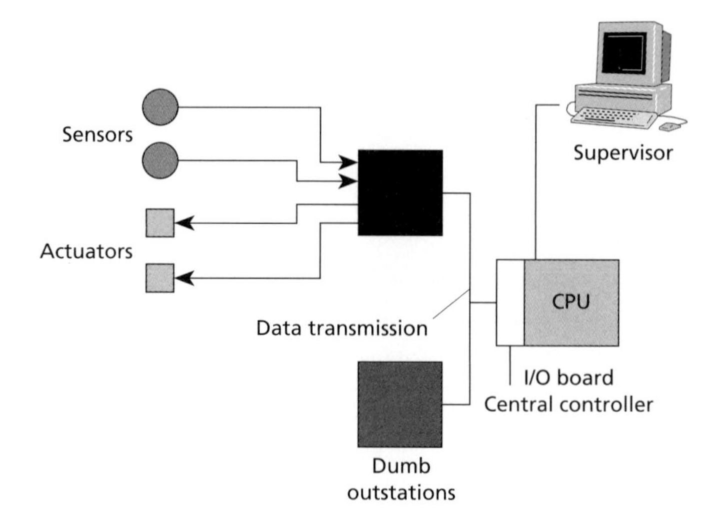

Figure 4.2 Centralised controller with dumb outstations

- Sensors and actuators are connected to local outstation.
- Outstation contains signal conditioning and communication circuits.
- Outstation performs no control function.
- CPU polls each point in turn.
- Outstations vulnerable to CPU failure.
- CPU has to handle large amounts of data.
- They have been superseded by intelligent outstations.

4.1.3 Programmable controllers

While the use of dumb outstations reduces the wiring requirements for a large BMS, it is vulnerable to a failure in the CPU. It is now the general rule to use programmable controllers or intelligent outstations (Figure 4.3) which incorporate their own microprocessor and carry out local control functions, which will operate in the event of a CPU problem. Communication with the central supervisor is for the purpose of supervision, fault reporting, data collection

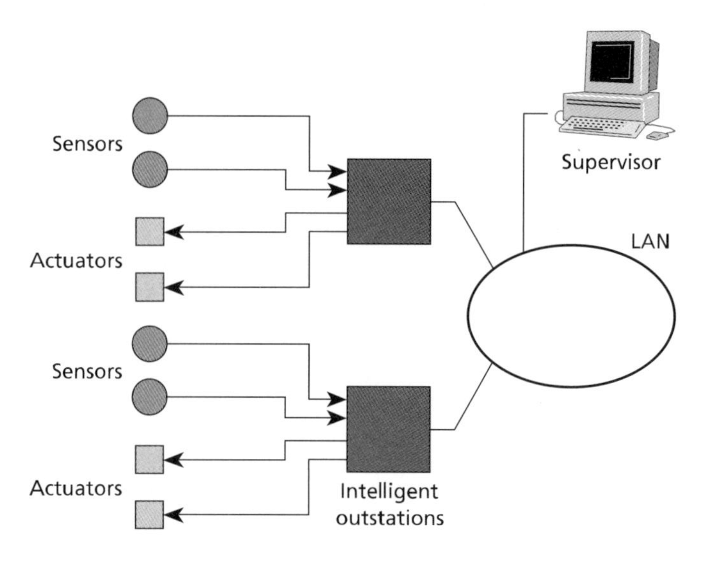

Figure 4.3 BMS with intelligent outstations

and the ability to change control parameters as necessary. Communication between outstation and supervisor is normally using some form of local area network; this may employ a proprietary protocol or one of the standard systems discussed further below. The intelligent outstation, also known as the universal or freely programmable controller, is the main building block of the modern BMS and is described in detail in 3.8. The main features are:

— Control function is incorporated in outstations.

— Outstations function independently of CPU.

— Central station accepts reports and fault alarms.

— Central station may alter set points and other parameters.

— Devices communicate via a local area network or bus.

— It is the basic component of a modern control system.

A unitary controller is distinguished from a general controller in that it is designed to control a specific type of plant, e.g. a VAV box or air handling unit. Sensors and actuators may be hardwired or communicate with the unitary controller over a field bus. Unitary controllers are found supplied with packaged systems; see 4.1.6.

4.1.4 Bus systems

Decentralisation is carried further using a bus system (Figure 4.4). Intelligent devices in the control system incorporate their own microprocessor and communications ability. Devices are connected to a bus which allows communication between all devices using a well-defined protocol. Each device is individually addressable and programmable. For instance, an intelligent temperature sensor at the field bus level will transmit a message onto the network identifying itself as well as including a temperature value. Any controller or other device on the bus may then use this message. In order to limit total traffic on a network, a bus network may be divided into segments connected by bridges, which can transmit only messages destined for outside the segment. Connection between buses at different levels is via gateways or protocol interface units, which can perform translation where the control levels employ different protocols, as well as filtering. Some of the major bus systems used in BMS are discussed below. The characteristics of a bus system are:

— All devices communicate using standard protocol.

— All devices are individually addressable.

— Devices communicate directly with each other as required.

— The system is decentralised.

— Interoperability between different manufacturers is a possibility.

4.1.5 Modular systems

A particular example of a bus system is based on the concept of modular controls. The system uses a range of modular units, each of which is programmed to control a specific item of plant. In addition, there is a range of zone controllers which perform the function of a local user interface. All the modules communicate on the same twisted-pair bus system, using the LonTalk protocol. Installation and configuration are simple; the control modules are pre-programmed and are self-tuning, so virtually all the installation that is required is hardwiring to the plant that is to be controlled and the setting of the module's address. Adjustment of set points and time schedules is done at the zone controller, which is normally mounted in the controlled space and available for adjustment by the user. The system is suitable for situations where the services of a controls specialist is not available for installation or commissioning. Systems can be built incorporating several hundred modules and remote supervision is possible.

4.1.6 Packaged systems

HVAC plant, such as multiple boilers or chillers, may be supplied as a packaged unit complete with factory installed unitary control system (Figure 4.5). In the example of a multiple boiler unit, the fitted controller would incorporate all safety cutouts, ignition procedures, boiler sequencing and the control of ancillary devices such as flue dilution fans. Such a unit will control itself independently. Typically, the required flow temperature is set directly at the local controller, but it may be desirable to connect it to the main BMS for the purpose of time scheduling and transmitting a general alarm to the supervisor. It will be necessary to check the compatibility of the manufacturer's controller with the main BMS. The main features are:

— packaged plant with factory-installed unitary controller

— often an economical solution

— limited interface with BMS

— sometimes compatibility problems.

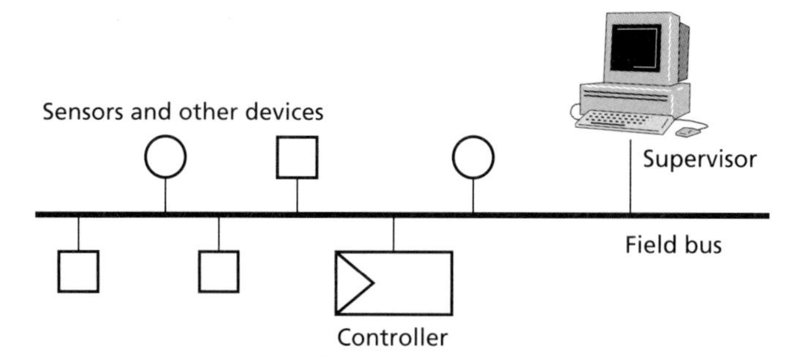

Figure 4.4 Bus system

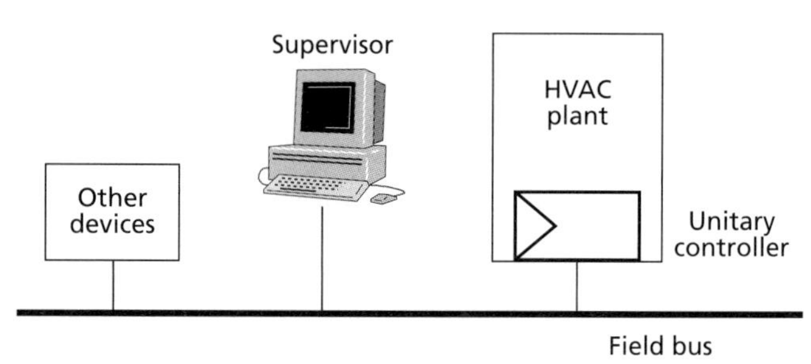

Figure 4.5 Factory-fitted unitary controller

4.1.7 Three-level model

The BMS of a large modern building may be arranged in a hierarchical structure. A useful concept is that of communication level. The European Committee for Standardisation (CEN) divides communications within a building into three levels[20] (Figure 4.6):

- management level, e.g. supervisor to supervisor
- automation level, e.g. DDC controllers
- field level, e.g. sensors and actuators.

Information which is needed locally is not passed upwards to higher levels, reducing the traffic on the system. Communication with sensors and actuators takes place at the field level. Intelligent controllers sit at the automation level, as does the head end supervisor, which has access to all systems at this level. The management level accesses information from the automation level and has access to a wide area network, so that it can exchange management information with independent systems in the same or other buildings. The control function at management level is limited. Interfaces between levels are used to:

- restrict low level traffic from passing to the level above
- translate from one protocol to another.

The three-layer structure has the advantage of organising network traffic efficiently, but requires the additional complication of linking the three levels. A single network carrying all communications may be sufficient for many applications.

4.1.8 Utilisation of IT network

Modern organisations usually have an extensive IT system, which necessarily incorporates a network covering the building. IT networks are designed to carry very large amounts of information and it may be possible to use the network for the purposes of the BMS (Figure 4.7). This can save infrastructure costs by the elimination of a separate BMS network, but places increased demands on the installation to ensure that the BMS is operational in time for building commissioning. Anyone with access to the IT system has potential access to the BMS, which could allow staff to monitor or reset local temperatures. A BMS requires real-time operation and there are potential problems over sharing the network with other IT traffic, which may involve transmission of large files with the possibility of delays. Using the IT system for BMS purposes:

- presupposes an extensive IT network
- reduces infrastructure cost
- high speed data transmission
- ready access to BMS by staff via workstations
- BMS may be disrupted during modifications to the IT system
- controls commissioning dependent on IT network.

The term 'systems integration' implies that several, or all, building services systems are brought together into a single cooperative management system and are able to communicate with each other and act accordingly. For example, information on occupation level derived from the access control system can be used as an input to the

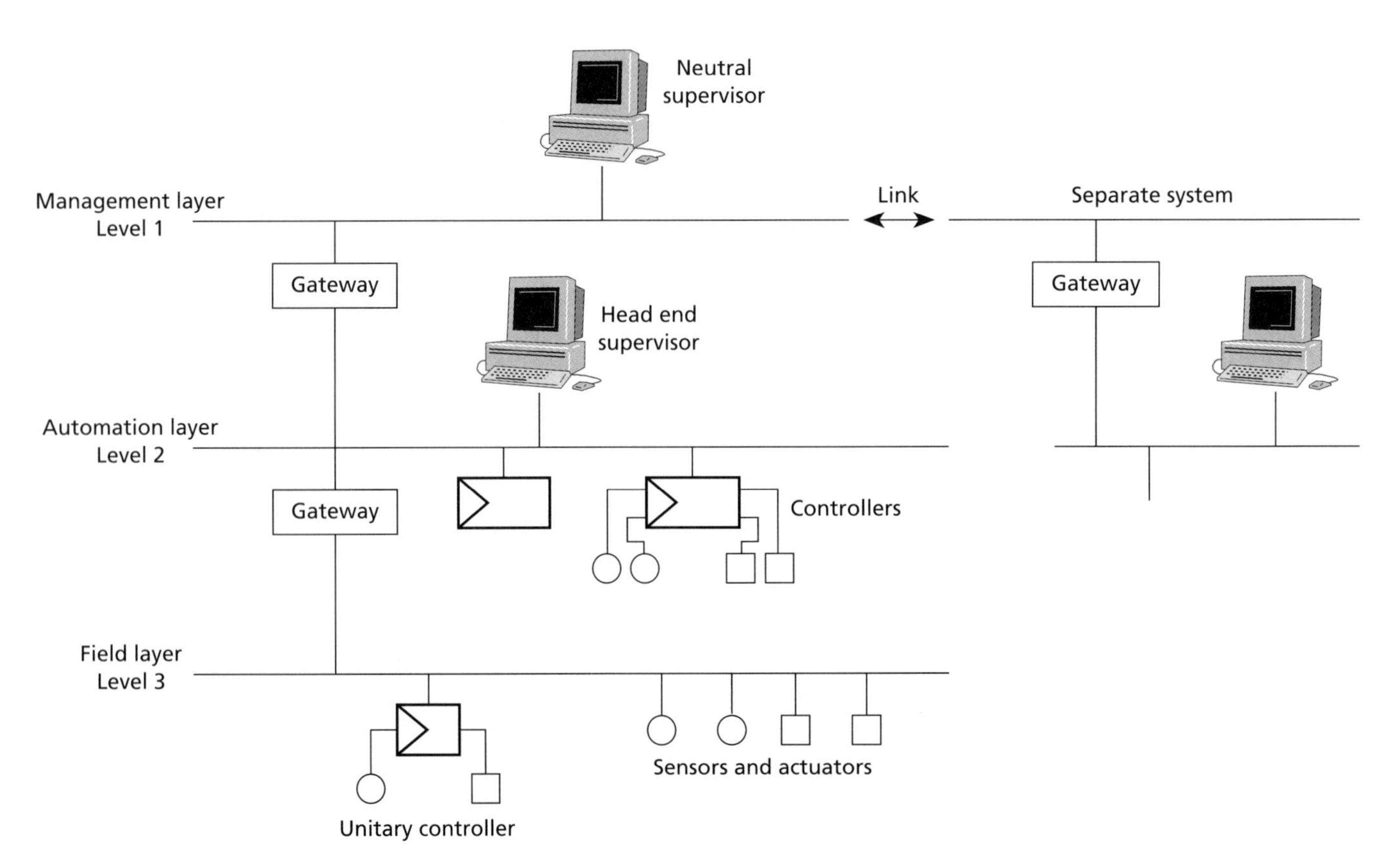

Figure 4.6 The three-layer model of BMS architecture

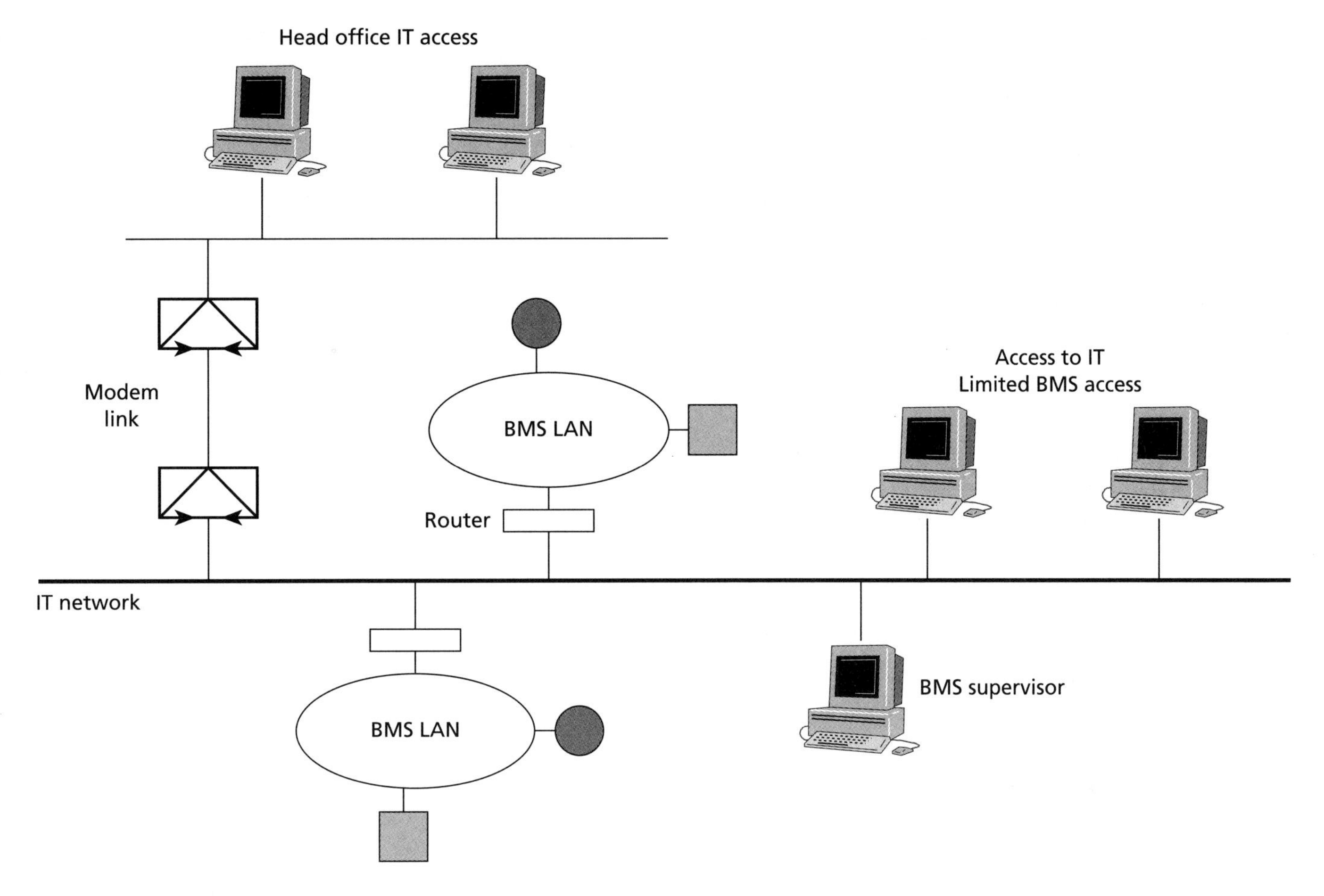

Figure 4.7 BMS and IT shared network

ventilation controller. The increasing use of standard protocols makes communication between systems easier; gateways can be used where it is necessary to bridge between different protocols. The concept is also known as the intelligent building system (IBS). All building services may be controlled from a single supervisor. Systems integration is discussed further in 4.4.

4.2 Networks

Network implies that a number of devices are connected with each other via a communications system. A communications network is characterised by two essential parts:

— physical medium, which is used to transport the signals, e.g. wire, optical fibre, radio

— protocol, which is the set of common language rules for the communication signals.

Devices communicating over a segment of the physical medium and using the same protocol form a local area network (LAN). Separate LANS, which may employ a range of physical media and protocols, can communicate with each other with the help of routers, forming a wide area network (WAN). The components of a WAN may be widely separated, communicating via satellite. An internetwork is an alternative name for a WAN. The manner in which networks are arranged and interconnected is termed the network topology, discussed further below.

4.2.1 Topology

The manner in which devices and their interconnections are laid out is termed the topology of the network. The chosen topology will affect the amount of cabling required and the flexibility of layout. It will also have an influence on the technical aspects of communication, such as signalling rate and reliability. Figure 4.8 illustrates the main topologies.

— *Point to point*. Devices are connected directly to each other. Features are:

  — Only one connection affected by a fault.

  — Simultaneous transmissions are possible.

  — Wiring cost is high.

  — There is no automatic interconnection between devices.

— *Star structure*. Each device is directly connected to a central unit. Features are:

  — Failure of one device does not affect others.

  — Failure of central unit gives system failure.

  — Cabling cost is high.

— *Ring structure*. Bus devices are interconnected in a ring structure. Each device communicates with its neighbours and passes messages on. It is possible to incorporate watchdog relays, which bypass individual nodes when a device is powered down; this retains the integrity of the ring. Features are:

  — Cabling cost is modest and expansion is easy.

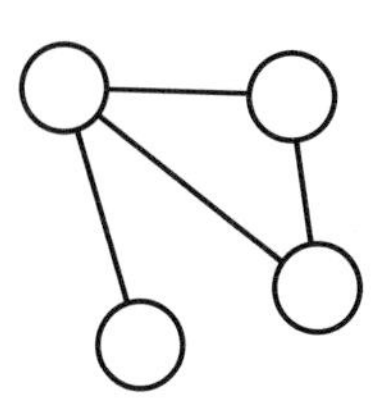

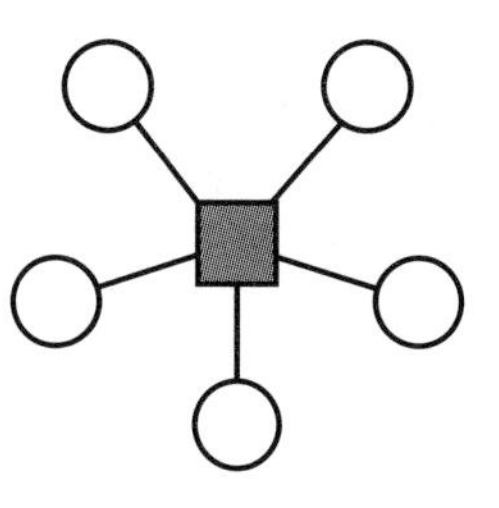

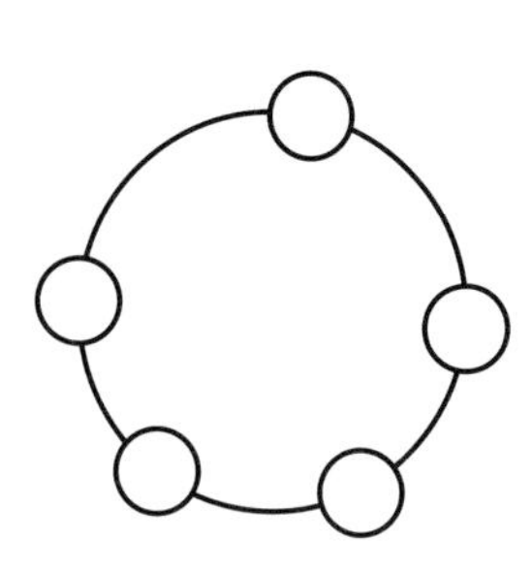

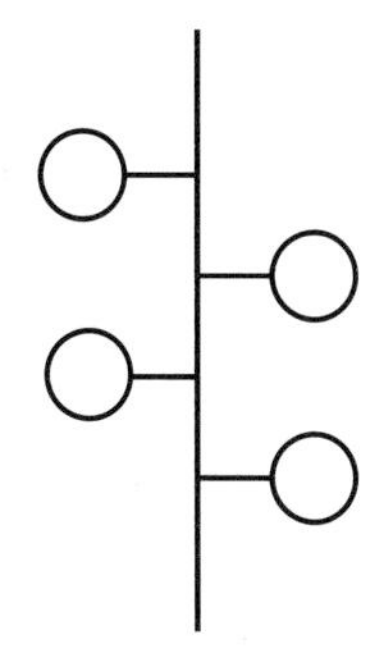

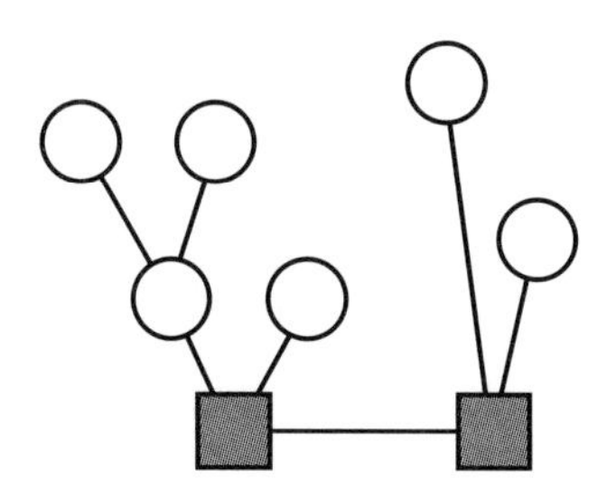

Figure 4.8 Network topologies

— Failure of one device leads to system failure.

Example: Arcnet.

— *Bus structure*. All bus devices communicate over the same transmission path. Features are:

— Any device can communicate with any other.

— Failure of one device does not affect communication between other devices.

— Cabling cost is low and expansion is simple.

— Bus can only handle one message at a time.

Example: Ethernet.

— *Tree structure*. The tree structure is a combination of the structures described above. Information sent to a root is passed upwards to all devices sitting on the relevant branches, which are connected as a bus structure. The devices at the roots of the network serve to amplify and send on messages received from other branches, or may act as gateways, converting messages from one transmission medium to another. Features are:

— Aids organisation of subsystems.

— No direct connection between all bus devices.

— Failure of the root leads to system failure.

Examples: LonWorks, EIBus.

The most common topology used in the major field network systems is a form of tree structure, whereby local bus topologies may be connected together using couplers or routers at each root. By this means, subsystems using different communication media may be coupled together and operate using the same protocol.

4.2.2 Communication media

Communication between devices takes place over a local area network (LAN). This has two aspects, the network protocol and the physical embodiment in the communications medium. Several communication media are available for BMS purposes (Table 4.1). Achievable transmission rates are steadily increasing.

Coaxial cable and fibre optic cable have very high data-carrying capability and can transmit data over long

Table 4.1 Communication media

Communication medium	Transmission rate (bits/s)	Application
Power line carrier	5 k	Limited data-carrying capacity. Application is primarily domestic or retrofit
Telephone line	56 k	Useful for communicating between physically separated LANs. Unsuitable for real-time continuous control
ISDN telephone line	128 k	High capacity
Unshielded twisted pair (UTP)	1 M	The most common medium for voice and data communication. Four-core cable may be used for digital signalling
Shielded twisted pair	16 M	Offers higher capacity than UTP but bulkier and more expensive
Coaxial cable	10 M	High capacity data transmission. Basis of the older Ethernet systems
Optical fibre	1000 M	The highest data transmission capacity. Immune from EMC problems. Unsuitable for multiple local connections
Radio	10 k	Short range licence-exempt radio provides flexible operation

distances. They require specialist installation and are used in many IT systems. While a BMS can well share the use of such a network, the high specification is not necessary for a BMS. The most widely used medium for a BMS bus is twisted-pair copper wire. Restricting the transmission rate to 10 kbit/s makes the medium resistant to interference and the cable may share trunking with power cables; no specialist installation is required. Where it is difficult to run cables, it is possible to use power line carrier or radio systems. Power lines are susceptible to much electrical noise, but the development of spread spectrum techniques has made power line carrier systems more resistance to interference and good signal rates are possible.

Low power radio systems operating in the UHF band may be used without a licence for telemetry and data transmission(1). They have the advantage of low infrastructure cost, low installation cost and flexibility in use. National authorities have issued specifications for licence-exempt radio transmission, which define the maximum permissible RF power, the frequency and bandwidth of the transmission and the number of channels. Standards differ between continental Europe and the UK. In the UK there are two relevant specifications:

— *MPT1340* defines a single-channel low power system with a range of tens of meters. Unit costs are low.

— *MPT1329* allows up to 31-channel operation and can achieve a range of several kilometres with good penetration within buildings. Data transmission rates of up to 10 kbit/s are possible.

Table 4.2 gives the classification of data cables(2). Category 6 cabling is the current standard for data cabling.

Table 4.2 Data cable classification

Category	Description
1	Basic telephone cable for voice
2	Data cable for use to 4 Mbit/s
3	LAN cable characterised up to 16 MHz (Class C link)
4	LAN cable characterised to 20 MHz
5	LAN cable characterised up to 100 MHz (Class D link)
6	LAN cable characterised up to 250 MHz

Several protocols have been developed, not all of which have been exploited for use in BMSs. Table 4.3 summarises some of the bus protocols that are in use.

4.2.3 Structured cabling

The increasing use of IT in business operations has required developments in the cabling infrastructure that interconnects the many terminal devices. Networks require sufficient capacity to deal with demands such as video conferencing; network loads of 1 Gbit/s have been reached and are projected to increase. The consequence is that the IT cabling in a building may rapidly become inadequate, with the consequent expense and disruption of upgrading.

The concept of structured cabling, also known as universal cabling, aims to install a cabling infrastructure in a building or group of buildings which will meet the requirements of all potential users of the building during its lifetime, without the need for re-cabling. PCs and other terminal devices, including BMS components, are plugged into local terminal outlets. The initial installation aims to provide sufficient terminal outlets to accommodate future internal rearrangements and variations of work requirements.

The cabling topology is illustrated in Figure 4.9. Three levels of installed cabling are involved, plus a patch cable to connect each device to the local terminal.

— *Horizontal cabling*. Each work area of the building, typically a floor or part floor, has a floor distributor, also known as the patch frame, mounted in the telecommunications closet. Direct cabling, normally Category 5, runs from the floor distributor to each of the local terminal outlets in the work area. The outlets are typically floor boxes containing an RJ45 socket, installed at sufficient density to allow for future needs. The length of horizontal cabling is limited to 90 m.

— *Building backbone cabling*. Within a building, each floor distributor is connected to the building distributor via building backbone cabling. This cabling is also termed building riser cabling, since it commonly interconnects floors of a building.

Table 4.3 Some network protocols

Protocol	Standard	Comment
Ethernet	*ISO 8802-3*	Data transmission up to 10 Mbit/s. The most common IT medium(3)
ARCNET	*ANSI 878.1*	High performance LAN with data exchange up to 2.5 Mbit/s. Used widely in industrial applications(4)
BACnet MS/TP		Master Slave/Token Passing serial communications using an *EIA-485* signal. Provides adequate performance at reasonable cost
Point to point	*EIA 232*	Serial communication between two devices. Typically used for dial-up communication using modems
LonWorks		The Lon network is based on the proprietary Neuron chip and can be applied to a wide variety of physical media
Batibus	*ISO JTC 1 SC25*	Uses twisted pair at 4.8 kbit/s
TCP/IP		Transmission Control Protocol/Internet Protocol. It encompasses media access, packet transport, session communications etc. TCP/IP is supported by a large number of hardware and software vendors. Several Internet protocols deal with aspects of TCP/IP. The most important have been collected into a three-volume set, the *DDN Protocol Handbook*(5)
European Installation Bus		Uses twisted pair at 9.6 kbit/s

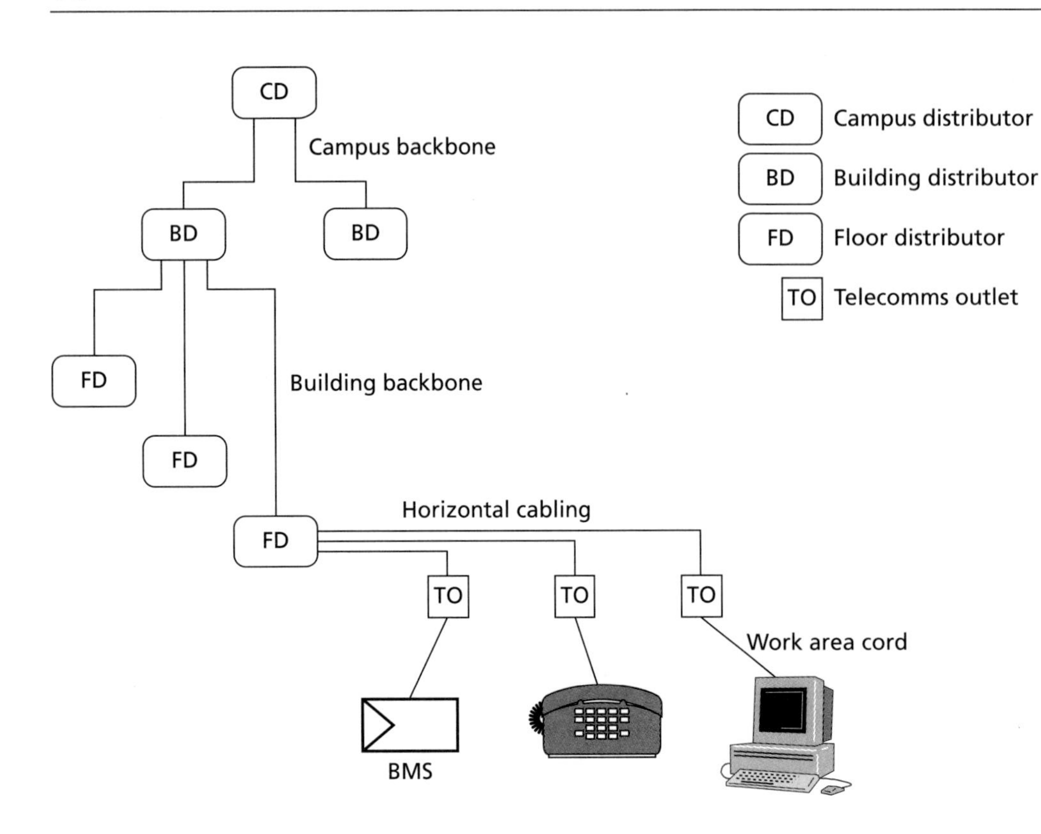

Figure 4.9 Topology of structured cabling installation

— *Campus backbone cabling*. Where several buildings share the same network, campus backbone cabling is used to connect building distributors with the campus distributor. Fibre optic cable is commonly used for this purpose.

— *Work area cabling*. Individual devices, such as PCs, printers, telephones and BMS components, are connected to the terminal outlets via work area cabling, also known as patch cabling. Each device which is to be connected to the network is provided with a jack plug compatible with the terminal outlets.

For BMS purposes, use of an 8-pin RJ45 plug and socket is recommended. Thus, where a BMS system is to share a structured cabling system, it would be specified that all control components, e.g. actuators, controllers, VAV boxes, would be provided with a fly lead terminated by a standard jack plug. The structured cabling system must be specified to provide a large number of sockets in suitable locations, in service ducts and plant rooms, which will be adequate for present and future needs.

Where structured cabling is specified, the cabling system is designed as a complete system to incorporate all data transmission needs. The wiring is then installed by a single contractor. This eliminates many problems of coordination and commissioning and is claimed to lead to cost savings. However, any decision to incorporate the BMS with a structured cabling system should be considered carefully. Flexibility is not normally a high priority with HVAC and BMS systems, since major repositioning of plant is an unusual event and is likely to be associated with a major refurbishment in the building. Sharing communications networks may introduce contractual problems over installation, commissioning and maintenance, which should be resolved at an early stage. For some BMS applications, the current-carrying capacity of the cable used in an IT network may be inadequate; again, this problem must be resolved in good time.

4.2.4 Standardisation

Components of a network communicate with each other using a common communication protocol. The communication medium itself may take a number of physical forms, but is typically a low voltage twisted-pair cable. Communication normally takes the form of a message or telegram. When required to do so, a device transmits a telegram onto the bus, containing such information as the identification of the sending device, the information to be transmitted, e.g. a temperature value, the address of the device for which the message is intended and additional information needed for error checking. Each device therefore requires a microprocessor to handle message coding and the transmission and reception of messages to the bus; this must include procedures to deal with message prioritisation and the avoidance of message collision when several devices are attempting to transmit at the same time. The rules for message formation form the protocol.

Historically, manufacturers have developed their own proprietary protocols, but there is now a strong move to the adoption of standard protocols. A major advantage of using a BMS network with a standard operating protocol is the degree of compatibility that may be achieved between different pieces of control equipment. Three levels of compatibility can be distinguished:

— *Compatibility or coexistence*. Devices from different manufacturers may use the same communications network without interfering with each other, but can only communicate within their own subgroup of devices.

— *Interoperability*. All devices on the network share the same communications protocol and so may communicate with each other, e.g. one maker's light switch will operate another's light fitting.

— *Interchangeability*. Devices are completely standardised, both physically and operationally, so that devices from different sources may be freely interchanged.

Substantial progress has been made towards the goal of producing networks with interoperability. This implies that devices from different manufacturers which conform to the required standard can be plugged into to the BMS network and operate with devices from other manufacturers. This development has depended on the development and acceptance of the necessary protocols, which are well-defined sets of rules by which devices identify themselves and transmit and receive messages. Several protocols have been developed by national and international bodies and by manufacturers. Full interchangeability is rare. It would benefit users in that more than one source of supply would be available, offering the benefits of competition. However, the requirement that the devices should be operationally identical would hold back developments.

4.2.4.1 Standardising bodies

A well-defined protocol is essential for a bus system if it is to find wide application and allow devices from different manufacturers to operate together. A bus system requires the following properties:

- a governing association
- a sufficient range of published standard protocols
- a testing procedure
- a certification procedure.

The European Committee for Standardisation (CEN) took on the task of defining norms for BMS protocols and started by classifying communications within a building into three levels (Table 4.4).

The management network exists to allow data to be collected for the management of facilities and energy and financial reporting. It can be used to connect complete systems together. The automation network operates at the level of controllers and user interfaces. This is the level at which most BMSs operate, e.g. the connection between an intelligent outstation and the head end supervisor. Interfaces to related services such as lighting, fire or security, take place at this level. FND (Firm Neutral Data Transmission) is a German DIN standard intended for linking BMS central stations and uptake outside Germany and Switzerland is low. At the automation net level, FIP (Field Instrumentation Protocol) is a French standard, which has not become established in building services applications. The field network is the low level network used to connect small devices such as unitary controllers, sensors and actuators. EHS (European Home Systems) was developed with EU funding for domestic applications; this is expected to become incorporated into a new common standard, mentioned in 4.2.5.5. The other major standards are considered in more detail below.

The International Organisation for Standardisation (ISO) has developed a design model which all protocol development should reference. It is called the Open Systems Interconnection (OSI) Basic Reference Model, *ISO 7498*[6], and is also known as the seven-layer model (Table 4.5). The model represents an effort to make the communication problem manageable by breaking it down into a series of smaller problems, each of which can then be tackled independently of the others. This has been done by defining a hierarchy of functions arranged one on top of the other in seven layers where each layer deals with one or more of the issues. The seven-layer model is independent of the CEN 3 level hierarchy; each of the CEN levels contains a protocol which can be analysed in terms of the OSI model.

Table 4.4 Levels of communication within a building (*CEN TC247*, WG4)

Level	Example	Proposed draft standards
Management	Supervisor to supervisor	BACnet, FND
Automation	DDC controllers	BACnet, FIP, Profibus
Field	Sensors, actuators	LON, EIB, Batibus, EHS

For example, the bottom two layers of the stack are the physical and data link layers. According to the OSI model, these layers concern the type of physical interconnection of computers, the electrical signalling, addressing, error detection scheme, and medium access method. The selection of these characteristics, taken together, constitutes what is commonly referred to as a local area network or LAN. The next layer up is called the network layer and describes the characteristics of protocols designed to allow multiple LANs to be connected together. The transport layer optimises the actions of the network level and represents a link between the lower and higher layers. The session and presentation layers deal with wide area networks. The presentation layer provides for a common language for the entire network, so that devices using different protocols may communicate with each other. Finally, at the top, is the application layer which addresses the communication requirements of specific applications, for example, building automation and control. An application layer protocol defines the specific format and content of the messages that two or more computers, each cooperating in a particular application, will use in conversing about their common activities.

Several bus networks are in widespread use for building management systems. Major systems are described in the following sections. The systems share many concepts, though may use different words to describe them. This guide uses the term device to describe a whole physical piece of equipment that is connected to the net; the device contains the necessary communications processing power as well as the hardware necessary for it to carry out its function. The term object may have slightly different meanings in different systems, but is normally defined in terms of the software function of a component or a device.

4.2.5 Standard systems

4.2.5.1 BACnet

BACnet, an acronym for Building Automation and Control Network, was developed under the aegis of the American

Table 4.5 Open Systems Interconnection Basic Reference Model

Layer 7	Application	User interface	
Layer 6	Presentation	Common language	User
Layer 5	Session	Opens and closes dialogue sessions	
Layer 4	Transport	Network optimisation	
Layer 3	Network	Data packet transport round network	
Layer 2	Data link	Message structure	Transport
Layer 1	Physical	Communication medium and signalling	

Society of Heating, Refrigerating and Air Conditioning Engineers (ASHRAE) to provide on open system specifically designed for the requirements of BMSs. It is now formally defined in *ANSI/ASHRAE standard 135-1995*[7]. BACnet focused on defining a method for communications for the functions commonly found in building automation. BACnet is administered by a standing committee and is designed as a high level protocol which may use a number of transmission protocols at the field bus level. Table 4.6 shows the organisation of BACnet in terms of the OSI model; the commonly used media are in bold type.

In terms of the OSI model, therefore, BACnet is a four-layer protocol stack consisting of an application layer protocol, a network layer protocol, and several data link and physical layer protocols. BACnet's simple network layer allows multiple BACnet networks to be connected in order to create a BACnet internetwork. This capability is most often used in order to allow networks employing different data link layer technologies to be linked together through routing devices. For example, a large building could have multiple programmable controllers and operator workstations connected by an Ethernet network, and each programmable controller could be connected to low cost application-specific controllers using a master–slave RS 485 bus or other low cost bus.

BACnet was defined from the start for use with building services. BACnet defines 18 standard object types, where an object is an abstract representation of a control system element. Object types are defined for analogue and binary inputs and outputs, control loops, schedules and so on, chosen to facilitate the modelling of DDC systems; further objects will be added to deal with other building management functions such as fire and security. Each object consists of a set of properties, where each property contains a value of some specified data. All objects have the properties of an object identifier, and object name and an object type. In addition, there are other properties such as present value, status and so on. BACnet carries out operations by invoking a set of standard services which govern the rules by which messages are constructed and tasks performed.

4.2.5.2 LonWorks

LonWorks is the general name for field bus systems based on the LonTalk protocol, which is physically implemented in the Neuron chip. This is a programmable integrated circuit developed to handle network and input/output functions. The aim of LonWorks is the establishment of interoperable systems, whereby devices using the LonTalk protocol, from different manufacturers, can freely communicate with each other over the LonWorks network. To this end, an independent association, the LonMark Interoperability Association, has been set up, which is responsible for establishing standard specifications for the functionality of devices which are to be connected to the network. In the language of LonWorks, a device such as a sensor, controller or actuator has an abstract representation as an object. Each object has a strictly defined functional profile, which sets out the inputs and outputs that the device exchanges with the rest of the network. As well as issuing functional profiles, the LonMark Interoperability Association offers means for conformance testing of products. Products conforming to the guidelines are issued with a LonMark logo as a warranty for interoperability.

Table 4.6 OSI model of BACnet protocols

Layer	Examples				
Application OSI 7	**BACnet application**				
Simple network OSI 3	**BACnet Network Layer**				
Data link OSI 2	***ISO 8802-2* Type 1**		MS/TP	**Dial-up**	**LON**
Physical OSI 1	**Ethernet**	Arcnet	RS 485	**RS 232**	**LON**

Layers 4–6 refer to wide area networks. Layers 1 and 2 constitute the LAN.

LonWorks is capable of employing several different communication media. It is possible to build a fully distributed HVAC control system in which the components of a control loop exist as physically separate objects, from different manufacturers, communicating with each other over the building's field bus network.

Figure 4.10 illustrates the components of a LonWorks network. Individual devices are connected to the bus at nodes. The term device is used to imply a physical device such as a controller or sensor. The protocol treats the system as a collection of objects. Objects are well-defined building blocks of the system, with a defined set of properties. For instance, a sensor object has properties which identify its type, its individual identity and its output, e.g. temperature, together with several other optional properties. A physical device, such as a controller, may consist of several objects. Each device must contain the processing power necessary for it to communicate according to the protocol. Each device must contain a transceiver, which sits between the Neuron chip and the bus. Transceivers are available for several media, including twisted pair, power line carrier, wireless and infra red. Figure 4.11 illustrates the components of a temperature sensor. The A/D converter conditions the temperature signal from the temperature sensor into a form that can be recognised by the processor. The function of the processor is to generate a signal which will follow the strict protocol of the network, identifying the sensor and transmitting a temperature value that can be accessed anywhere else on the network. This signal is then converted to a form appropriate to the local communication medium. Depending on the application, the device may be powered from the network or from an external supply.

The nature of the system is such that each object performs a well-defined function. For instance, a sensor, controller and actuator driver may be physically separate devices and attached to the bus at different points, but bound together by addressable software and operating as a single controller. The various protocols define a large number of objects which together cover most requirements of a BMS. It is not necessary that the operation of a device be transparent to the network and in many cases it is more practical to have a complex device, such as a VAV controller, built in a conventional way with hardwired sensors and actuators. Such a unitary device may be connected to the net using a limited protocol which allows the supervisor to know the operational status of the controller and alter the set point if necessary; however, the supervisor has no knowledge of the inner workings of the controller.

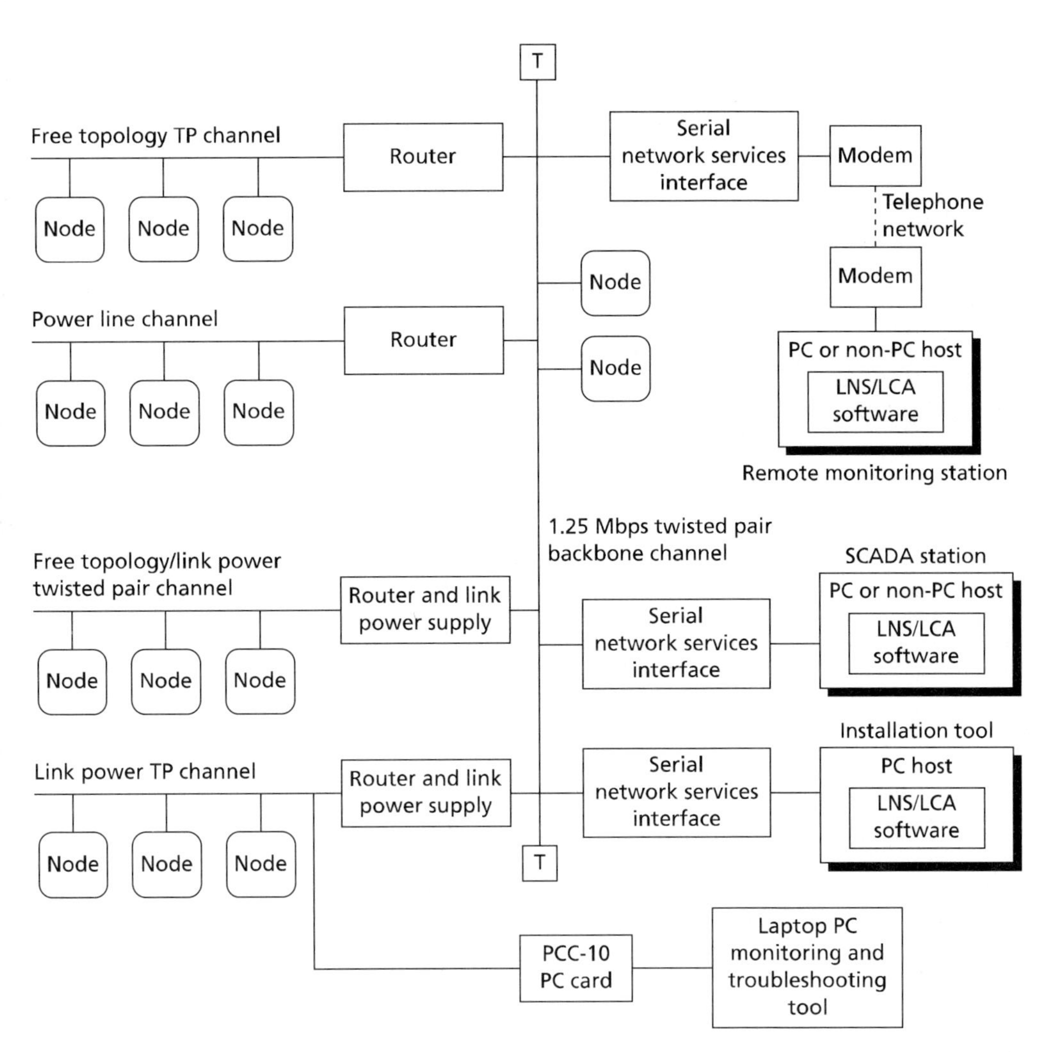

Figure 4.10 Components of a LonWorks network

Different communication media may be used within the same network by the application of routers. Routers also serve the function of partitioning sections of the network from traffic in another area, so that information which is only required within one area is confined to that area and not transmitted over the entire network. In addition, interfaces are available which allow a LonWorks application to run on a non-Neuron host. Thus a PC can become a node on the network. The use of an interoperable network for BMS allows information to be exchanged between subsystems that might otherwise present problems of intercommunication. For instance, an IR presence detector could send output to lighting, security and HVAC subsystems over the network.

The application of fully interoperable LonWorks systems is limited by the range of functional profiles which have been defined and their incorporation into products. Even where profiles have been defined, there are situations where it may be more practical to use a conventional intelligent controller hardwired into the unit which it controls; this is likely to be the case for factory-fitted controllers. It is possible to design an interface with a LonWorks field network so that the controller may communicate with the rest of the system, allowing monitoring of performance and the resetting of parameters at the head end. This represents a compromise with full interoperability, but may often be a practical solution.

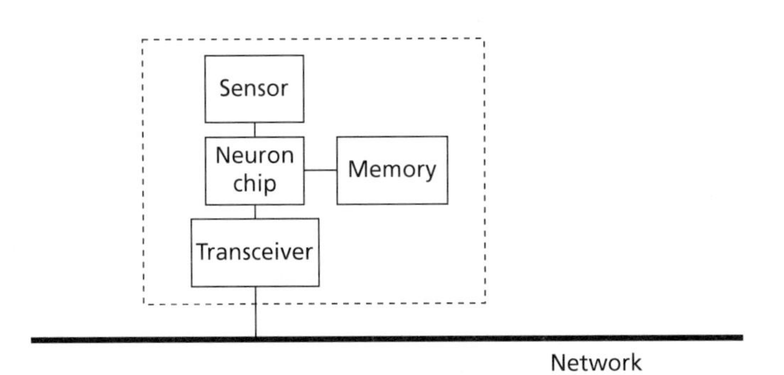

Figure 4.11 Components of a LonWorks temperature sensor

4.2.5.3 EIBus

EIBus is a European standard administered by the European Installation Bus Association, with its headquarters in Brussels. The EIBA has members drawn from the major manufacturers of electrical installation equipment in Europe and has as its aim of guiding development of a highly reliable bus system for BMS which will be available all over Europe. The EIBA is independent of individual companies; it sets out the technical directives for the EIBus system and lays down quality requirements. Members who pass its tests may use the EIB mark. The EIBus has been developed as a simple and reliable system which may be installed without special skills by an electrical fitter. The bus itself consists of an unshielded twisted pair, which is normally installed adjacent to the power supply cables, sharing the same trunking. Devices sitting on the bus communicate by sending event-driven telegrams. A relatively low information rate of 9.6 kbit/s is used. The choice of frequency makes installation less critical. There are no restrictions on the layout; branching and looping are permitted.

Each device on the system comprises a bus coupling unit (BCU) and an application unit (AU). The AU may be some form of sensor or actuator. In the case of a sensor, the AU sends information to the BCU, which encodes it and transmits a message onto the bus. In the case of an actuator, the BCU receives a message destined for the AU, decodes it and instructs the AU on the action to perform. The BCU is programmed during the set-up phase of the installation. This is done using a PC plugged into the system using EIB Tool Software.

The smallest EIBus system (Figure 4.12) consists of a bus line, some bus devices and a power supply, which provides 28 V DC to the line. Up to 64 bus devices may be connected to a single bus line, which has a maximum length of 1000 m. Up to 12 lines may be interconnected using line couplers to form an area. If further expansion is required, up to 15 areas may be connected to a backbone line. Each device is characterised by its physical address of area, line and device number.

Line couplers act to control traffic. When setting up the system, devices are bound together by giving the source and target address. The coupler will pass line-crossing telegrams but all other messages are confined within the line boundary, thus reducing the amount of traffic on the network. A PC may be connected to the bus by a gateway, to provide head end supervision. A PC is also used during the commissioning period with appropriate software.

4.2.5.4 Batibus

Batibus originated in France as an open field bus system designed specifically for use in buildings. The protocol is defined in French and international standards and administered by the Batibus Club International, based in France. It is designed to be an interoperable system and product certification is carried out by an independent company. Batibus finds its main applications in small to medium buildings and has been incorporated in products from several manufacturers.

The communication medium is an unshielded twisted pair, operating at 4.8 kbit/s; it may be installed next to power cables if a shielded insulated twisted pair is used. The cable may be installed in a free topology and carries the power supply for the devices. Devices are connected to the bus via a communications module. The address of each device is set physically at the device by setting a drum wheel or dip switches; in some cases the device carries a display and keypad for the purpose.

Each network can accommodate 240 separate addresses. The actual number of devices may be greater, since devices of the same type which are required to operate together may share the same address. Depending on the cable size used, the total network length may extend over 2 km. The bus cable operates at 15 V and network devices may be powered directly from the bus up to a total consumption of 150 mA, typically 75 devices. Up to 1000 externally powered devices may be connected to the bus.

4.2.5.5 HBES

At the time of writing, the EIB Association, Batibus Club International and the European Home Systems Association have announced a programme of convergence to produce a common standard to be known as Home and Building Electronic Systems (HBES).

4.3 Electromagnetic compatibility (EMC)

The extensive wiring of a control system makes it susceptible to interference from the many sources of electrical noise present in and around buildings. The effects

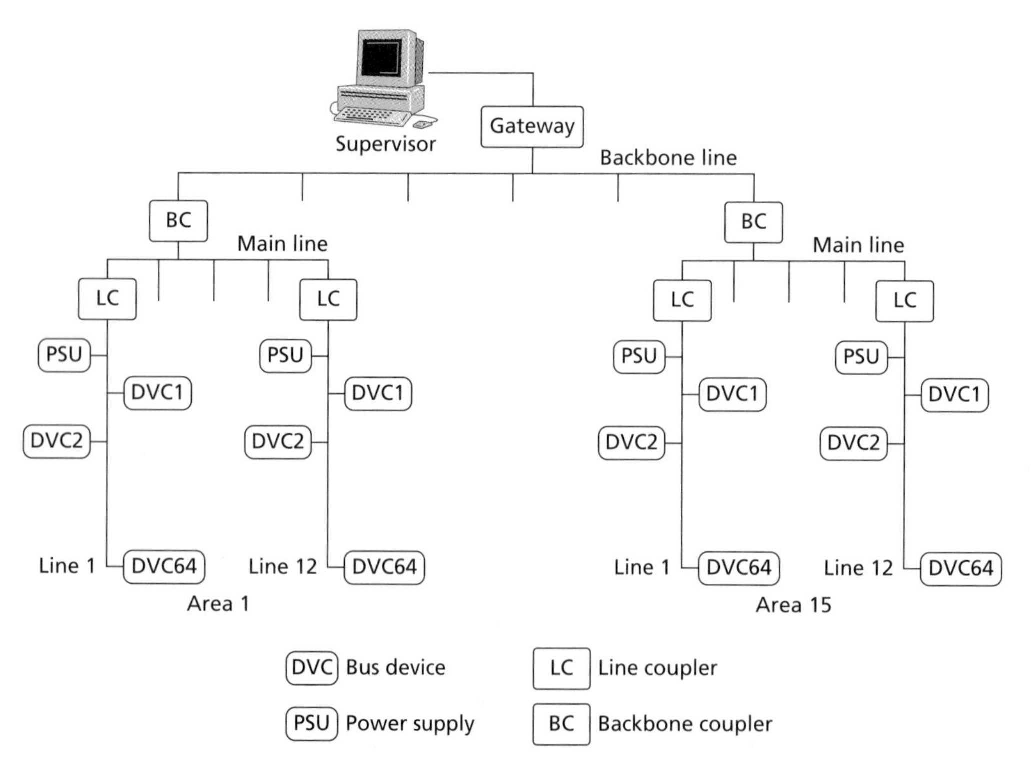

Figure 4.12 Topology of EIBus

of interference can range from minor errors in a temperature reading to complete system failure and equipment damage. Since the control system operates continuously in real time, it is vulnerable to interference. A malfunction that may be tolerable if it involves a temporary deviation in HVAC performance, becomes intolerable if it represents a threat to safety or security. The requirement that a system has a satisfactory EMC standard may be specified using a standard clause from Section A64 of the *National Engineering Specification*[8]. Two alternatives are given: Clause 7080 states the general requirements of compatibility and is recommended for general use; where it is essential to identify specific EMC standards to be met, Clause 7090 may be used.

The chance that a BMS will suffer from interference depends on:

— the levels of conducted and radiated electromagnetic disturbances in the building

— the immunity that has been designed into the individual components that make up the BMS

— the way that the BMS, as a networked system of controllers, has been installed.

Ambient electromagnetic noise levels can be high in modern commercial buildings with their high densities of electrical and electronic equipment, and in HVAC plant rooms containing heavy electrical plant. Industrial processes can produce substantial interference and there may be sources of noise external to the building. As far as possible, these problems should be tackled at source at an early stage. The prime responsibility of the controls specialist is to select equipment that complies with the immunity and emissions limits in the EMC product standards and to install them in accordance with good practice. The BMS installation should then have satisfactory reliability and comply with the UK regulations. The following draws on *BRE Digest 424*[9], which should be referred to for more detailed advice.

4.3.1 The Electromagnetic Compatibility Directive

The EMC Directive[10] sets out the two requirements, that equipment should neither suffer from, nor cause, electrical interference. Compliant equipment carries a CE mark and it is now a criminal offence to sell or bring into service non-compliant equipment. It should therefore be possible to assume that control products meet the essential requirements of the EMC Directive. At the moment the requirements are generic; product-specific standards for building control products are in preparation. Most products are self-certified by the manufacturer, based on tests in their own or third-party test house. For a bespoke product, such as a control panel, where the expense of testing would not be justified, it is possible to meet the regulations by the preparation of a technical construction file, where an approved third party certifies that the construction complies with the objectives of the Directive.

4.3.2 Installation

The use of approved components is not sufficient to guarantee freedom from interference. The EMC of an installation depends also on the degree of coupling to noise sources; BMS, with their extensive signal cabling, are especially vulnerable. There is a balance to be made between the costs of providing protection by using high immunity components or by using lower immunity products which may require special installation measures to prevent interference.

4.3.2.1 Location

EMC should be a major consideration, along with ease of access, in the location of controllers and other components of a control system. While placing control components close to the equipment being controlled has advantages, it may expose them to high levels of disturbance. Controllers and cabling should be placed as far away as practicable from sources of electrical noise; these include switched loads such as relays and contactors, lifts, air-handling units, chillers and variable-speed drives. Transformers, busbars and lift control equipment can be strong sources of 50 Hz magnetic fields, which can cause disturbance to computer display screens; a separation of up to 5 m may be necessary to avoid wobbling images.

There may be sources of interference external to the building. It is advisable to carry out an EMC survey in advance of construction if the planned site is near to radars, radio transmitters, electrified railway lines or heavy industrial plant. It may be necessary to install special mains-conditioning units to provide clean power to the BMS. In extreme situations, it is possible to incorporate architectural screening to screen rooms or even the entire building against interference[11].

4.3.2.2 Power supply and earthing

Power supplies in HVAC plant rooms should normally be adequate for BMS controllers that comply with the relevant immunity standards. However, power supplies may suffer from a wide range of disturbances and the risk of interference will be minimised if supplies from electrically noisy circuits are avoided. If practicable, field controllers should be supplied via a separate ring or spur circuit from the local distribution board.

The earthing of electronic equipment and signal cabling can introduce problems. Earth systems in a building provide the functions of:

— safety

— lightning protection

— EMC.

Safety earthing provides protection against electric shock by connecting exposed metalwork to the earth lead in mains wiring. Lightning protection for a structure is provided by lightning conductors mounted at the top of the structure, connected by down leads to earth electrodes. Transients induced in external cables, lightning conductors and other metalwork can propagate throughout a building; remote strikes to power and data cables can generate transients that propagate by conduction down the cable before entering the building. Lightning protection of buildings is covered by *BS 6651* and is discussed by Johansson[12]. With a properly designed lightning protection system, the risk to components conforming to the EMC Directive should be low. BMS cables running outside the building should be buried; screened twisted-pair cable in metal conduit may be used for short distances. To improve reliability, surge protection devices may be installed where external cables enter the building. For advice on earthing practice, see *Earthing Practice*[13].

4.3.2.3 Cabling

The cabling that connects components of a control system is susceptible to interference from adjacent power cables and from radio frequency interference. Techniques for avoiding interference include the use of screened cables, care with routing cables, segregation of power and signal cables and correct earthing practices.

Many types of screened cable are available. A comprehensive treatment of data and other types of cabling is given in *Electric Cables Handbook*[14]. Table 4.7 gives an indication of the performance of different types of cable in attenuating magnetic and electric fields. It is important to remember that screening with a conductive material, such as copper, is effective at attenuating high frequency electromagnetic fields, but has little effect on power frequency magnetic fields. To attenuate 50 Hz magnetic fields it is necessary to shield with a magnetic material such as steel, or to use twisted-pair cables.

In practice, a low cost, foil-screened twisted pair cable is suitable for most BMS applications for frequencies below 1 MHz. For IT and data communication circuits operating at frequencies above 1 MHz, such as Ethernet, the cable type will normally be specified by the equipment supplier. It may be coaxial cable or suitable twisted pair. The use of coaxial cable is not recommended for low frequency use as noise induced in the screen will be added to the signal.

BMS signal cables should be kept as far away as practicable from sources of interference. In particular, untwisted cables should not be exposed to magnetic fields from high current equipment such as transformers. The *IEE Wiring Regulations* permit signal and power cables to share the same conduit, providing the signal cable has adequate insulation. The regulations are written from the point of view of electrical safety, and do not concern interference. Sharing a conduit or trunking makes for economical installation and some field bus systems, e.g. EIBus, are designed so that this will not introduce any EMC problems. However, in general it is recommended that signal cables without screening should be separated from power cables by a minimum distance of 150 mm. Ideally signal and power cables should be routed in separate trays or trunking, and cross at right angles where they meet. Table 4.8 presents general recommendations for different types of screened signal and power cables; manufacturers' recommendations should be followed where appropriate. The control system specification should lay down what standard of installation is required.

Table 4.7 Screening cables against 50 Hz magnetic and radio frequency (RF) electromagnetic fields

Code	Cable type	Noise reduction	
		50 Hz	RF
Plain	Plain, no screen, no twists	None	None
	Plain in steel conduit or trunking	Good	Good
	Plain with braid or metal tape screen	None	Good
UTP	Plain two-core unscreened twisted pair	Good	None
STP	Braid-screened twisted pair	Good	Good
FTP	Foil-screened twisted pair	Good	Good
MICC	Mineral-insulated copper clad	None	Good
	Twisted-pair MICC	Good	Good
SWA	Steel wire armoured	Good	Good

Table 4.8 Minimum separation (mm) between BMS signal cables and power cables[15]

Signal cable	Power cable		
	Twin and earth	SWA	MICC
Plain untwisted	150	125	Touching
UTP	75	50	Touching
All screened	Touching	Touching	Touching

The earthing of circuits and cable screens can present problems. Correct earthing depends on the type of cable, the signal frequency and whether the signal is single ended or differential. In a differential circuit, neither of the pair of signal wires is at earth potential; induced potentials from interference are induced onto both wires, and the input of a differential amplifier is able to reject common-mode disturbances. Most BMS analogue and digital circuits are low frequency and single ended, with the ground line at earth potential. For such cables, both the ground line and the screen of the cable should be connected together and earthed at the controller end of the cable only. Earthing the screen at both ends of the cable creates an earth loop, which can cause large 50 Hz currents to be induced, which in turn can induce unwanted noise in the signal cable. Some communication circuits employ differential amplifiers. Neither signal lead is earthed and the screen should be connected to earth at one end only. Cabling for a standard system should follow manufacturer's advice. High frequency cables above 1 MHz behave rather differently and the outer screen is earthed at both ends and perhaps at several points along its length.

4.3.2.4 Control panels

Control panels must comply with the EMC Directive. This may be achieved by constructing the panel entirely from CE-marked components, in which case each component must be installed in accordance with instructions supplied with the component. Alternatively, the complete panel may be tested as a unit for compliance with the relevant standards, in which case the panel itself will have a CE mark. Unauthorised modifications made after installation could invalidate the CE mark.

BRE Digest 424[9] summarises good practice in the design of control panels, including the following points:

— Sources of broadband electrical noise, such as contactor solenoid coils and switch contacts for external inductive loads, may need suppressors.

— Metal cabinets with low impedance bonding between separate parts provide good screening against radiated electromagnetic disturbances. Mild steel cabinets give good protection, while metal-coated plastic enclosures may be inadequate against low frequency magnetic fields. Conductive gaskets may be required where a high level of interference is expected, such as near a transmitter or radar installation.

— Use a Form 4 control panel cabinet, which is divided into cubicles with each motor drive in a separate cubicle.

— Interfacing relays may be mounted in either the drive or controller section. If in the controller section, they should be in a separate area from the controller or separated by a metal screen.

— Cables entering the panel should preferably be led to a termination strip mounted on the cabinet, rather than to the controller itself. Connections within the cabinet should be routed wherever possible along a metal surface.

— Variable-speed motor drives should be carefully installed in accordance with the manufacturer's instructions, which may require the use of a designated VSD enclosure and screened power cables.

4.4 Systems integration

A modern building incorporates a range of building services other than HVAC. Such systems have in the past been developed by different manufacturers and in operation they worked completely independently of each other. There are great potential benefits if the different systems can be brought together so that they can communicate with each other and work cooperatively. The term systems integration is used to describe the bringing together of equipment from different sources and the combined working of different building services systems, bringing about some or all of the following benefits:

— *Savings in installation costs*. The integrated systems may share the same physical network for communication.

— *Greater choice for the designer*. Devices from different manufacturers use the same protocols and can be used with the same network.

— *Simpler building operation*. All building services systems can be supervised from the same central interface. Where appropriate, systems can be accessed from other points in the network.

— *Integrated operation*. Where fully implemented, systems integration allows systems to interact, giving intelligent operation of the whole building.

From the point of view of the user, a major benefit of systems integration is that the entire BMS may be operated from a single head end PC, so that operating staff have only to learn one set of operating software. This benefit can be achieved by the use of a neutral head end; see below.

4.4.1 Implementation

Systems integration can be implemented at different levels:

— *Within a system*. The term systems integration is sometimes used to describe the ability of devices from different manufacturers to use the same communications network. This is better described as interoperability rather than as full systems integration.

— *At the head end supervisor*. The separate systems act independently of each other, but are all brought to a common supervisor, known as a neutral head end, which incorporates gateways allowing it to communicate with all the systems. Thus all systems may be controlled using a common interface with a consistent look and feel. Information from the different systems may be combined into reports and operating schedules may be set in common. It differs from full integration in that there is no direct communication between different systems; if data is to be passed from one system to another, it has to be sent via the supervisor.

— *Full integration*. Information is exchanged between all systems, which are programmed to respond appropriately. This produces the intelligent building.

A modern building will have many networked systems running in parallel:

— IT network

— telephone network, paging and fax

— security and access control

— fire and emergency systems

— closed circuit television (CCTV)

— lifts

— lighting

— HVAC

— energy management.

It is possible for all systems to share a common IT network for communication with evident savings in cabling costs; the practical considerations are dealt with further in 4.1.8. Intelligent building operation comes when the different systems are able to communicate with each other directly and take appropriate control decisions. Some examples will make this clear (Table 4.9).

The availability of comprehensive information allows powerful reporting applications to be provided. Energy metering may be combined with temperature and occupancy information for M&T analysis, as well as providing the information for tenant utility billing. Run time or fault data may be fed to a maintenance scheduling program, which will provide the facilities manager with notice of required actions.

Table 4.9 Examples of integration between systems

Integrated systems	Features
Access and HVAC	Access control system informs the HVAC system of the number of occupants in an auditorium, which adjusts ventilation rate accordingly
Security and CCTV	If a visitor is denied access, CCTV and PA systems are activated, allowing the supervisor to see and speak to the person and take appropriate action
Energy management and HVAC	The energy metering system reports that maximum demand limit may be exceeded, so that the HVAC controls may shed load
Lighting and security	Occupancy detectors in the lighting control system inform the security system of the position of occupants during out-of-hours working

4.4.2 Problems with integration

A fully integrated building management system offers considerable advantages to the building user. However, there are difficulties in achieving this, which may be classified as technical and contractual.

4.4.2.1 Technical

The technical difficulties are real, but being overcome. At the field level, interoperability requires that all devices conform to the protocol standard being used. An example is LonWorks set of definitions of Standard Network Variable Types, which defines the protocol for devices ranging from temperature sensor to VAV controller. In situations where the required device either does not match the standard, or the standard has not yet been defined, it will be necessary to provide a customised solution, both for the device in question and any other devices on the network that need to receive the message.

Network technicalities are discussed elsewhere, but there may be problems with widely different systems sharing the same communications medium. Most building services operate in real time, sending relatively small amounts of information at frequent intervals. The building user's IT system, however, is designed for the needs of the business and is likely to transmit very large files at irregular intervals. There are difficulties in ensuring that all systems achieve their required traffic access. This situation can be improved by confining control traffic to a segment of the network, or sub-LAN, by the use of a bridge or router to connect it to the rest of the internetwork. Local traffic is confined to the segment and only information which is destined elsewhere is allowed out via the bridge.

4.4.2.2 Fire protection system (FPS)

While at first sight the fire alarm system would seem to be an ideal candidate for integration, there are a number of obstacles to allowing the fire system to share the same network as other building services. Fire protection requires the monitoring of fire and smoke detectors throughout the building, the transmission of information to and from the fire alarm system, and the initiation of appropriate control actions.

Integration of the FPSs and BMSs for monitoring purposes is fairly straightforward. Autonomous subsystems, e.g. fire, security and BMSs, are linked together at the management level: see Figure 4.6. Each of the systems is capable of operation on its own. The fire detection system is complete and contains all the fire alarm components such as fire panels. The fire detection system can pass information upwards to the central supervisor, e.g. alarm annunciations and associated display of fire locations. The central supervisor may or may not be able to pass commands to the fire alarm system. If so, the commands would be for non-critical functions, which would not affect operation in an emergency. Many parts of the fire alarm network must be capable of prolonged operation during a fire, which in practice means wiring in MICC (mineral-insulated copper-sheathed cable). MICC is not suitable for general purpose data cabling, on account of its expense and limited bandwidth.

In the event of a fire being raised, the BMS is required to take appropriate actions, and the control strategy for each plant should contain settings for a fire alarm mode (see 5.1.5). In discussion with the Building Control and Fire Officers, the functions may be divided into primary or supplementary safety operations. Primary, safety-critical operations are normally controlled directly from the FPS, leaving the supplementary operations to be controlled via the BMS. Fire and smoke control dampers are generally classed as safety-critical and are not part of the BMS.

Code of Practice BS 7807[16] sets requirements for design, installation and servicing of integrated fire detection systems. It does in principle allow a considerable degree of integration between fire and other systems, providing the requirements of *BS 5389* are met[17]. In practice, the limiting factor for integrating systems is the high cost of suitable cabling. Any integrated system requires the approval of the Building Control Officer and the Fire Prevention Officer, with the likely involvement of the building insurers. Where it is desired to achieve the flexibility and sophistication of an integrated fire detection system, the potential difficulties of obtaining approval and any possible contractual problems should be tackled at the earliest possible stage. The general of fire engineering and the provision of safety systems is covered by CIBSE Guide E[18].

4.4.2.3 Contractual

Systems integration requires the bringing together of equipment and systems which have in the past been handled as independent contracts. Where an integrated network is planned care must be taken to ensure that clear lines of responsibility are laid down. It may be that a single company, possibly a systems integrator or one of the suppliers, should be responsible for providing the communications network and coordinating all other companies that are to use it. Not only must there be clear responsibility for design, installation and commissioning, but there must be a responsibility that extends throughout the operating life of the integrated system. Some standards require a central database to hold all information on the network topology and linking software. There must be clear ownership of the database and arrangements for its upkeep.

There are technical advantages to be gained by the integration of security systems with the BMS; for instance the presence detectors used for lighting control can feed information to the security system. Such integration may contravene security standards and not be acceptable to the insurance company involved. Potential conflicts of responsibility may be avoided as far as possible by creating sub-networks for primary systems so that each can be fully installed and tested for independent functionality before adding the interoperable functions. Any problems occurring on integration will overlap two or more networks and be the responsibility of the systems integrator to resolve.

4.4.3 Integration with IT systems

Systems integration moves the BMS closer to the IT system. The term information technology is used to cover a number of meanings. The CIBSE Applications Manual *Information Technology and Buildings*[19] classifies information systems as belonging to three generic areas:

— user information systems: voice, text, image and data networks used by the occupants of the building for business purposes

— building information systems: for the control and supervision of building management, energy, fire and security systems

— miscellaneous information systems: public address, closed circuit TV, paging and signage, which span or fall between the first two categories.

The spread of IT types from user IT to building information systems is illustrated in Figure 4.13. The more fully integrated the building systems, the greater the interaction between types, e.g. the telephone network may by used to allow building occupants to change temperature set points, or the paging system employed to give an automatic callout to a maintenance engineer in the event of a malfunction.

4.5 User interface

Modern building management systems are controlled by means of software operations. Control is no longer by knobs and switches, hardwired to the device to be operated. The use of computers as central supervisors gives the designer almost total freedom to design the operator interface. The ability to control and monitor every aspect of the building's operation has in turn increased the amount of information to be handled by the supervisor with a corresponding increase in complexity. Since a building service installation will be operated by a range of users with disparate responsibilities, qualifications and experience, it is vital to design operator controls which can be used at the appropriate level by a wide range of users.

4.5.1 Levels of operation

Interaction with the BMS may take place at all levels of the system and at each level there may be different requirements for different classes of operator (Table 4.10). Large systems contain all levels of operation. Management level activities are primarily information gathering and data analysis, without active control of the system. The management level computer can communicate with several remote sites or even countries. In many system, the management and operations level functions are housed within the same computer; some supervisory programs will carry out both management and control functions.

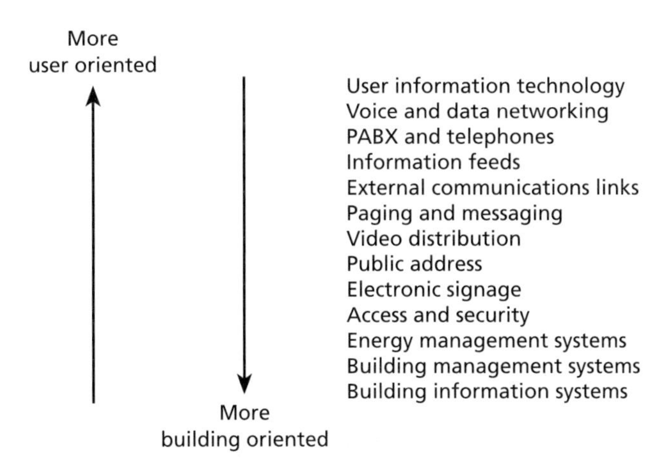

Figure 4.13 The range of information types found in a modern building

Table 4.10 User interaction with BMS

Level	Operator	Function
Management level	Facilities manager System administrator	Reporting Energy M&T Off-line data analysis
Operations level central supervisor	Non-technical personnel (security, caretaker) System operator	Response to alarm messages and instructions Rescheduling, parameter adjustment, monitoring
	Specialist engineer	Reprogramming, fault finding, expansion
Service tools	Specialist engineer	Monitoring, reconfiguration, fault finding
System level outstation	Non-technical personnel	Some local control of conditions
	Specialist engineer	Parameter adjustment, reprogramming, fault finding
Zone level local control	Occupants	Set point adjustment

4.5.2 Hardware

4.5.2.1 Zone level: stand-alone controllers

Small HVAC systems may operate without a BMS, using a number of dedicated controllers which are designed to perform a particular function. Each controller is hardwired to the relevant plant and does not communicate with other controllers. Examples are compensators, optimisers and three-term temperature controllers. Such controllers are small in size and incorporate their own user interface. Typically, this consists of a liquid crystal display plus a keypad or a number of multifunction knobs and buttons. They are capable of sophisticated control functions and require careful setting up by a trained operator. It may be possible to disable the set-up functions, leaving the main set point, e.g. room temperature, available for change by unskilled staff. Alternatively, the setting knob may be mounted remotely from the controller in the occupied space, with the controller itself in a protected space such as the plant room.

The limited size of the display puts great demands on the ergonomic design of the interface. Much use is made of symbols and menus. However, it is rarely possible to make the full range of user adjustments self-explanatory. It is advantageous if abbreviated instructions be kept with the controller behind a drop-down cover or elsewhere. It is essential that the instruction book is kept in a place where it will be found when required at a later date.

4.5.2.2 Operations level: plug-in keypad

A simple keypad may be plugged into a BMS outstation to review point values, receive alarms and provide some parameter adjustment. The keypads have a small LCD display and a limited number of keys. The device may be permanently mounted on an outstation or be used as a portable readout. Alarms may be directed to the keypad, which contains a built-in buzzer. Some types of keypad may be mounted on a motor control panel and be used to view data and make changes to parameters for all the controls mounted within the panel.

4.5.2.3 Systems level: touchscreen

A touchscreen can sit on the network, providing reduced input and output capacity compared with the main supervisor. However, subject to access control, it can have access to the entire network. The user communicates with the display by touching the screen display, which may consist of pictures, graphs, knobs, dials and text. The screen takes up less space than a PC. It has no movable parts and is suitable for wall or panel mounting. A typical use is to provide limited control of local plant operation by a minimally trained user. The display screens can be specially designed for the intended function.

4.5.2.4 Operations level: supervisor

The central supervisor, also known as the head end supervisor, commonly consists of a PC together with a printer. By the use of appropriate gateways and interfaces, the PC can communicate with control and bus systems employing a variety of communications protocols, and if desired, with other building management systems such as fire, lighting, security etc. The head end supervisor is the prime interface between the system operator and the BMS. Where the BMS operates over a LAN or bus system, it will normally be possible to access the system from anywhere on the network using a PC and appropriate software. The use of a portable supervisor is of great value when performing commissioning or maintenance. Use is of course dependent on access control.

4.5.3 Users

The enormous power and flexibility of the PC-based supervisor means that the software must be carefully chosen to allow appropriate operation by different levels of user. There are three levels of operation:

— *Operation by unskilled personnel*. When the building is unoccupied, or the operator is absent, unskilled personnel such as a caretaker or security guard can monitor critical alarms from the BMS. The supervisor should display or forward only alarms requiring action and also give instructions to allow the action to be carried out.

— *Operation by trained personnel*. In a large system, the supervisor will be operated by a person trained in its use, but who will not necessarily be a building services engineer. The operator's responsibilities include monitoring the plant for satisfactory operation, resetting parameters or time schedules as required. All alarms must be responded to; a well-designed supervisor requires confirmation of the action taken before allowing an alarm to be turned off.

— *Operation by the controls engineer*. The supervisor may incorporate software which allows reconfiguring of the control system, and uploading and down-loading of software to outstations.

4.5.4 Software

Several software packages for supervisors are available and may be expected to include some or all of the following features:

— *Graphics*. Graphic displays of plant operation contain diagrams of the plant with live point values displayed, giving on-screen displays of temperatures, flows etc. plus the operating state of items of plant. Set points may be adjusted directly and plant items switched on and off.

— *Data logging*. Point values, both analogue and digital, are routinely recorded in the system database. Data may be displayed graphically, either as a real-time trend graph of the present situation or a review of historical data.

— *Alarms*. Certain points may be assigned alarm levels and alarm priorities. When an alarm event occurs, the supervisor software takes appropriate action, typically:

  — displaying an alarm message on screen

  — sending an alarm message to another node on the BMS

  — relaying an alarm message via email, pager or fax

  — updating the alarm log, which includes acknowledgement and action taken.

— *Reports*. Reports are predefined summaries of historical data, which may be printed automatically or on demand. They may also be transferred to a spreadsheet for off-line analysis.

— *Time scheduling*. Time schedules determine the operation of building services on working days, weekends, holidays and maintenance periods. The supervisor software incorporates a calendar function which allows advance time scheduling in a user friendly manner.

— *Security*. Access to the supervisor must be controlled to prevent unauthorised use. Each user is assigned a password which has an access level associated with it. Typical access levels are:

  (1) *Alarms only*. Allows viewing and printing of alarms.

  (2) *View*. Allows viewing of all graphics.

  (3) *Read/write*. As view, but allows changing set points and time schedules.

  (4) *Supervisor*. As read/write, plus changing system parameters and passwords.

For additional security, all accesses to the supervisor are logged. In addition, the supervisor may contain additional software relevant to the operation of the BMS, such as monitoring and targeting or maintenance management. A suitable drawing package will allow the production of the system graphics.

4.6 Summary

Conventional controllers are hardwired to the systems they control and do not communicate with each other. They can provide satisfactory and economical control of small buildings. The modern programmable controller contains a microprocessor and performs local control of a piece of plant or subsystem and exchanges information with the rest of the building management system. The programmable controller is the workhorse of most BMSs and can be configured to perform a wide range of control functions. Packaged plant is usually supplied with its own hardwired dedicated controller, which has limited communications with the main BMS.

Communications within a BMS take place at three levels: the management level for supervisors, automation level for controllers and field level for sensors and actuators. Local area networks may be organised in a variety of topologies. The most common is the tree structure. Devices on the same branch communicate freely with each other over a field bus. Gateways at the root of the branch filter messages and translate to a different protocol or communications medium as required. A BMS network may therefore include different communications media and protocols.

The most common medium used by a BMS is the unshielded twisted pair, which is cheap and simple to install. Where the BMS shares a network with IT and high data transmission rates are required, coaxial cable is used. Telephone lines or radio links may be used to link physically separated parts of a wide area network.

The use of standard communications protocols allows devices from different manufacturers to communicate with each other, and the major protocols are described, including BACnet and LonTalk. The use of extensive networks risks problems of electromagnetic compatibility. Components and installation practices should meet the requirements of the EMC Directive to prevent problems arising.

The philosophy of systems integration brings building services systems together, so that such systems as HVAC, lifts, lighting, security may share information and be controlled from a common supervisor. This leads to improved efficiency of operation, but integration requires careful planning to ensure clear lines of responsibility.

The complexity of a large system places demands on the user interface design. The supervisor should be arranged so that access is possible at several controlled levels, from simple supervision to fault location and reconfiguration.

References

1 Simms R How far will it go? *Electronics World* (September) 780–781 (1998)

2 *BS EN 50173* Information technology — generic cabling systems (London: Stationery Office) (1996)

3 *ISO/IEC 8802-3 Information processing systems — Local area networks — Part 3 Carrier sense multiple access with collision detection (CSMA/CD) access method and physical layer specifications* (Geneva: ISO) (1993)

4 *ARCNET local area network standard ANSI/ATA 878.1* (New York: American National Standards Institute) (1992)

5 *DDN protocol handbook* (Menlo Park CA: SRI International) (1989)

6 ISO 7498 *Information processing systems. Open systems interconnection Basic reference model* (Geneva: ISO) (1984)

7 *BACnet — A data communication protocol for building automation and control networks* ANSI/ASHRAE 135-1995 (Atlanta, GA: American Society of Heating, Refrigeration and Air Conditioning Engineers) (1995)

8 NES *National engineering specification* (London: Chartered Institution of Building Services Engineers) (1997)

9 *Installing BMS to meet electromagnetic compatibility requirements* BRE Digest 424 (Watford: Building Research Establishment) (1997)

10 EMC Directive *EEC Statutory Instrument No 2372 89/336* (Brussels: EEC) (1989)

11 Bromley A K R Electromagnetic screening of offices for protection against electrical interference and security against electronic eavesdropping *CIBSE 1991 National Conference*, Canterbury, 7–9 April 1991 (London: Chartered Institution of Building Services Engineers) (1991)

12 Johansson M *Lightning protection of buildings and their contents: a review of current practice* TN 1/94 (Bracknell: Building Services Research and Information Association) (1994)

13 *Earthing practice* (Potters Bar: Copper Development Association) (1997)

14 BICC *Electric cables handbook* 3rd ed (Oxford: Blackwell Science) (1997)

15 ECA *Recommended cable separations to achieve EMC in buildings* (London: Electrical Contractors' Association) (1993)

16 BS 7807 *Code of practice for design, installation and servicing of integrated systems incorporating fire detection and alarm systems and/or other security systems for buildings other than dwellings* (London: British Standards Institution) (1995)

17 *BS 5839: Fire detection and alarm systems for buildings: Part 1 1988. Code of practice for system design, installation and servicing* (London: British Standards Institution) (1988)

18 *Fire engineering* CIBSE Guide E (London: Chartered Institution of Building Services Engineers) (1997)

19 *Information technology and buildings* Applications Manual AM7 (London: Chartered Institution of Building Services Engineers) (1992)

20 ISO *Building control systems* ISO 16484 (Geneva: ISO) (2000)

5 Control strategies for subsystems

5.1 Safety
5.2 Boilers
5.3 Chillers
5.4 Control of hydraulic circuits
5.5 Central air handling plant
5.6 Energy recovery
5.7 Mechanical ventilation
5.8 Variable air volume
5.9 Constant-volume room terminal units
5.10 Fan coil units
5.11 Dual duct systems
5.12 Chilled ceilings
5.13 Heat pumps
5.14 Natural ventilation
5.15 Lighting controls
5.16 Summary

A large HVAC system may be broken down into subsystems, each with its own control strategy for effective operation. This section sets out the major strategies for the basic HVAC subsystems, which can subsequently be put together to form a complete application.

5.1 Safety

A control system must ensure safe operation of the system it controls, give adequate warning of any malfunction and if necessary take appropriate action in the event of equipment failure. A BMS employs a variety of alarms, interlocks and control strategies to do this. These strategies are found over a wide range of different HVAC systems and are treated together below to avoid repetition.

The operations considered may be classified as:

- alarms
- interlocks
- safety strategies.

5.1.1 Alarms

Alarms are installed to indicate to the operator that a system variable has exceeded a pre-determined limit. This may simply be a warning that some maintenance action is required, e.g. a filter is becoming blocked or a piece of plant has exceeded its run-time limit, or be a statement that a failure has occurred requiring immediate attention. At the design stage it must be decided how the alarm is to be displayed and what action must be taken. Alarm enunciators are available which give visual and audible indication that a fault has occurred; they may be placed adjacent to the plant in question or where they will be seen by the operator. Otherwise they are displayed on the central supervisor. Alarms may be given priority grades which determine how the alarm is to be handled. To ensure that the alarm is not ignored or cancelled without action being taken, the alarms should 'lock on' until cleared; the appropriate operator identification should be required before the clearance is accepted, together with an acknowledgement of the action taken.

Alarms may be classified as:

- digital alarms, assigned to the open or closed position of a contact
- analogue alarms, initiated when a value falls outside the high or low alarm limits
- status alarm, initiated when the status of a contact differs from the expected value, e.g. a device has not responded appropriately to a control signal
- totaliser alarm, initiated when the accumulated value of a variable exceeds its high limit value
- hours run alarm, initiated when the total hours run of an item of plant exceeds the set limit.

5.1.2 Interlocks

Interlocks ensure that particular items of plant may only operate together, or are prohibited from operating simultaneously. While this could be programmed into the BMS, the use of the term interlock implies that correct operation is sufficiently important for the functions to be hardwired. An example would be the requirement that a minimum water flow through a boiler exists before the boiler can be fired. Hardwired interlocks may be provided for the following:

— Boilers, interlocked to:
  — pumps and primary loop flow
  — plant room ventilation and flow dilution proving
  — gas detection and gas valve
  — panic buttons.
— Chillers interlocked to:
  — primary pumps and chiller flow
  — motorised valve end switches
  — dedicated high and low pressure switches.
— Plant shutdown interlocked to fire alarm.
— Start-up and shut-down sequences may be hardwired where this is important for safety or plant protection

5.1.3 Safety strategies

Safety switches override normal plant operation to prevent harm to plant or personnel and are normally incorporated in the BMS software. This does not necessarily imply any malfunction. For a large system, the logical hierarchy of the many interlinked safety switches becomes complicated and is best set out as a decision tree. The use of complex computer systems may serve to conceal potential problems or unexpected control actions in alarm situations. Advice is given by the Health and Safety Executive(1).

Hot or chilled water systems

For systems dependent on the circulation of hot or chilled water, the presence of water circulation enables the whole HVAC system and:

— In the event of flow failure, the system is shut down and the boilers or chillers are turned off.
— Frost protection is normally provided during off periods. If the outside temperature or the interior of an exposed boiler house falls below say 3°C, the circulation pumps are turned on to prevent freezing in exposed pipework. If the water circulation temperature falls below say 20°C, the heat source is enabled to provide warm water. In addition, the space temperature in the most exposed part of the building should be maintained above 10°C.
— If the system is operating under optimum start control, the compensator is disabled and the flow temperature set to maximum.

Air handling systems

A minimum value of differential pressure across the supply fan is normally required to enable the whole system. If the pressure fails, the controller shuts down supply and exhaust fans and sets dampers to required positions. The following actions should be carried out:

— Interlock supply and extract fans so that both come on together.
— In the event of fire or smoke signal, stop supply and extract fans, close intake and recirculation dampers and open extract damper; details to be agreed with fire officer.
— Frost protection is required to prevent cold outside air freezing the coil. Operate water pumps to provide flow though coils when the outside air falls below 2°C. Open valves to ensure flow through coils. Frost precautions may include closing outdoor air dampers and starting air recirculation fans.
— The measurement of differential pressure across filters will give warning of a blocked filter requiring replacement. This is displayed as an alarm.

Boilers

— Protection against catastrophic failure will normally be provided by the inbuilt controls, e.g. flame failure and over-temperature cutouts and low water level provided as part of boiler package. Boiler firing is inhibited unless flow is proved.
— Boilers may suffer damage from thermal shock if return water is supplied at too low a temperature to a hot boiler. A minimum return temperature to a hot boiler is required to prevent damage from boiler shock and condensation within the boiler. Temperature limits are advised by the boiler manufacturer.

Package controls. Items of plant such as boilers and chillers are factory fitted with safety switches such as over-temperature cutouts or flame failure devices, which operate through the unitary controller. In most cases, these may be wired to the BMS which will display an alarm and indicate the failure mode to the operator.

Many of the system variables monitored for alarm and safety reasons are part of the control system. However, it is bad practice to use the same sensor and signal processor for both control and alarm functions. The two should be physically distinct.

A related function of a BMS is to ensure correct sequencing of operations during start-up and shut-down. During start-up, plant may be switched on in sequence to ensure that maximum demand is not exceeded. During the run-up time of fans and pumps, it will be necessary to inhibit the action of various alarms and switches, so that they do not interpret the initial period as a fan or pump failure. The alarm inhibit time has to be set to a suitable value. On close-down of the system, whether routine or due to a failure warning, valves, actuators and other devices should be driven to appropriate positions. Typically for an air system with recirculation, the intake and exhaust dampers are closed and the recirculation damper is opened.

5.1.4 Selection switches

It is normal practice for each item of plant to be provided with a Hand/Off/Auto selection switch. This switch is operated manually and allows the plant item to be switched off or operated independently of the BMS for such tasks as maintenance or commissioning. When the switch is in the

Off position, the plant is shut down regardless of the BMS operating signal. The Hand position activates the plant as if it were receiving a 'normal operation' signal from the BMS. Safety features and interlocks remain in operation. The status of the selection switch is monitored by the BMS.

5.1.5 Fire and smoke control

The design, installation and operation of the fire control system is separate from that of the HVAC and BMS. The fire system is subject to approval by the fire control officer and it is unusual for approval to be given to integrate the fire systems with other building systems. The operation of fire and smoke control dampers is therefore not the function of the BMS. The BMS receives fire alarm signals and must act accordingly. The control strategy for each subsystem of the HVAC plant should set out the control action to be taken in the event of receiving a fire alarm signal. For much of the plant, the response to a fire alarm is plant shut-down. Where there is a central air handling unit, the response to a fire alarm may be either of:

— plant shut-down, including supply and extract fans with inlet and exhaust dampers closed

— plant shut-down with the extract fan continuing to run with the exhaust damper open

In all cases a fireman's override switch must be available allowing the fire officer to manually control the ventilation plant when arriving on site. CIBSE Guide E[2] discusses the fire precautions relevant to a BMS and CIBSE Technical Memorandum TM16[3] gives the legislative background.

5.2 Boilers

A great range of boilers for the production of steam and hot water is available. Most burn gas or oil and available sizes of boilers range from small domestic units with simple on/off control to multiple boiler units delivering over 10 MW of heat. Electrode boilers, which pass electric current directly through the water to be heated, are available in sizes up to 2 MW and can be used for hot water or steam generation, often taking advantage of off-peak supplies. The principles of boiler control remain the same for all sizes. Boiler controls may be classified as:

— Operating controls, which undertake routine operations, such as firing and shutting down the burners and controlling the water leaving temperature. In multiple boiler systems, the operating controls sequence the firing of the separate units.

— Limiting controls, which ensure that system variables do not exceed any preset limits, which may have safety implications. The limiting controls may link to safety interlocks, which shut down boiler operation in the event of a malfunction of boiler or related equipment. Safety controls and operating controls should be separate.

There is a wide range of available boilers and installation arrangements and this is reflected in the variety of installed circuits and control strategies. Many are unsatisfactory, giving inefficient or unstable operation in practice. Recommendations for good practice have been brought together by BRECSU[4] which sets out a number of recommended circuits and control strategies (Table 5.1). The following sets out the principles of control of the boiler control and primary water circuit. Control of secondary circuits is dealt with in 5.3.3.

5.2.1 Types of boiler

New boilers must comply with the *Boiler (Efficiency) Regulations 1993*[5] and have some or all of the following characteristics:

— low water content and low thermal mass

— high part load efficiencies

— improved heat exchangers and insulation

— low standing losses

— packaged modular arrangements for larger outputs.

One result of the improved design is that it is not usually worthwhile to isolate boilers when heat output is not required, leading to simpler and more reliable multiple boiler installations, compared with earlier installations which required isolation of unused boilers.

Condensing boilers achieve high efficiency by condensing water vapour in the flue gases and so extracting additional latent heat.

Table 5.1 Classification of boiler controls

Control type	Description
Operating controls:	
— Leaving water thermostat	Controls water temperature by burner on/off, high/low or modulation
— Pump overrun	Pump runs after burner turns off to dissipate heat and prevent boiler overheating and locking out
— Flue gas	Monitor oxygen in flue gas and regulate air supply for maximum efficiency
— Ignition and shut down	Automatic burner ignition sequence
— Sequencer	Controls the operation of multiple boilers
— Sequence selection	Rotates the order of operation of multiple boilers
Limiting controls:	
— High limit stat	Locks out boiler if leaving water temperature exceeds limit
— Water pressure	High and low limits in pressurised systems
— Water flow	Check water flow in boiler
Safety interlocks:	
— Flame safeguard	Check before and after ignition. Shut fuel supply and purge if no flame
— Burner fan	Check for forced draught burners
— Gas detection	Check for gas in plant room and close gas valve if detected
— Flue gas dilution	Check fan operation
— Water level	Checks water level
— Fuel-oil temperature	Closes fuel supply valve if heavy oil outside temperature limits
— Fusible link above boiler	Closes fuel supply valve if high temperature detected

Condensing boilers:

- Have an additional heat exchanger in the flue gases.
- Are efficient when not condensing, but require a water return temperature of less than 55°C for condensing operation. Maximum efficiency is achieved with a return temperature of 35°C.
- Can achieve efficiencies in the range 85–95%.
- Are generally higher first cost than non-condensing.

Space heating requirements vary over the season and there are advantages in using multiple boilers operating under control of a sequencer. As the heat load increases, individual boilers are progressively turned on. Each boiler performs at or near design conditions, giving high efficiency and reduced wear. The multiple boiler system may comprise independent boilers or an integrated unit consisting of several identical modules. Some modular boilers have several burners firing into a single combustion chamber. The use of multiple boilers provides safety in the event of a boiler failure; if two boilers are installed, each with an output of one half of the design maximum, one of them will be capable of heating the building for more than 90% of the heating season. Boilers are now available which maintain high efficiency when operating at low loads and the potential improvement in seasonal efficiency from using multiple boilers must be carefully weighed against the increased complexity of the installation.

Hot water service requirements, i.e. hot water for washing, catering and other purposes, may be supplied in four ways:

- central calorifier, supplied by main heating boiler
- central dedicated HWS boiler
- local storage, heated by gas or electricity
- local point of use, usually electric.

In general, it is more efficient to separate space heating and hot water service systems. A self-contained central hot water system is normally much more efficient than using the main central heating boiler to provide hot water, particularly in summer. One solution is to use electric immersion heating combined with off peak storage in summer and retaining use of the main boiler in winter. Small decentralised gas-fired storage water heaters close to the point of use can significantly improve efficiency, especially in an extensive building where distribution losses associated with a central plant would be high. Where hot water demand is low, local instantaneous point of use water heating is economical.

5.2.2 Boiler controls

Boilers are supplied by the manufacturer with built-in controls. These controls must meet the provisions of HSE PM5[6], which sets out safety requirements for boiler controls and these will normally be met by the manufacturer's control panel. Many manufacturers offer a full controls package controlling all the relevant aspects of boiler operation set out in Table 5.1, which the manufacturer will supply and commission. The control unit will normally interface with the BMS, giving the BMS access to alarms and status information and allowing adjustment of set points.

5.2.2.1 Burner controls

The controls on the burner are linked to a thermostat which senses the leaving water temperature. The burner control may be:

- on/off
- two-stage (high/low), giving typically 40% output at the low setting
- modulating, giving an output range from 20 to 100% output at good efficiency.

5.2.2.2 Boiler protection

Some boilers risk corrosion damage if condensation occurs on the heat exchanger and so the system design and operation must ensure that the return temperature to the boiler must exceed a minimum value, typically between 25 and 55°C. This is termed back end protection. The boiler may also require a minimum flow rate during firing. A typical weather compensation circuit using a mixing valve, shown in Figure 2.12, results in a constant flow through the secondary circuit at the variable compensated temperature. The variable return flow to the boiler will be at a low temperature during mild weather. It may be necessary to employ a boiler protection circuit, which uses an additional pump to maintain flow temperature in the boiler (an example is shown in Figure 5.22).

A particular example is a condensing boiler with split heat exchanger. This requires a low water temperature flowing through the secondary heat exchanger to produce condensation in the flue gas, while maintaining back end protection in the primary heat exchanger. This may result in connecting the two heat exchanger sections to different parts of the system. An example is shown in Figure 5.1 for a typical weather-compensated circuit. The cool return water from the secondary circuit passes through the secondary heat exchanger and a standard boiler protection circuit is provided in the primary.

The boiler protection circuit finds application for older, high water content boilers operating with variable flow in the primary circuit. The control of heating circuits is discussed in BRECSU General Information Report GIR 40[4], which strongly recommends that with modern high efficiency boilers a constant primary flow circuit should be

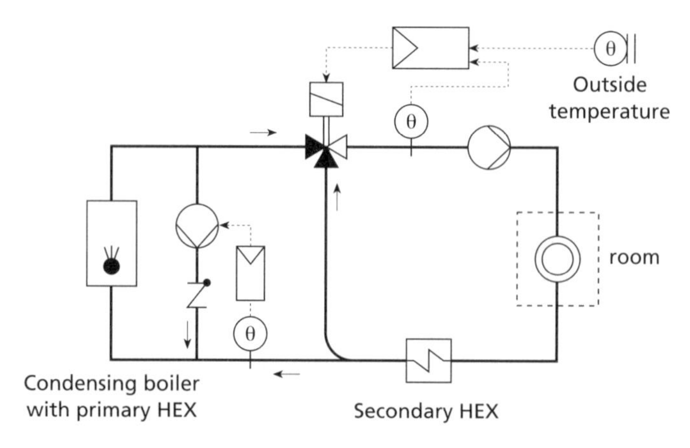

Figure 5.1 Condensing boiler fitted with a two-part heat exchanger in a compensation circuit. Boiler protection for the primary heat exchanger is provided

used, with a system header to decouple the flows in primary and secondary circuits. This obviates the need for back end protection and gives simple effective control (see Figure 5.23).

If the burner and primary circulation pump are switched off simultaneously in a lightweight boiler, there may be sufficient residual heat to cause local boiling, with the risk of damage to the boiler. It is therefore usual to provide a pump overrun function; the pump runs for a short period, up to 5 minutes, after the burner switches off, to dissipate the residual heat. The situation is different for high water content boilers. Here, the residual heat represents a worthwhile amount of energy. The function of the pump overrun function is to move this useful heat into the secondary heating circuit, rather than allow it to contribute to the boiler standing loss. Circulation is maintained for a short period until flow and return temperatures are roughly equal.

5.2.2.3 Water heating plant

A recommended water heating plant for general use utilises separate primary and secondary water circuits, linked by a common header which acts as a hydraulic decoupler. A primary pump maintains water flow round the primary circuit and through the boiler. The primary pump is normally positioned in the return pipework to the boiler, particularly when open non-pressurised systems are used. Hot water is drawn from the common header to supply secondary circuits and HWS calorifiers as required. With this layout, compensation control adjusts the flow temperature of the secondary water flow temperature and does not influence the boiler. It is important to ensure compatible flows where the primary and secondary circuits interact; see 5.4.1. The use of a hydraulic decoupler is recommended as the constant flow of water through the boiler allows stable control to be achieved. Figure 5.2 shows a simple primary circuit; the boiler is protected by a high limit thermostat on the flow water and a flow switch on the primary return. Other boiler protection controls were discussed above. There are many variations of water circuit, the most important of which are discussed further in 5.4.

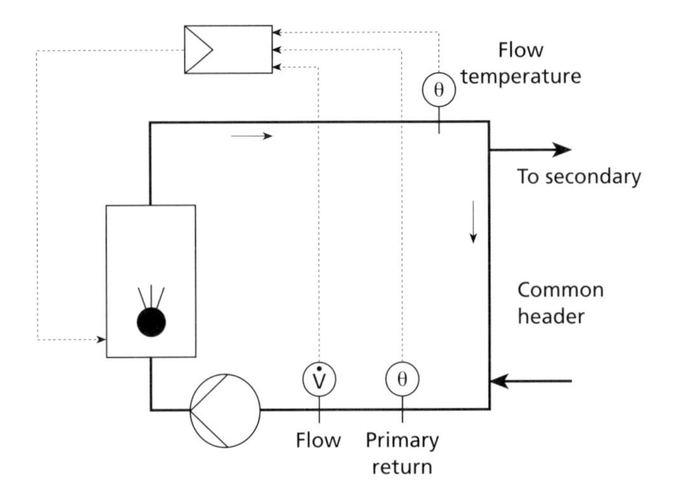

Figure 5.2 Simple boiler primary circuit. The boiler feeds a common header and primary flow is maintained by a pump. The boiler firing is controlled to give the desired flow temperature; a low primary return signal will enable the boiler

For smaller systems where there is only one secondary circuit it is possible to use the same circulation pump to drive both primary and secondary circuits. Figure 2.12 illustrates a simple weather compensation circuit where the primary flow is compensated. This has the disadvantage of reducing the return temperature to the boiler at low loads, with the attendant risk of condensation and corrosion.

5.2.2.4 Multiple boilers

Multiple boilers are used with some form of sequence control, which switches units on or off to match the heat demand. Separate boilers, not necessarily identical, may be controlled by a sequencer. A modular boiler can be used, consisting of a group of similar modules that are brought into use as required. The unit may have separate combustion chambers or a common combustion chamber. In order to share operating duty equally between all components of a multiple boiler system, boiler sequence selection is employed. This changes the order of operation of the boilers in rotation, either on a time or usage basis. Control is simplest where all components of the boiler system are identical. Circumstances where non-identical boilers are used include:

— The use of a small boiler for summer hot water service.

— The use of a single condensing boiler, where the cost of specifying all condensing units is not justified. The condensing boiler should always be the lead boiler to ensure maximum utilisation.

Poor control of boiler operation can add 15 to 30% to fuel consumption[4] compared to a well-controlled system. The use of modern boilers, with good efficiency and low standing losses, has simplified some aspects of boiler control. With a multiple boiler installation, heat losses from an unused boiler are reduced if that boiler is cold or cooling; when the heat output from a unit is not required it is more efficient to isolate the unit and prevent hot water flowing through it. However, experience has shown that the use of individual boiler pumps and isolation valves can cause many problems. The low standing losses of modern boilers allow simple sequencing control of multiple boilers without the necessity for isolation.

These low losses also enable a multiple boiler unit to operate with constant flow through all boilers without significant heat loss. This allows a simplified hydraulic layout, giving effective control without the use of individual boiler pumps or isolation valves. The use of separate primary and secondary circulation pumps combined with a low loss header ensures that a constant flow is maintained through the boilers, irrespective of any control action in the secondary circuit; this enables effective boiler sequencing control based on the return temperature.

Figure 5.3 shows the control principle for multiple parallel connected low water content boilers. This is recommended by BRECSU[4] for simple effective and reliable control. The lead boiler is enabled at all times heating is required. As the heating demand in the secondary circuit increases, the return water temperature drops and subsequent boilers are enabled according to the change in return temperature. Individual boiler temperature is controlled by thermostats on the boiler leaving water temperature. Integral control should not be used, since the fall in return temperature with

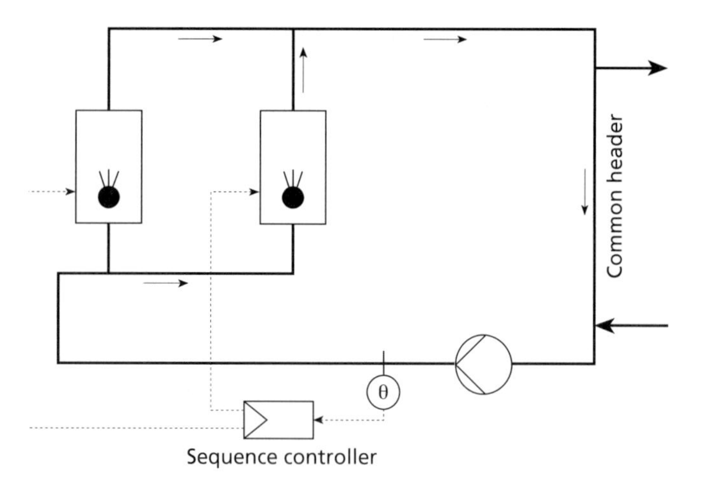

Figure 5.3 Multiple parallel boilers operating under control of a sequencer. Individual boiler firing is controlled by boiler thermostat

increased heat load is part of the strategy and is not to be treated as a load error.

In conditions of low load, the return water temperature entering the boilers is higher than at design conditions. The return temperature controls the number of boilers in service, and takes boilers off-line as the return temperature rises. The leaving water temperature of the operating boilers will be higher than the design temperature; however, this flow is then diluted by water flowing though the non-operating boilers; this is the phenomenon of flow dilution. The flow temperature of primary water entering the common header is the mean of the leaving temperatures of the multiple boilers. Individual boiler thermostats must be set higher than the design flow temperature to obviate the risk of lockout at low loads. Due regard must be paid to the provisions of PM5[6] on allowable temperatures and pressures.

Some designs of boilers, particularly older models with large water content, have an unacceptably high standing loss when warm. Overall efficiency in a multiple boiler system is therefore improved if boilers can be isolated from the primary water circulation and allowed to cool when not required. This adds to the complexity of the control system and can lead to control problems; in addition, some manufacturers do not recommend isolation valves because of concern about the effects of repeated heating and cooling of the boiler. If isolation is to be used, it must be engineered properly, as inadequate isolation gives rise to inefficiencies and control problems. Tight shut-off is required, as even a residual flow of 1% of nominal flow rate can give a heat loss equal to 40% of that obtaining at 100% flow[7].

Figure 5.4 shows a suitable control circuit for multiple parallel connected large boilers which need to be isolated. The lead boiler is enabled at all times heating is required and subsequent boilers and pumps are enabled according to the total heat load of the system, which is measured as the product of the primary circulation flow and the difference between flow and return temperature in and out of the common header. During boiler warm-up, the valve is in the recirculation position to prevent corrosion and cold slugs of water being delivered to the header. A separate control is required during system start-up, since there is then no temperature differential across the header. Outside air temperature can be used as an indication of system load. Individual boiler firing is controlled as normal by a thermostat on the boiler outlet. Control by heat load requires more instrumentation than control by temperature differential, but is required since the use of separate boiler pumps makes it difficult to ensure a constant primary flow. This circuit is more expensive to install than the simpler system shown in Figure 5.3 and has to be justified by the reduction in heat loss from off-line boilers.

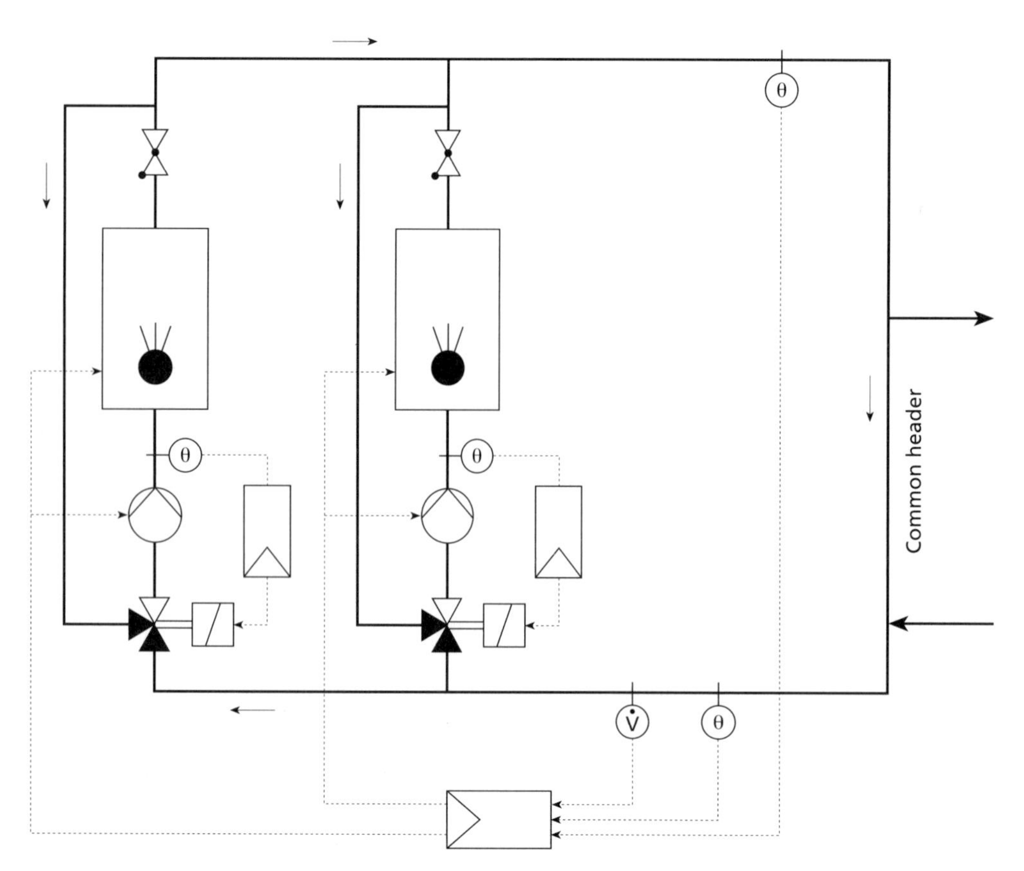

Figure 5.4 Multiple parallel high water content boilers. The sequence control enables additional boilers according to measured heat load

Series connection of boilers has the practical disadvantage that it is more difficult to isolate a unit for maintenance than with parallel connection. The resistance to flow in the primary circuit is also greater, resulting in higher pumping head. Nevertheless, this system gives simple effective and reliable control, without any of the possible problems of high boiler leaving temperatures caused by flow dilution. Figure 5.5 shows series-connected boilers feeding a common header. Constant flow in the primary circuit is maintained at all times. The lead boiler is enabled at all times heating is required. Subsequent boilers are enabled as the return temperature drops in a manner proportional to temperature. Individual boiler firing is controlled by the boiler's own thermostat.

5.2.3 System pressurisation

Pressurising a hot water system allows higher operating temperatures. Simple systems rely on the static head provided by a feed and expansion (F&E) tank; note that the relevant measurement when calculating pressure and allowable working temperature is the height of the feed and expansion tank above the highest point of the circulating system. The F&E tank is open to atmosphere and so will absorb oxygen and may also become contaminated. The use of a sealed system eliminates absorption of air and the attendant corrosion risk; higher working pressures may be obtained in buildings where it is impossible to mount an F&E tank at sufficient height. Pressurisation may also be applied to chilled water systems. A pressurised system must meet the requirements of PM5[6]. This details a number of control interlocks to ensure safe working and specifies that the temperature of the water leaving the boiler must be at least 17 K below the temperature of saturated steam at a pressure equal to the pressure at the highest point of the circulating system. It is necessary to monitor pressure to ensure that systems do not over- or under-pressurise.

5.2.3.1 Sealed expansion vessel pressurisation

This system uses a sealed expansion vessel incorporating a flexible diaphragm. The space above the diaphragm is charged with air or nitrogen to the required pressure and provides system pressurisation during normal operation. The movement of the diaphragm accommodates expansion of the water in heating. If system pressure falls due to water loss, the pressurisation pump is activated; this adds water to the vessel to bring the pressure back to the working level. An anti-gravity tank acts as a buffer between the hot water circulation and the expansion vessel to prevent hot water reaching the vessel during expansion (see Figure 5.6).

In addition to the pressure sensor which controls the pressurisation pump, high and low pressure switches are installed to act as interlocks. Activation of the high pressure switch disables the boiler and closes the fuel supply; circulation pumps continue for the run-on period. Activation of the low pressure switch disables the boiler and turns off circulation pumps. These interlocks are normally hardwired and require a manual reset.

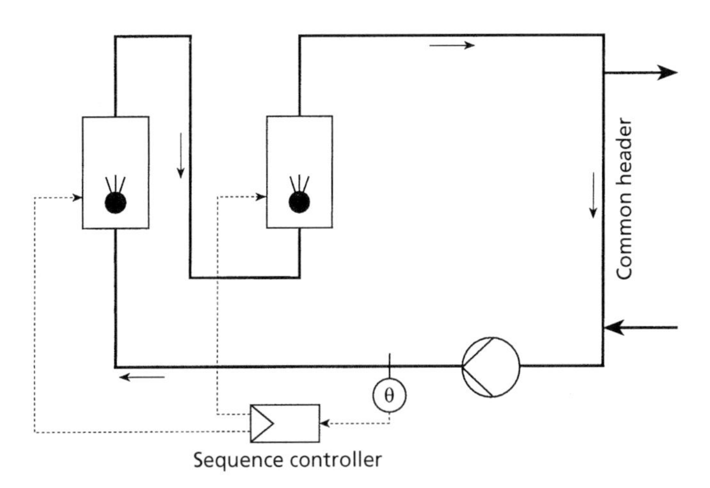

Figure 5.5 Series-connected boilers operating under control of a sequencer

5.2.3.2 Pump pressurisation

An alternative system uses a pump to maintain system pressure. Figure 5.7 illustrates the system. When the system is below pressure, the pressurisation pumps operate to restore pressure. If the system pressure is high, a spill valve opens and relieves any excess pressure by allowing water to pass to the expansion vessel. The pumps do not run continuously, but operate more often than is the case with the sealed system above and it is therefore common practice to provide duplicate pumps. The system can be provided as a packaged pressurisation unit with controls incorporated. High and low pressure interlocks should be provided, as for the sealed system.

5.3 Chillers

The basic components of an air conditioning system using chilled water as a distribution medium are:

- chiller, which contains the refrigeration unit
- heat rejection system, e.g. air cooled condenser or cooling tower
- chilled water distribution system.

The control of the distribution system is considered in detail in other sections. The chiller is the heart of the chilled water system and is classified by the type of refrigeration cycle:

- *Vapour compression*. This is the most common type, using an electrically driven compressor
- *Absorption*. Absorption chillers use heat to drive the refrigeration cycle. They are used where heat is cheaply available; they have some advantages in large systems.

Vapour compression units are classified by the type of compressor, which are shown in Table 5.2 together with typical applications and capacity control methods.

Table 5.2 Compressor types

Compressor	Application	Capacity control
Centrifugal		Inlet vane and VSD
Reciprocating	All sizes	Cylinder unloading
Screw	> 200 kW	Slide valve and VSD
Scroll	>50 kW and < 200 kW	VSD

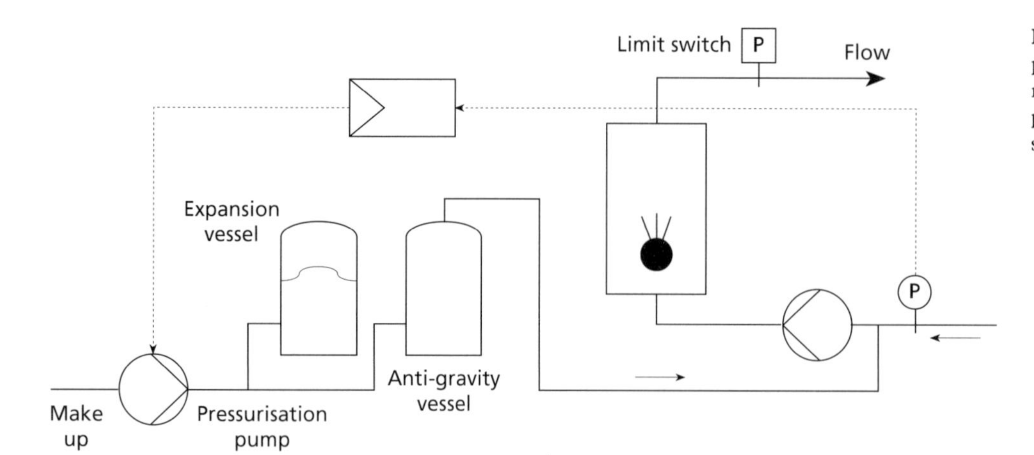

Figure 5.6 Sealed expansion vessel pressurisation. Pump runs when required to replace water and make up pressure. High and low limit pressure switch fitted

5.3.1 Chiller control

All types of compressor are capable of some form of capacity control. Reciprocating chillers are still the most commonly used, and consist of multi-cylinder units. Capacity control is by unloading one or more cylinders, i.e. by letting the suction valve open so that the unloaded cylinder does not pump refrigerant. Depending on the number of cylinders, this gives four or more stage capacity control. Rotary compressors using a combination of variable-speed drives (VSD) and mechanical control can give smoothly modulating control of chilling capacity; in most cases the output may be modulated down to about 10% of full output. The efficiency of operation is strongly affected by the operating temperatures and pressure of the chiller unit. The primary factor affecting compressor efficiency is the pressure difference across the compressor, known as the refrigerant pressure. Reducing the refrigerant pressure increases efficiency. Refrigerant pressure is reduced by the following:

- reducing condenser water temperature (but there is a minimum temperature for safe operation)
- raising the chilled water temperature
- reducing load (but overall chiller efficiency falls off at loads below about 30% full load).

Improved efficiency will be achieved by resetting the chilled water flow temperature upwards when the building cooling demand is light. Where variable-speed pumping of the chilled water is employed, the necessary flow rate and hence pumping energy consumption will be increased if the chilled water flow temperature is increased, compared with the alternative of maintaining flow temperature and reducing flow rate. It is possible that this will negate any improvement in chiller efficiency. Each system has to be considered independently to decide whether chilled water temperature reset is energy efficient.

Heat rejection from the chiller is commonly either by using a separate cooling tower or an integral air-cooled condenser. Air-cooled condensers are used in most small self-contained or split package systems up to about 1 MW capacity. Such units have the advantage of being self-contained, requiring no additional water supply and obviating any risk of Legionnaires' disease. The refrigerant gas leaving the compressor is circulated through a heat exchanger which is cooled by ambient air drawn over it by a fan. Fan speed is controlled as a function of load and ambient conditions to minimise overall energy consumption. Fan control for a packaged unit containing an air-cooled condenser is provided as part of the unitary controller provided with the unit and is not regarded as part of the BMS. The control of cooling towers is considered separately below.

It is universal practice that the chiller manufacturer provides a factory-fitted controller which is programmed to ensure safe and efficient operation over a wide range of loads. The dedicated controller looks after capacity control and condenser water temperature, plus other items such as thermal expansion valves or hot gas bypass. For multiple chillers, it will take care of sequencing. The controls specialist is therefore not required to configure the chiller controls for optimum efficiency. For many systems, the chilled water temperature is set at a constant value at the

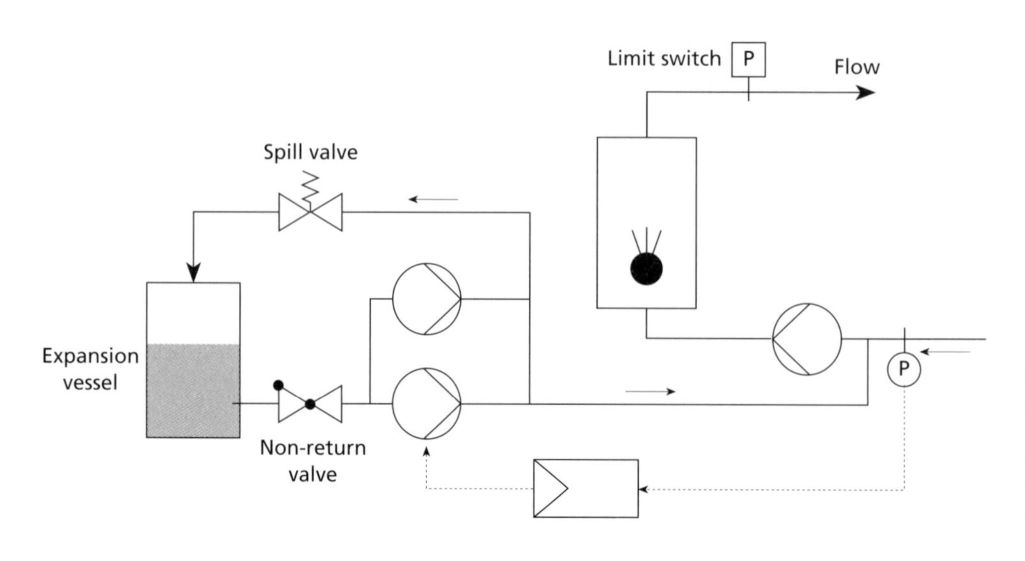

Figure 5.7 Pump pressurisation. The pump runs to maintain system pressure on receipt of low pressure signal. Excess water passes to the expansion vessel via the spill valve. High and low pressure limit switch is fitted

chiller controller, leaving the only function of the BMS to turn the chiller on and off at appropriate times. However, the more comprehensive chiller controllers are designed to communicate with a BMS. This provides for control of the supply of chilled water and monitoring of operation, while leaving the detailed control and optimisation of the working of the chiller to the fitted controller.

Table 5.3 summarises an interface between a chiller controller and the BMS. The example shown is for a reciprocating chiller and the controller allows the BMS direct control of capacity by loading or unloading the compressor cylinders. In general, it is sufficient for the BMS to supply the chilled water temperature set point and allow the chiller controller to optimise operation. Several points are available to the BMS which may be used for condition monitoring and maintenance.

Table 5.3 Chiller control interface for reciprocating chiller

Point	Type
Controlled points:	
— Water temperature set point	AO
— % load limit (10–100%)	AO
— Lead unit No	AO
— Start/stop command	DO
— Load the chiller	DO
— Unload the chiller	DO
Monitored points:	
— Chilled water temperature	AI
— Condenser water temperature	AI
— Compressor oil temperature	AI
— Compressor oil pressure	AI
— Compressor suction pressure	AI
— Compressor discharge temperature	AI
— Compressor full load current	AI
— Compressor run hours	AI
— Compressor number of starts	AI
— Compressor anti-recycle timer	AI

AO: analogue output; AI: analogue input; DO: digital output.

5.3.2 Multiple chillers

The use of multiple chillers offers standby capacity and improved efficiency at part loads. A typical arrangement is shown in Figure 5.8, where two multi-cylinder compressors are mounted in parallel. The lead compressor runs preferentially; the lag compressor is activated when the lead compressor can no longer meet the load.

This arrangement can provide good control of the leaving temperature of the chilled water. In the example, the chillers have a design capacity to produce a leaving temperature of 6°C when the return temperature is 10.5°C. In the example, each chiller control is set at 4.5°C. This set point has to be lower than the leaving temperature set point since if only one chiller is in operation, the water leaving the operating machine will be mixed with the warmer water passing through the inactive machine; this is termed flow dilution. If the chilled water temperature leaving a chiller falls below the lower limit of the operating differential, that chiller is unloaded one step. The sequence control sensor is mounted in the mixed chilled water flow. If the mixed flow falls below the lower limit of the operating differential, a chiller step is unloaded. The sequencer unloads the lag chiller in preference to the lead. It is also necessary to set a time delay, known as the load/unload time; this is the time allowed for the chiller to bring the water temperature under control after a load change. If the water temperature has not been brought within the differential band within this time, a further step is loaded or unloaded. There can be circumstances where one chiller has sufficient capacity to meet the building load, but would need to cool to a flow temperature below its set point; this is a consequence of flow dilution. In this situation, both chillers run at part load. The operating differential of the sequence controller must be greater than the change in leaving temperature produce by loading or unloading one step; this is necessary for stable operation. This method of step control can give good control of the mixed leaving temperature.

Multiple chillers are not necessarily of the same size or type. There are advantages in using a mixture of sizes, e.g.

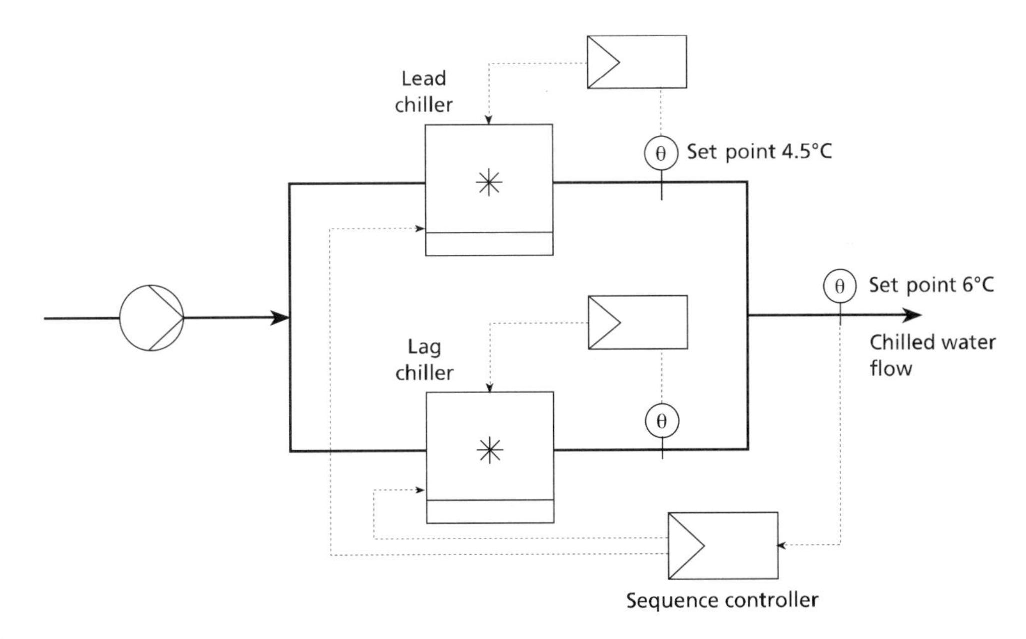

Figure 5.8 Two chillers in parallel operating under sequence controller

chillers in the ratio 30/50/70. This allows better matching of capacity to load than the use of equal-sized chillers; the chiller management controller assesses the load and selects the combination of chillers which best matches it. In some situations, a combination of vapour compression and absorption chillers may be economical, either because of the availability of a source of cheap heat or a requirement to avoid peak electrical loads. The optimisation of a multiple chiller system requires knowledge of the efficiency curves of each chiller at all temperature differentials. A typical calculation sequence, starting at minimum load, would be to select and start the most efficient chiller for that load. As the load increases, the data is constantly checked. When the optimum load for adding a second chiller occurs, based on current load, temperature differences and efficiency curves, the selected chiller is started and the loading of each chiller is adjusted for maximum system efficiency. If the decision is made to add or drop a chiller, minimum on and off time considerations must be considered.

5.3.3 Cooling towers

A cooling tower (Figure 5.9) employs evaporative cooling and can cool the condenser water to within 3 K of the ambient wet bulb temperature. By comparison, an air-cooled heat exchanger can achieve cooling of condenser water to within about 11 K of ambient dry bulb temperature. The use of poorly maintained cooling towers has been associated with outbreaks of legionellosis and this has restricted their application in recent years[8,9]. However, the greater efficiency of the evaporative cooling tower ensures its continued use in appropriate situations.

The performance of a cooling tower is characterised by two temperature differences:

— *range*: the difference between entering and leaving water temperatures

— *approach*: the temperature difference between the water leaving the cooling tower and the ambient wet bulb temperature (ambient dry bulb temperature in the case of a 'dry' cooling system).

A schematic diagram of an open cooling tower and its controls is shown in Figure 5.9. In an open cooling tower the condenser water itself is sprayed in the tower to achieve maximum evaporative cooling. As shown above, the maximum working efficiency of the chiller is achieved at the lowest possible condensing water temperature, subject to a minimum temperature, typically 18°C, below which there is a risk of damage to the compressor. Figure 5.10 shows the lowest possible condenser water temperature leaving a cooling tower as a function of ambient wet bulb temperature and this should give the maximum possible chiller efficiency. The condenser water is subject to low and high temperature limits to avoid damage to the chiller. As shown in Figure 5.9, control of condenser water temperature is by a combination of fan speed control and variable bypass of the flow of condenser water though the tower. Where multiple fans are employed, sequential switching of fans may be used instead of fan speed control. The range, i.e. the amount of cooling produced in the tower, may be increased by increasing the cooling tower fan speed and reduced by turning the fan off and using the bypass valve to reduce the flow of condenser water through the tower. The bypass valve is always fully open to the tower when the fan is on.

There are three major control strategies for cooling tower operation:

— fixed set point

— variable set point (fixed approach)

— near optimal control.

With fixed set point control, the condenser temperature is controlled at a constant set point, typically about 26 to 28°C, by the use of the modulating bypass valve and the cooling tower fan. For simple systems, the fan control is on/off. This has the advantage of simplicity, but in part load conditions the condensing water temperature is unnecessarily high, resulting in high chiller power consumption. Variable set point control adjusts the set point of the condenser water temperature in accordance with a characteristic curve of the form shown in Figure

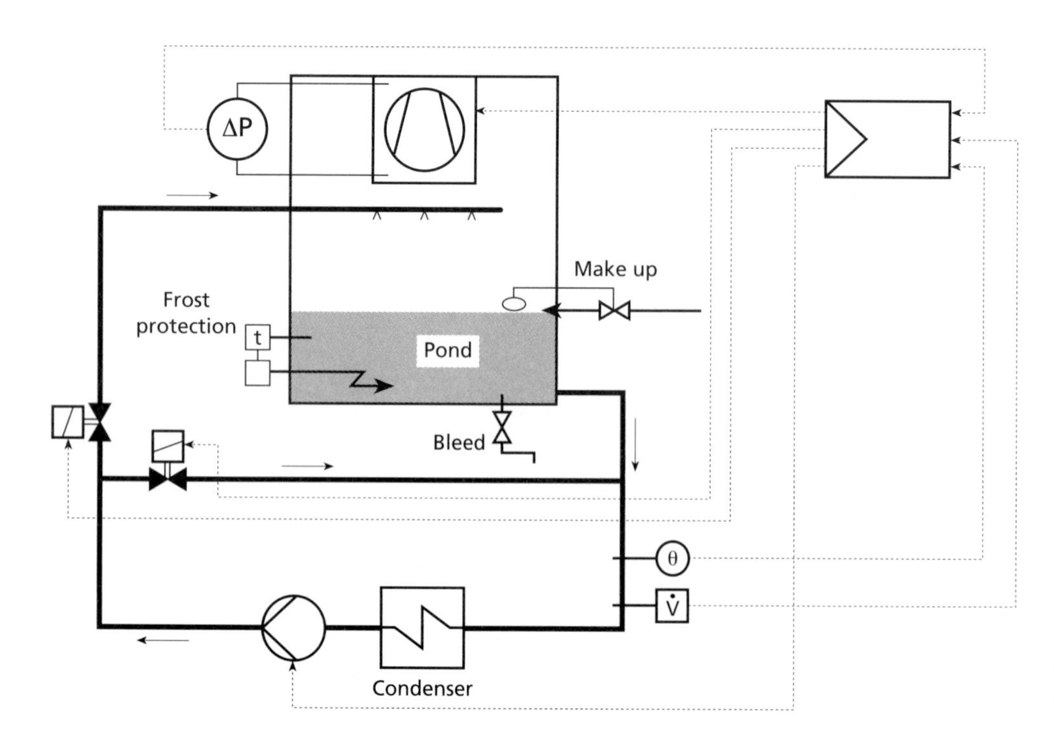

Figure 5.9 Open circuit cooling tower with bypass circuit

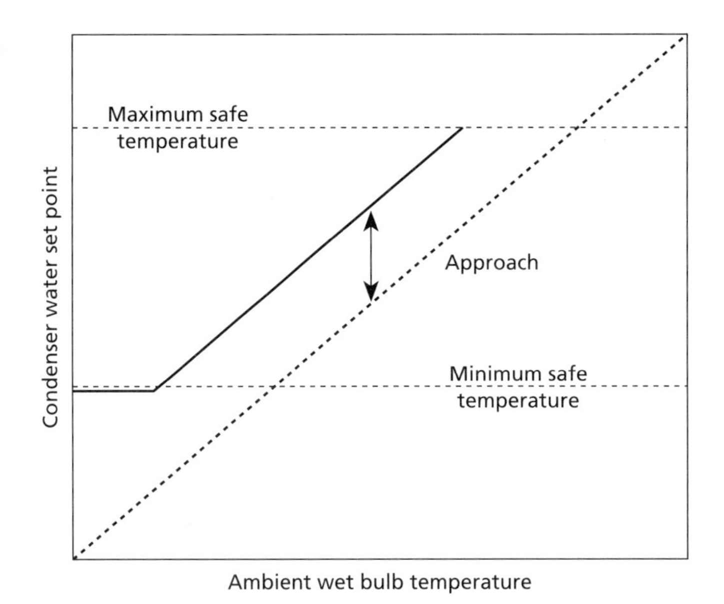

Figure 5.10 Minimum condenser water temperature for an open circuit cooling tower

5.10. Restricting the set point to the minimum approach value means that the fan speed will not be increased in a vain attempt to reach an impossibly low temperature.

Variable set point control does not necessarily result in minimum energy consumption, since the saving in chiller power consumption is to some extent offset by an increased cooling tower fan consumption. A control strategy developed in the USA(10) and since applied to the UK situation(11) achieves low combined energy consumptions. The strategy is termed near optimal control and consists of a simple control algorithm that regulates the fan speed in proportion to the part load ratio, i.e. the ratio of the actual load to the maximum system load. Upper and lower condenser water temperature limits are incorporated to prevent damage to the chiller. This is an open loop control and it is necessary to calculate in advance the system parameters of the control algorithm from performance data of the cooling tower and chiller. The controller requires three inputs: chilled water flow and return temperatures and the condenser water supply temperature. A value for chilled water flow is also necessary to calculate the chiller load; this value can be supplied from commissioning data or measured directly. The control output is a signal to run the tower fans at a speed proportional to the cooling load. Near optimal control is particularly suited to installations with large centrifugal compressors that are operated throughout the year over a range of ambient temperatures.

As well as control of condensing water temperature, a cooling tower requires other controls for safe operation:

— *Freeze protection*. Water in the tower is at risk of freezing in cold weather; the danger is increased by the use of a bypass valve or when the chiller is out of action. An immersion heater controlled by a thermostat in the tower pond is used to prevent freezing. Alternatively, the pond may be drained to a sump located in a warm space when a risk is present.

— *Bleeding*. The concentration of dissolved and suspended solids increases over time due to evaporation and the collection of airborne dirt. The pond requires to be bled periodically and topped up with fresh water. This may be on a time schedule or controlled by monitoring water conductivity. The bleeding program may be combined with chemical treatment of the water. A water meter is often installed in the supply pipe to the pond, which may be used to control chemical dosing equipment.

— *Interlocks*. A flow switch is positioned in the condenser water circuit. If flow is not proved, both chiller and cooling tower fan are switched off. A differential pressure switch may be used to prove cooling tower fan operation.

In a closed circuit (indirect) cooling tower, the condenser water is circulated though a heat exchanger within the tower. Condenser water and cooling tower water are thus kept separate, which has the advantage of preventing possibly dirty water passing through the condenser. Control strategies are the same for the open tower. The closed circuit tower still involves the use of sprayed water with the attendant risk of the dispersal of contaminated droplets. Where it is necessary to use a dry system, it may be desirable to use a dry cooling tower. Condenser water is circulated through a water–air heat exchanger which is cooled by fans. This differs from an air-cooled condenser where the refrigerant itself is circulated through an air-cooled heat exchanger. The use of a dry cooling tower avoids the use of long refrigerant lines where the heat rejection unit is remote from the compressor. It has the disadvantage compared with an air-cooled condenser of introducing an extra heat exchanger, with the associated loss of efficiency. The minimum approach for the cooled water is about 11 K of the ambient dry bulb temperature. Otherwise, the control principles are the same as for an evaporative cooling tower.

5.3.4 Free cooling

For systems which depend on the circulation of chilled water, it may be possible to dispense with the chiller when the outdoor conditions are such that the condenser water temperature from the cooling tower is lower than the required chilled water temperature. Three methods are available for the provision of free cooling:

— *Refrigerant vapour migration*. A special chiller design incorporates extra refrigerant paths which bypass the compressor. Heat exchange between condenser and evaporator takes place by gravity circulation of refrigerant. This is also known as the thermo-syphonic method of free cooling.

— *Auxiliary heat exchanger*. The condenser water and chilled water circuits are diverted to an auxiliary heat exchanger, where the cool condenser water extracts heat from the chilled water circuit.

— *Interconnected circuits*. The two circuits are directly connected, so that the condenser water flows in the chilled water circuit. This risks circulating dirty water to fan coil units or other terminals with risk of fouling and blockage. Water treatment and filtration assume great importance.

The term free cooling in this context is to be distinguished from the situation in air handling systems using economiser control, where cold outdoor air is circulated directly to the conditioned space. This is also termed free cooling and is treated in 5.7.5.

5.3.5 Ice storage

Ice storage systems run chillers overnight to produce ice; during the subsequent day the ice is melted to provide chilled water for cooling and dehumidification. While it is possible to size the ice store to be of sufficient size to cope with the next day demand, it is more common to opt for partial storage, where the ice store is not sufficient in itself to provide all cooling needs on a design day. Peak loads are met by discharging the ice store and running the chiller at the same time. Ice storage has several advantages over a non-storage system:

— *Smaller chiller size*. The chiller does not have to meet the instantaneous maximum design load, potentially giving reduced capital costs.

— *Lower running costs*. It may be possible to take advantage of reduced electricity prices at night or exploit the potential for reducing maximum demand charges.

— *More efficient chiller operation*. The chiller runs steadily at design load for extended periods and benefits from lower ambient temperatures, which results in reduced condenser water temperatures.

The two major types of ice store are classified as:

— *Direct storage*. This type of store is also known as an ice builder. Refrigerant coils are immersed in the storage tank and ice forms on the outside of the coils. When cooling is required in the main system, chilled water is passed through the tank and this melts the ice.

— *Indirect storage*. The most common form of indirect store is the ice bank. The primary chilled circuit uses a glycol mix. During charging, this is chilled and circulated through a cooling coil immersed in the ice store to freeze it. During discharge the same glycol mix is circulated through the coil in the now frozen ice store and then to a heat exchanger to cool the secondary chilled water. It is possible to dispense with the heat exchanger and circulate the glycol mix throughout the secondary system.

5.3.5.1 System configuration

Partial storage systems can be configured in different ways, the names reflecting the relative positions of the chiller and ice store. Chiller upstream configuration represents the great majority of installed systems.

— *Chiller upstream*. The chiller, ice store and load are piped in series. Chilled coolant leaving the chiller flows directly to the coil in the ice store. When both chiller and store discharge are operating simultaneously, the chiller receives water from the load at relatively warm temperature, allowing the chiller to operate at a high COP.

— *Chiller downstream*. The chiller receives water leaving the ice store and it is possible to modulate the flow and bypass round the store to achieve almost constant operating conditions for the chiller. This arrangement allows low coolant flow temperatures to the load if required, but requires that the chiller operate at low temperatures with a lower COP.

— *Parallel system*. The piping arrangement allows the chiller or ice store to operate independently of each other, or in a variety of arrangements. It is potentially more efficient, but requires more pipework, valves and controls.

Figure 5.11 shows a chiller upstream ice storage system. The system is arranged so that it can provide some cooling to the building during the night-time charging period. The circuit shown is suitable for a night-time cooling load of up to about 20% of the peak load. Larger night-time cooling loads require a separate chiller which is available 24 h per day but does not contribute to ice store charging. The figure shows an injection circuit used to modulate the supply of coolant to the secondary heat exchanger. If simultaneous building cooling and store charging is required during the night, use of the injection circuit avoids supplying coolant at below zero temperature to the heat exchanger. If simultaneous cooling and charging is never required, the extra pump and three-port valve may be dispensed with. The various operating modes are tabulated in the figure. The daytime cooling load is met by chiller, ice store or a combination of the two. Temperature control of the coolant flow to the load is modulated using the three-way mixing at the exit of the ice store. The temperature drop across the secondary heat exchanger must equal the sum of the temperature drops across the chiller and ice store. This may require a larger temperature differential than usual with consequent attention to design. The design and control of ice storage systems is dealt with in CIBSE TM18[12].

5.3.5.2 Control techniques

It is important that the controller be able to monitor the amount of ice in the store, both to terminate charging when the maximum capacity has been reached and to estimate the

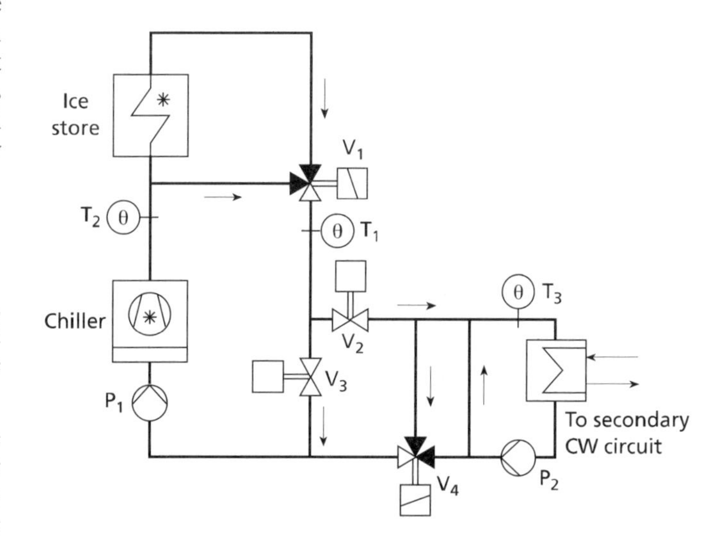

Mode	Pumps		Valves				Set points		
	P_1	P_2	V_1	V_2	V_3	V_4	T_1	T_2	T_3
Charge	On	Off	Open	Close	Open	Close		–5	
Chiller alone	On	On	Close	Open	Close	Open		4	
Discharge alone	On	On	Mod	Open	Closed	Mod	3		6
Discharge + chiller	On	On	Mod	Open	Open	Mod	3	8	6
Cooling + charge	On	On	Open	Open	Close	Mod		–5	6

Figure 5.11 Ice storage: series configuration chiller, upstream. If simultaneous charging of the store and building cooling is not required, the injection circuit with P_1 and V_4 may be dispensed with. The table shows settings for chiller priority operation. Set points are illustrative only

available capacity remaining during discharge. Several types of ice inventory estimation are used:

— *Volumetric devices*. These detect the pressure change produced by the increase in depth of the store caused by water expansion on freezing.

— *Heat flow monitoring*. These methods calculate the heat entering and leaving the store by measuring flow rate and temperature difference across the coil and so estimate the quantity of ice melted. This method estimates the quantity of ice remaining during discharge and requires that the store start each day fully charged to give a datum.

— *Store exit temperature*. During charging, coolant leaves the store at an initial temperature of about –0.5°C. When charging is complete, this temperature falls rapidly and the chiller is switched off when the exit temperature reaches –4°C. In practice, the fall in temperature may not give a clear endpoint, since the layer of ice on the coil inhibits heat transfer.

Charging strategy for a partial storage system. The ice store is charged each night during the low rate tariff period. Three levels of control are possible:

— *Charge to full*. The charge period is initiated by a time switch or signal that the low rate tariff period has started. The chiller then operates at full load until the charge control device indicates that the store is fully charged, irrespective of the level of daytime demand. The control is reset to prevent cycling.

— *Skipped charge*. If only a small amount of the ice store has been depleted during the day, the subsequent night's charge may be omitted.

— *Optimum charge*. Each night, the controller must calculate the total expected cooling required for the following day, together with an estimate of the ice remaining in the store after the previous day's operation. The difference, plus an allowance of 10% or so, is the required overnight charge.

Standing losses from an ice store are small and the energy penalty from charging fully each night may not be enough to justify a complex control system. For many buildings during the British winter, free cooling will be sufficient to provide enough chilled water and it will not be necessary to run the ice store. As the weather warms up in spring, the first day that chiller operation is required during the day marks the start of the ice storage season. The more sophisticated controllers use an estimate of the cooling demand profile during the day, both to estimate the total charge required and to optimise the discharge pattern. This is done by averaging the hourly figures over the previous three days, taking due account of weekends or holidays.

Control of the discharge cycle. Two main methods of control are available:

— *Store priority*. This treats the ice store as an extra chiller and modulates the supply of chilled water from the store to the rate that it can supply for the rest of the occupied period. If this is insufficient to meet the instantaneous demand, the chiller is run as well.

— *Chiller priority*. The chiller provides the base load, with the ice store being used to provide additional capacity as required. The chiller is turned off when the ice store is sufficient to meet the remaining demand for the day.

The store priority control may result in inadequate cooling ability to meet peak demands towards the end of the working day and the potential reduction in operating cost by exploiting lower cost stored energy may be nullified by increased capital cost and inefficient part load operation of the chiller. The chiller priority method results in both smaller chiller and ice store sizes and allows full advantage of the ability of the ice store to provide high cooling power when required for peak demands at the end of the day or for demand management. Figure 5.12 compares discharge pattern of the two strategies.

Achieving optimum discharge control is complicated by the fact that the available rate of chilled water supply from store falls off as the store is depleted. The aim of the discharge control is to run the chiller at maximum capacity during the day until the remaining demand can be met by the ice store alone. Maximum cooling demand occurs during the afternoon and the controller makes use of an estimate of demand profile obtained by averaging the previous three days demand. At hourly intervals during the day, the controller compares the estimated demand for the rest of the day with the remaining store capacity. If the store capacity exceeds the demand, the store is used alone. Otherwise, the chiller runs at full rate for 1 h, after which the comparison is repeated[13].

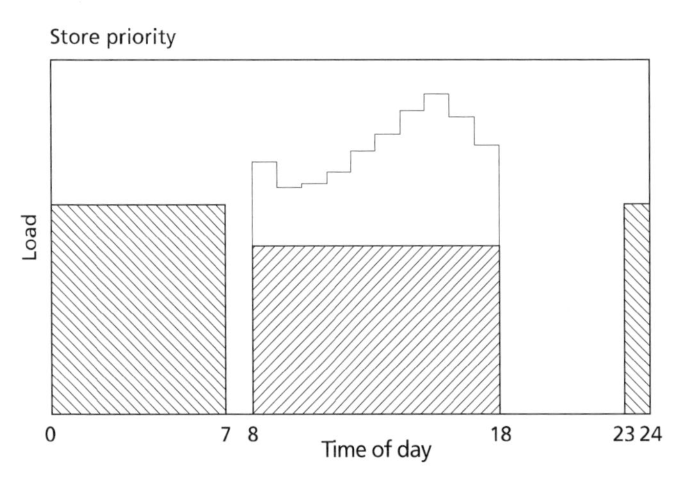

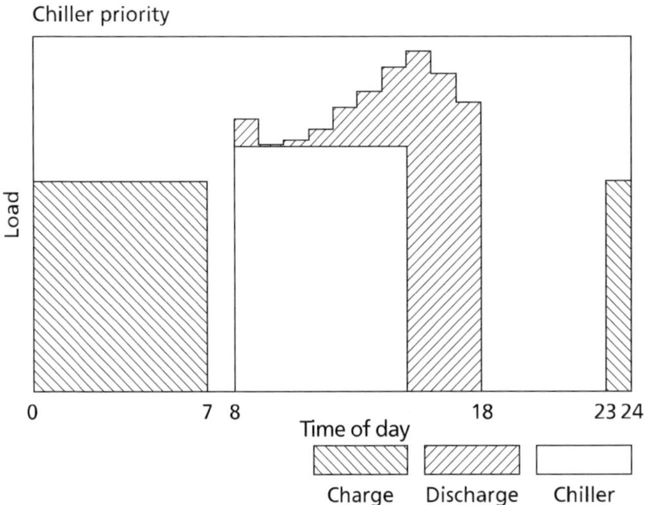

Figure 5.12 Comparison of store priority and chiller priority operation at part load

5.4 Control of hydraulic circuits

The variety of hydraulic circuits used in heating and cooling applications is very great. The following sets out some of the basic hydraulic circuits which are used for both primary and secondary systems and gives the main characteristics which are relevant to their control. Primary circuit implies the main distribution circuit of hot or chilled water, which supplies secondary circuits. Secondary circuit implies the hydraulic circuit which circulates fluid through a heat exchanger, which is typically a coil in a duct or fan coil unit, but may be a radiator, chilled ceiling or other device. In addition, there may be a circuit with its own pump ensuring the required circulation through the boiler or chiller. The discussion below is set out using boilers, but most applies equally to chilled water circuits.

5.4.1 Network design

It is essential that hydraulic circuits are properly designed and balanced if the control system is to be able to operate in a stable and effective manner. Sufficient flow of water must be available at each terminal and it must be controllable under all operating conditions; it is not sufficient to achieve design conditions alone. The problem of achieving controllability under part load conditions is greater when two-port variable flow circuits are used, compared with circuits using three-port valves. Three-port valves used in mixing circuits give more or less constant flow conditions, so that modulation of one valve does not disturb flow rates elsewhere in the system. In contrast, modulation of a two-port valve causes variations of flow and pressure elsewhere, which can lead to overflow or starvation of other terminals. This is discussed in more detail in 5.4.5.

While the detailed theory of hydraulic circuits is outside the scope of this Guide, three simple principles may be stated for good controllability[7]:

- The design flow must be available at all terminals.
- The control valves must have sufficient authority.
- Flows must be compatible at system interfaces.

These conditions may be achieved by a combination of satisfactory circuit design in the first place and proper system balancing on commissioning.

Availability of design flow

Consider a system which has a number of heating terminals controlled by two-port valves. Under steady operating conditions the valves modulate to provide the required flow through each terminal and the need for balancing valves may not be apparent. However, in boost or optimum start mode, all valves open fully. If the system is not properly balanced, some circuits will receive an overflow, with a corresponding underflow in less favoured circuits. The less favoured circuits will only receive their required flow when the favoured zones have reached their set temperature and their valves start closing. Since the output from a heat emitter is not a linear function of flow, the increased flow in the favoured circuits does not produce a corresponding increase in heat emission, with the result that the full output of the boiler is not available during start-up, when it is most needed. Mistaken attempts to cure the under-heating by increasing the pump pressure are unsatisfactory. Correct balancing is required to achieve design flow in all secondary circuits with the control valves fully open.

Valve authority

Valve authority was defined in 3.3.4.1. An alternative definition, which is more readily applicable to systems with two-port valves is

$$N = \frac{\Delta P_{min}}{\Delta P_{max}}$$

where ΔP_{min} is the minimum pressure drop across the control valve in its fully open position and ΔP_{max} is the pressure drop across the valve in its fully closed position. The authority can therefore be envisaged as a measure of the variation of the differential pressure across the control valve as the valve modulates. An authority of $N = 1$ implies no variation of pressure; an authority of $N = 0.25$ means a pressure variation over a range of 4:1 as the valve modulates from closed to open. An authority of $N \geq 0.25$ should be aimed at. A valve with a low authority will give poor control and may give rise to hunting operation. Satisfactory authority is achieved by selecting a valve with the required pressure drop at design flow rate and balancing the system to achieve the design flow rates. In some systems using two-port valves it is difficult to achieve the required authority and a differential control valve may be used to give a constant pressure drop across the control valve. This achieves an authority of $N = 1$ and is discussed further in 5.4.5.5.

Compatibility of flow

Where two circuits share a common section, the system must be balanced to ensure that the two flows do not interact in such a way as to prevent the desired performance being achieved. Figure 5.13 shows a boiler feeding a secondary circuit though a hydraulic decoupler; decouplers are discussed further in 5.4.3.3. If the flow in the secondary circuit is greater than that in the primary circuit, water will flow upward in the decoupler, with the result that cool return water mixes with the hot boiler flow, reducing the flow temperature in the secondary circuit. This can prevent the design boiler heat output being delivered to the heat emitters. The total pump flow on the primary side must be at least equal to that on the secondary and this must be established during balancing. The diagram shows a very simple layout. In practice, problems of incompatible flow are likely to arise with multiple boiler or chiller installations feeding a variable-flow secondary circuit.

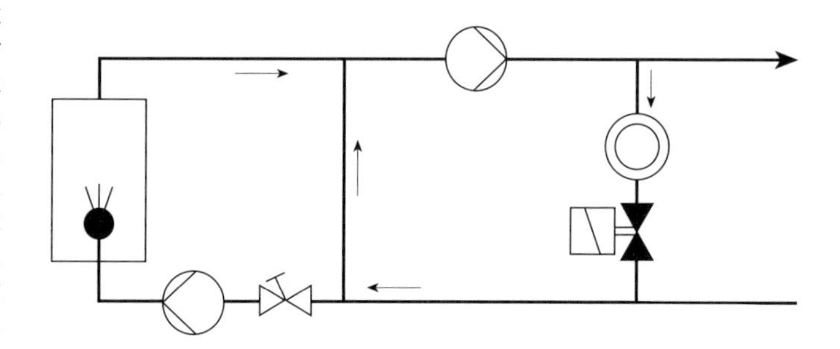

Figure 5.13 Incompatible flows. Reverse flow in the decoupler dilutes the flow temperature

Hydraulic circuits may be designed and analysed using the principles of network analysis[14]. Resistances of pipework and components may be estimated from dimensions or manufacturer's data; see CIBSE Guide C4[15]. Flow rates are calculated from the fundamental equation

$$\Delta P = RM^2$$

where:

M is the mass flow rate (kg/s)

R is the resistance (Pa s^2/kg^2)

ΔP is the pressure differential (Pa).

This differs from the familiar Ohm's law. In a hydraulic circuit the pressure drop is proportional to the square of the mass flow; this has the consequence that resistances in series are simply added, but two resistances in parallel are combined according to the relation

$$R = \frac{R_1 R_2}{R_1 + R_2 + 2\sqrt{R_1 R_2}}$$

Thus if two resistances of value R are connected in parallel, the combined resistance is only $0.25R$. Using the basic equation and the requirement that mass flow is conserved going round the circuit, it is possible to calculate flows and pressure drops at all parts of a circuit. Various tools are available to perform this[14]. A network analysis should be carried out at the design stage whenever possible, paying attention to the behaviour of the system at part load. As will be seen in later sections, modulation of individual two-port valves may result in over or underflow in other sections which cannot be corrected.

5.4.2 Basic circuits

Four basic types of secondary circuit are described in the following sections. A fuller analysis may be found in Petitjean[16] and Seyfert[7]. Valve selection and sizing follow the general principles set out above. In general, several secondary circuits may be fed off a single primary circuit, each of which is independently controlled by the demands of the zone or space being conditioned. The manner in which the circuits interact under part load conditions is of great importance to the design and successful operation of the system. Variable-flow systems present their own difficulties, which are considered separately.

5.4.2.1 Diverting circuit

Figure 5.14 shows a mixing valve installed in a diverting circuit. Several similar secondary circuits may be fed from the primary circuit, and the flow of fluid through each of heat exchangers under full flow conditions may be balanced by positioning a regulating valve as shown. The use of a three-port valve ensures that the total flow through the secondary circuit, i.e. heat exchanger and bypass, remains approximately constant at all valve positions. Note that the regulating valve is in a constant-flow section of the circuit. The control valve is selected and sized as described in 3.3.4 above, aiming at a valve authority of 0.5. The diverting circuit has the following characteristics:

— Flow is constant in the primary circuit.

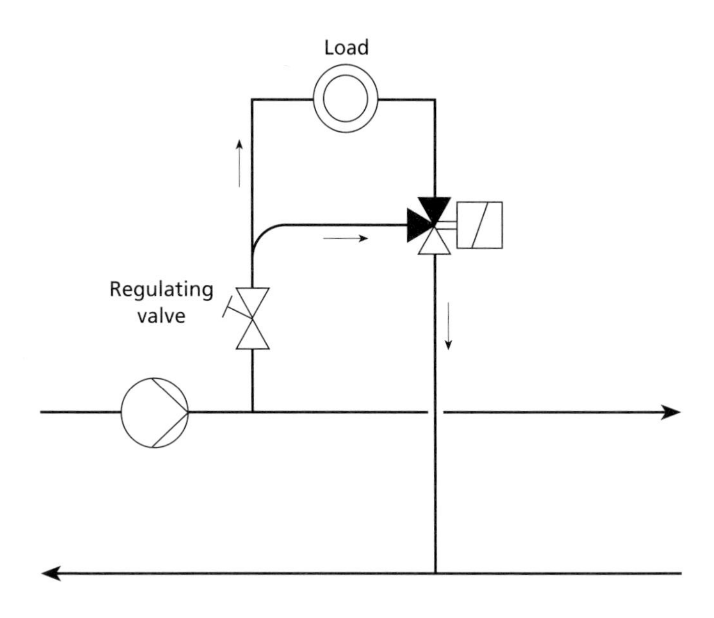

Figure 5.14 Diverting circuit with mixing valve

— There is constant temperature, variable flow in the heat exchanger.

— Variable flow in the heat exchanger may give poor operation, e.g. stratification.

— System remains balanced as individual loads vary.

— Full pumping load is maintained at low demand.

— Use of a diverting valve possible: a mixing valve in diverting application is usually preferred.

— A secondary pump is not always required.

5.4.2.2 Mixing circuit

In this circuit, the entire volume of water in the secondary circuit is pumped through the heat exchanger, and control occurs by mixing the water to modulate the temperature in the heat exchanger. The advantage of a mixing circuit over a diverting circuit is the even emission of heat over the entire surface of the heat exchanger, resulting in a small temperature gradient in the secondary medium; the continuous movement also reduces the risk of freezing in an air coil. The simple circuit shown in Figure 5.15 works best when there is little or no pressure drop between the flow and return pipes of the mixing circuit. A primary pump is not always necessary on small systems, provided that the requirements for minimum flow in the boiler are met; boiler protection can be provided by switching off the boiler when heating demand is satisfied, in conjunction with pump overrun and temperature interlocks. If a regulating valve is required to balance the secondary circuits it should be placed in the constant flow section as shown. The simple mixing circuit has the following characteristics:

— There is variable flow in the primary circuit.

— Constant flow through heat exchanger gives even temperature distribution.

— Can provide secondary circuits operating at different temperatures.

— Primary pump may not be necessary in small systems.

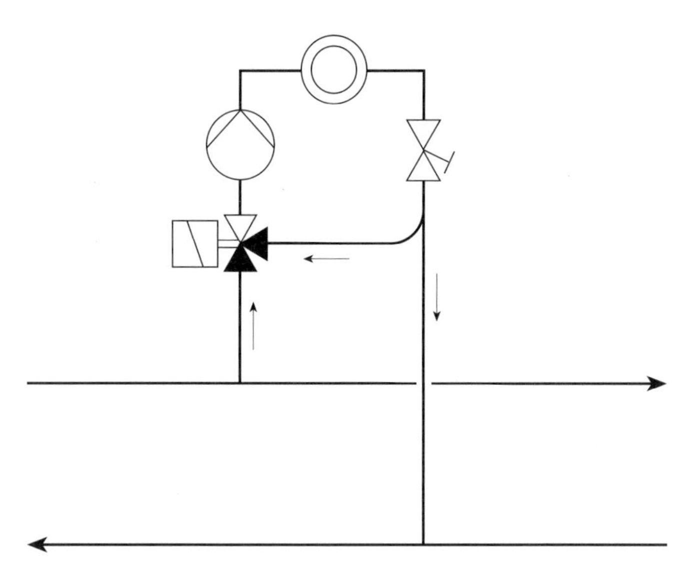

Figure 5.15 Mixing circuit

- There is full pumping load at part demand.
- There is an extra cost of secondary circulating pumps.
- Boiler protection at low demand is required.

In some heating systems, there is a requirement that the maximum flow temperature in the secondary circuit should be limited to a value below that of the primary flow temperature. This can be done by installing an internal bypass to the mixing circuit, as shown in Figure 5.16; this type of circuit is sometimes known as a dual mixing circuit. The regulating valve in the secondary circuit limits the flow through the heat emitter. The valve in the internal bypass is adjusted with the three-port mixing valve fully open to provide the required mixing of primary and secondary bypass flows which will provide the desired maximum secondary flow temperature. An application of this circuit is given in 5.4.6 on underfloor heating.

The mixing circuit works best where the primary flow and return pipes are at approximately the same pressure. Where a primary circulation pump is used, this situation may be achieved by the use of a hydraulic decoupler. If a pumped primary loop with a pressure drop between flow and return is required, then an injection circuit may be used, as described in 5.4.2.3 below. Otherwise, a local low-loss header can be incorporated as shown in Figure 5.17. This results in a constant flow in the primary circuit. Another approach is to use a differential pressure control valve (DPCV) across the secondary circuit to limit the pressure drop seen by the three-port mixing valve. The action of DPCVs is discussed in more detail below in 5.4.5.5.

5.4.2.3 Injection circuit

The use of an injection circuit gives greater separation between primary and secondary circuits (Figure 5.18). It requires sufficient pressure difference between the primary flow and return pipes to provide flow through the control valve. The secondary circulation pump pumps water continuously through the load. When there is no demand for heat, the control valve is closed and primary water flows via the bypass circuit through the control valve and back to the connection point on the primary return. When the control valve opens, a volume of water is injected into the secondary circuit and the equivalent amount leaves the secondary circuit via the control valve. Both primary and secondary circuits have constant flow rates. A regulating valve should be placed at every branching point in the primary circuit; a regulating valve may also be used in the secondary circuit to achieve the desired flow rate. The size of the control valve is determined from the nominal flow in the secondary circuit, calculated from the nominal heat output of the heat exchanger and the temperature drop across it, and the available differential pressure between the connection points on the primary flow and return legs. The variable-flow section, which is normally relevant to the calculation of valve authority, is the section between the primary and secondary bypasses; these resistances are likely to be negligible and valve authority will be high. With this form of circuit, correct hydraulic layout is important to avoid parasitic or single pipe circulation in the connection between primary and secondary circuits. Injection circuits are not suitable where it is important to avoid cross-contamination between primary and secondary

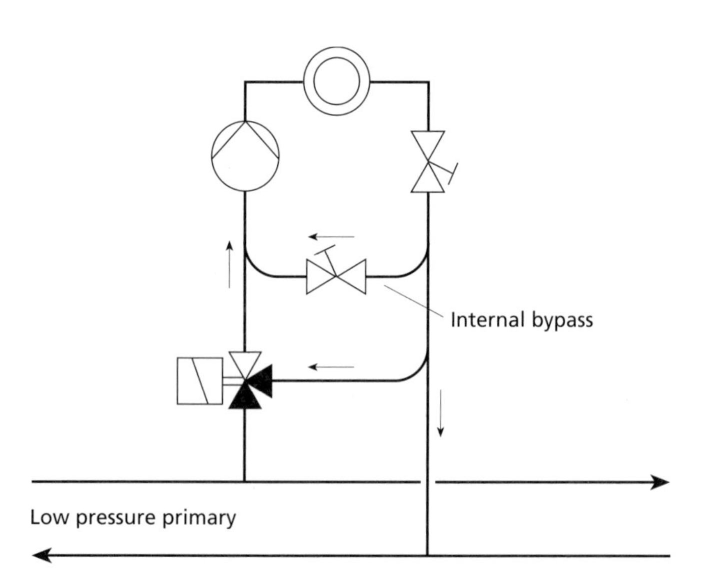

Figure 5.16 Mixing circuit with internal bypass. Used to provide temperature gradient between primary and secondary circuits

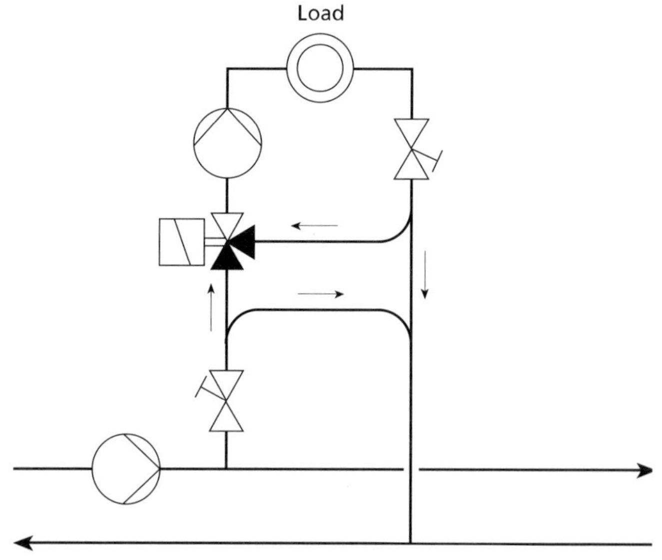

Figure 5.17 Mixing circuit incorporating low loss header. Suitable for use with constant flow pumped primary circuit

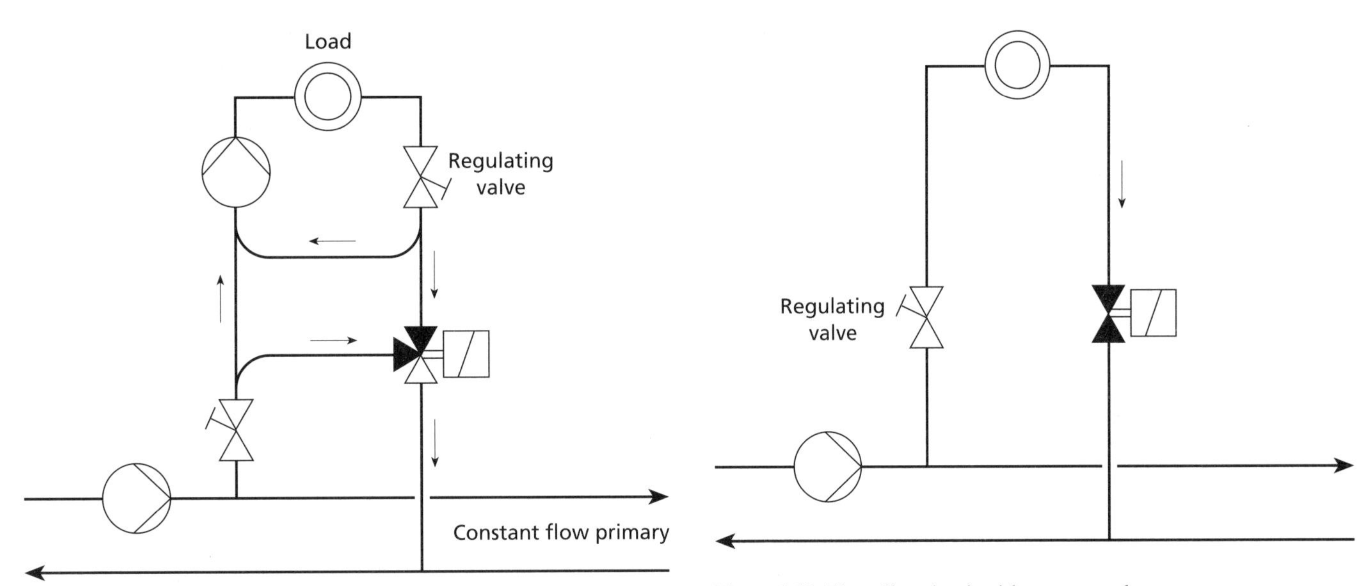

Figure 5.18 Injection circuit

Figure 5.19 Throttling circuit with two-port valve

circuits, or where the primary circuit uses medium temperature hot water (MTHW) or high temperature hot water (HTHW) and the consequences of valve failure would be unacceptable. In these situations a heat exchanger should be used to separate primary and secondary water.

Injection circuits have the following characteristics:

— Flow in both primary and secondary circuits is constant.

— Primary and secondary circuits are independent.

— Require sufficient pressure differential between primary flow and return.

— Suitable for district heating.

5.4.2.4 Throttling circuit

The use of two-port valves offers a cost saving over diverting or mixing valves. In Figure 5.19, the two-port valve controls the flow of water in the secondary circuit. Flow in the heat exchanger is at constant temperature, variable flow, with the possible problems of stratification. There are some problems in providing suitable valve authorities in throttling circuits. Figure 5.20 shows a ladder arrangement of multiple secondary circuits controlled by two-port valves, with illustrative pressures shown to demonstrate the regulation problem. The figures are based on a pump producing 100 units of pressure, loads requiring a pressure drop of 10 units at design flow and connecting pipework producing a pressure drop of 10 units between legs. In order to equalise conditions between the secondary circuits it is necessary to add regulating valves as shown. The regulating valve in the index leg, i.e. the leg furthest from the source, remains fully open while the others are adjusted to give equal flow in all legs. Valve authority is calculated as the ratio of pressure drop across the control valve (10 units) to the pressure drop across the part of the circuit with variable flow (100 units). In this case, the entire primary circuit flow is variable, resulting in low valve authorities of 0.1, well below that recommended for good control. The reverse return arrangement shown in Figure 5.21 has the advantage of using the flow and return pipework itself to balance the flows, but results in similar low authorities. Authorities could be increased by installing smaller valves, but with a consequent rise in pump head and energy consumption. Since the flow in the primary circuit is controlled by the throttling valves, provision must be made to ensure that adequate flow through the boiler is maintained or some other form of boiler protection provided. The use of variable-speed pumping overcomes some of the problems associated with the use of two-port valves and is considered separately in 5.4.5 below.

The characteristics of throttling circuits are:

— Flow in both primary and secondary circuits is variable.

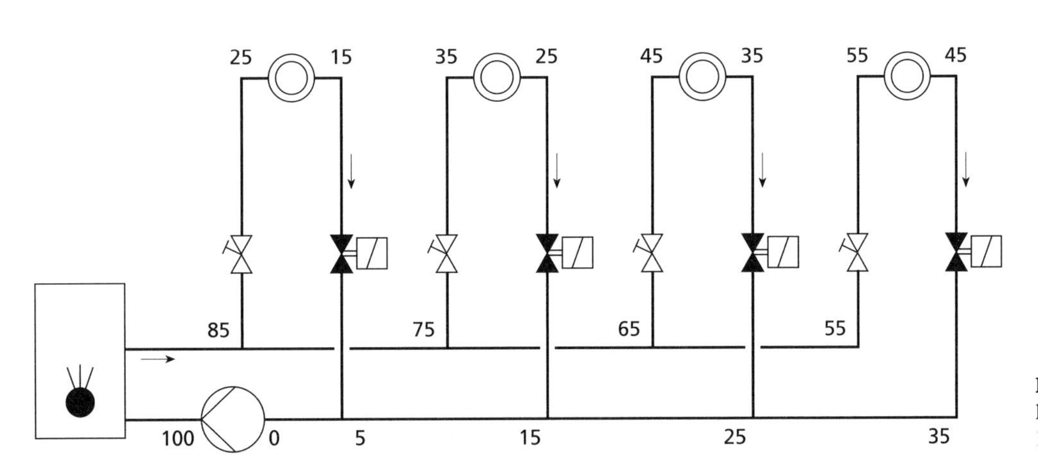

Figure 5.20 Throttling circuits in ladder configuration. Pump develops 100 units of pressure

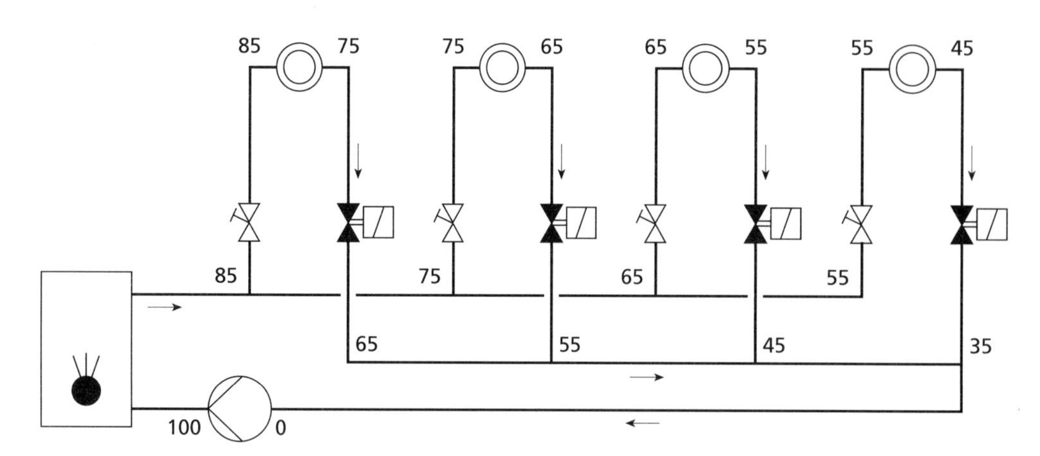

Figure 5.21 Throttling circuits in reverse return configuration

— Hydraulic balancing of circuits is required.

— Secondary circulation pumps not necessary.

— There is low valve authority.

— Pumping cost is possibly high.

— Cost of valves is lower compared with three-port.

— Boiler protection is required.

— Consider variable-speed pumping.

5.4.3 Primary circuits: boiler and chiller connections

A hydraulic heating or cooling system typically consists of a central source of hot or chilled water which is circulated in a primary system. A number of separately controlled secondary circuits obtain the necessary hot or chilled water from the primary circulation main. Secondary circuits usually fall into one of the main categories described above. Here we deal with some aspects of the primary circuits, which are described in terms of boilers. Chillers follow the same basic principles.

5.4.3.1 Central heating without primary pump

A central heating system employing secondary mixing circuits fitted with their own circulation pumps can dispense with the need for a primary pump; see Figure 5.22. The secondary pumps provide the pressure required to circulate water through the boiler circuit. This layout produces a variable flow in the boiler circuit and at times of low demand the water returned to the boiler will fall in volume and temperature. Some boilers require a minimum boiler return temperature to prevent condensation. This is known as back-end protection and can be achieved by the incorporation of a boiler circulation pump. A non-return valve is incorporated to prevent reverse flow when the boiler circulation pump is off.

5.4.3.2 Low pressure header system

All secondary circuits are connected to a distribution header. The header may be at some distance from the boiler but the header itself is of sufficient size to ensure that there is little pressure drop over its length; there should be negligible pressure difference (<500 Pa) between the connection points of the secondary circuits. The flow and return headers are connected at the far end by a hydraulic short circuit. A secondary circulation pump is required for each secondary circuit. Figure 5.23 shows connections for a radiator zone and a calorifier. The low resistance of the header means that only the resistance of the secondary circuits beyond the connection points need to be taken into account when calculating valve authorities; it is not necessary to include the resistance of the boiler and primary piping. The resistance in the primary circuit must be matched to the capacity of the primary pump by installing a regulating valve close to the primary pump; alternatively a variable-speed drive may be used for regulation. The flow rates in the primary and secondary loops must be set to avoid problems with incompatible flows; see 5.4.1 above. An example of a potential incompatibility is illustrated in Figure 5.24[17]. If the right-hand secondary circuit has too great a flow rate, there will be a reverse circulation between the connection points B and A, with the results that the right hand secondary circuit is supplied by its own return.

5.4.3.3 Hydraulic decoupling

Higher operating efficiencies may be obtained by distributing heating loads over several boilers; this allows

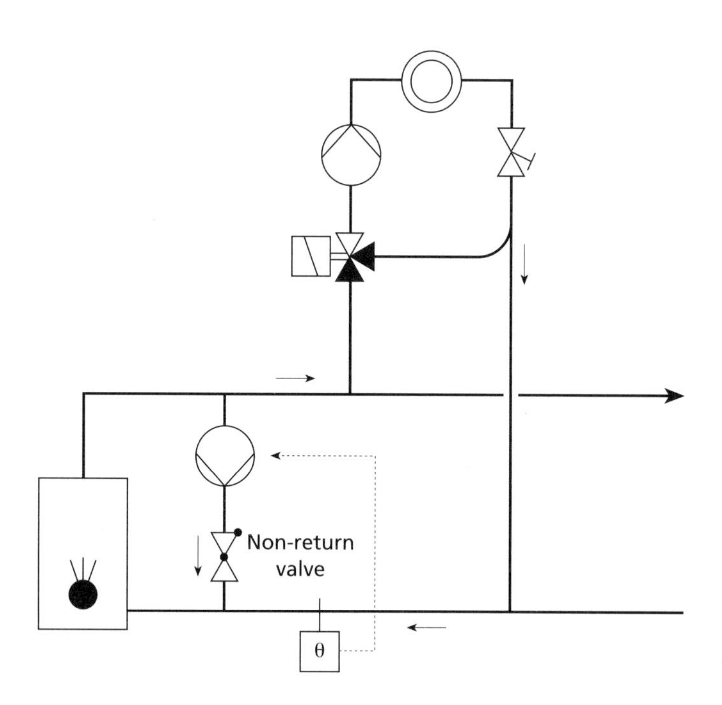

Figure 5.22 Boiler protection circuit

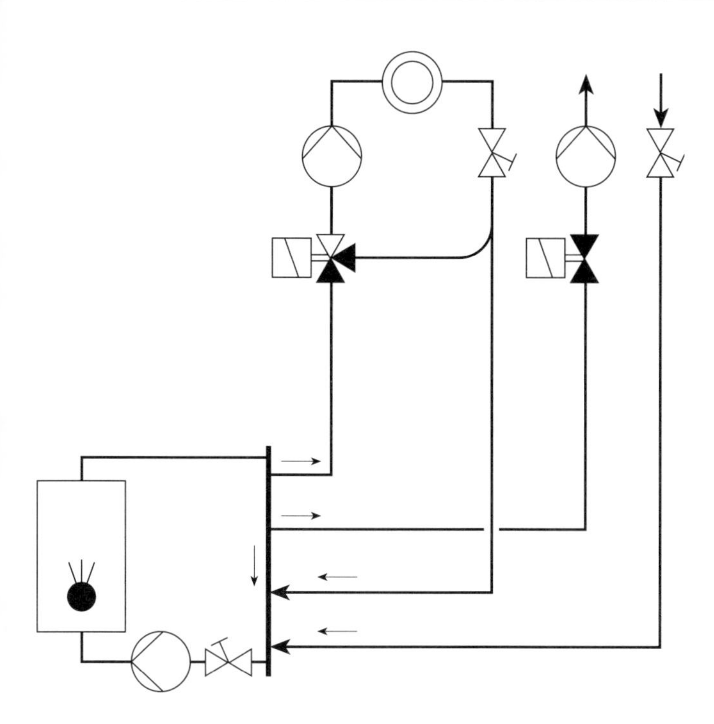

Figure 5.23 Low pressure header system

some boilers to be switched off at part load operation. Boiler sequencing is considered in more detail in 5.2.2. The use of multiple boilers with a variable-volume primary circuit produces a need to decouple primary and secondary circuits. This is achieved with a hydraulic decoupler, or buffer vessel, which is basically an upright pipe with four connections. The boiler pumps deliver water to the decoupler; in the absence of heating demand the entire volume passes down through the decoupler, which acts as a bypass. The decoupler consists of an upright pipe, whose bore should be about three times that of the connected supply pipes. This ensures a low flow rate and negligible pressure drop in the decoupler; the upright mounting, with flow entering at the top, prevents internal thermal circulation. The design should be such as to minimise any turbulence which might induce mixing between flow and return. To prevent dilution of the flow by return water the primary flow must be greater than the secondary flow; see 5.4.1 on compatible flow rates. Boiler operation is controlled from a temperature sensor which is located near the top of the decoupler. A variety of secondary circuits may be used. Figure 5.25 shows simple mixing circuits on the secondary. The decoupler is applicable to primary circuits with variable-speed pumping (see 5.2.2.4).

5.4.4 Calorifiers

Calorifiers are devices by which secondary hot water is heated by higher temperature primary water or steam. Storage calorifiers are in widespread domestic and commercial use for hot water services. A non-storage calorifier acts as a heat exchanger and is used to transfer heat from a high temperature circuit to a lower temperature secondary water circuit. Injection circuits provide an alternative to the use of a non-storage calorifier and are increasingly used for this purpose. A non-storage calorifier is essential where the high temperature source is steam or where it is necessary to separate the primary and secondary water circuits, such as swimming pool applications.

The circuit for a non-storage calorifier is shown in Figure 5.26. The secondary circuit may be for the provision of hot water services or an LTHW heating circuit. Where a non-storage calorifier is used for HWS, the secondary circuit will always be of the pumped circulation type, where hot water is pumped continuously round the distribution circuit, to make hot water instantaneously available at the taps. The temperature of the secondary flow leaving the calorifier is controlled by modulating the primary flow in the heat exchanger using a three-port mixing valve in diverting application; where the primary circuit uses steam, a two-port valve in the primary flow is used. An additional two-port isolating valve is fitted to the primary flow to provide tight shut off of flow to the calorifier. This ensures that there is no overheating or boiling in the calorifier, which might otherwise be caused by a control valve which did not close tightly.

The calorifier control is initiated by a demand signal from the secondary circuit. The origin of this signal depends on the application. It is likely to be a time schedule signal for HWS, or a heating demand for LTHW application. On receipt of the signal, the secondary pump is started and when flow has been proved, the primary circuit isolating valve opens and the control valve modulates to control the secondary water flow temperature at the required set point. If the high limit thermostat is activated, the three-port valve bypasses

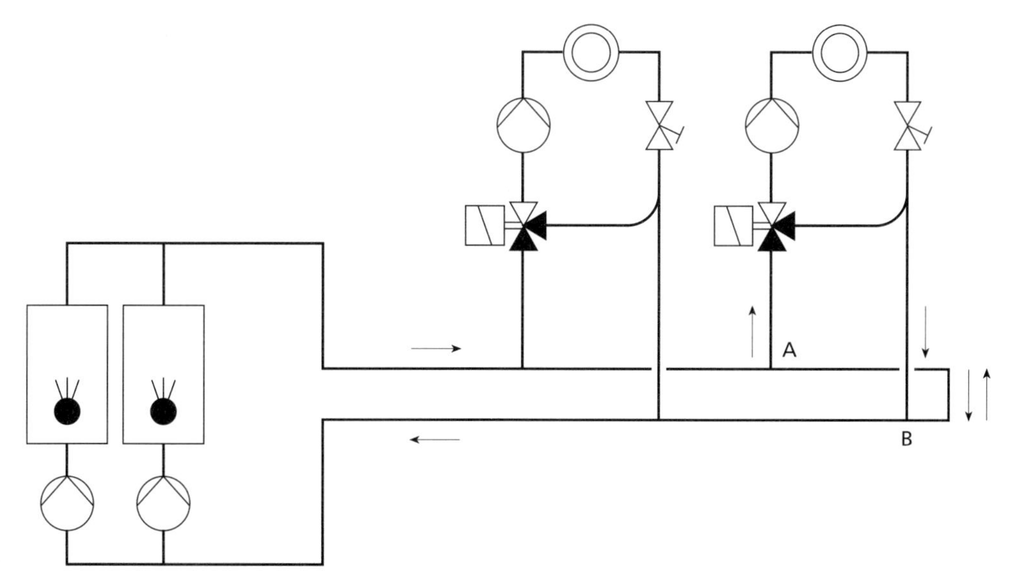

Figure 5.24 Incompatibility in a low loss header. Mismatch of flows may result in circulation from B to A

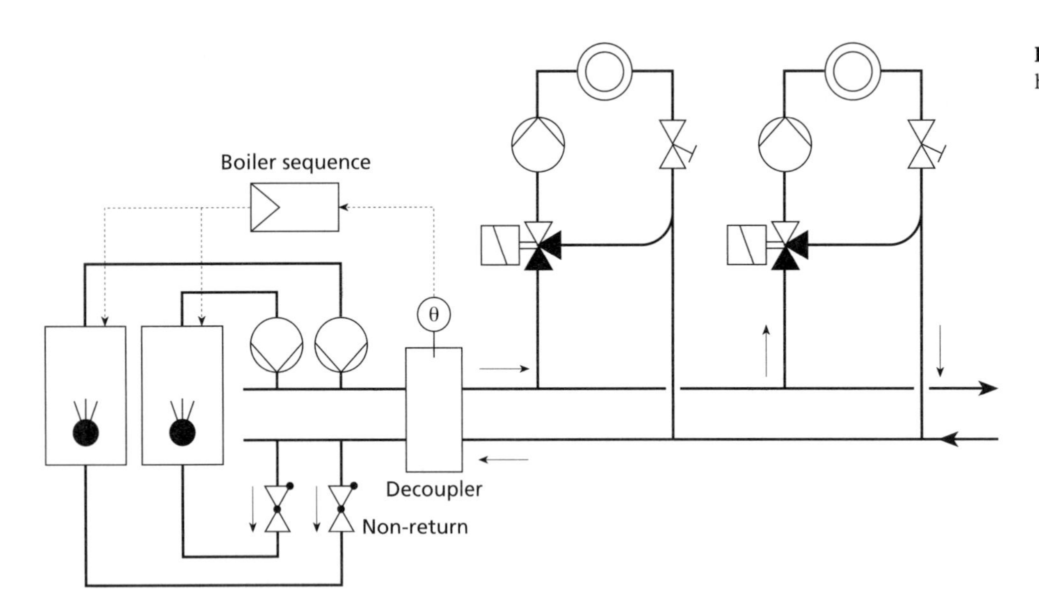

Figure 5.25 Two-boiler system with hydraulic decoupling

the calorifier and the isolating valve is shut. The high limit thermostat requires manual resetting. The same actions are taken by the control system if a failure signal is received from the secondary pump. When the demand signal ceases, the isolating valve is closed and the three-port control valve set to bypass the calorifier. The secondary pump continues to run during the pump overrun period and then is turned off; this minimises any risk of boiling due to residual heat.

The primary heat generator which supplies the primary hot water may also supply heat to other systems in the building. The secondary demand signal enables the heat generator. During start-up conditions, the heat load on the secondary circuit may be such that the calorifier remains cool, resulting in a return of cool water in the primary circuit; this may pose a risk to the boiler. In this case, it is possible to control valves in the circuits supplied by the secondary loop to remain in full bypass until the primary return temperature has reached an adequate temperature, whereupon the system switches to normal operation. This allows the primary circuit and boiler to reach operating temperature as quickly as possible.

Storage calorifiers are generally used for the provision of hot water services and a typical circuit is shown Figure 5.26, using LPHW in the primary circuit. Control of the stored

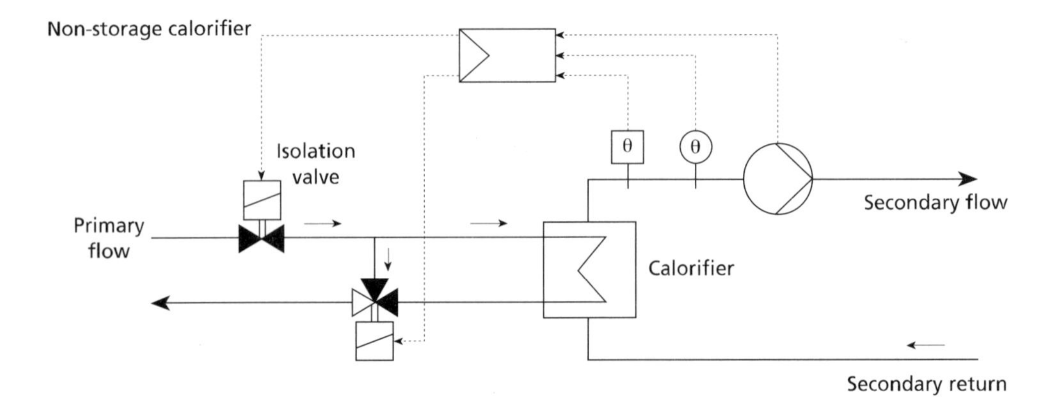

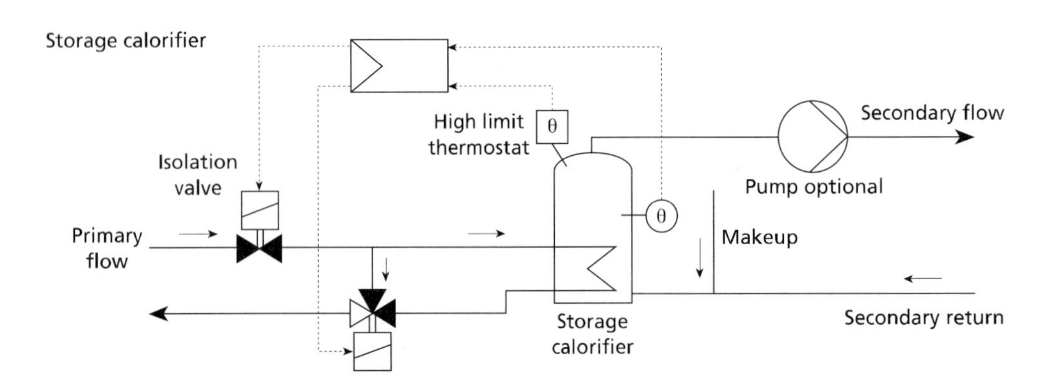

Figure 5.26 Control of calorifiers

water temperature uses on/off control, with a wide differential of up to 10 K. The heating demand is activated when the stored water temperature is below the lower temperature limit and the HWS time schedule is on. The isolating valve opens and the three-port valve opens fully to send primary water though the heating coil. This continues until the stored water has been heated to the upper temperature limit, when the valves close. The three-port valve does not modulate and operates fully open or closed. A high limit thermostat is fitted at high level in the stored water and is typically set to 90°C. If this temperature is exceeded, the isolating valve is closed to prevent primary water entering the calorifier. If a secondary circulation pump is fitted, it should overrun for 5 minutes. The thermostat requires manual resetting. Stored water systems should always be designed to reduce the risk of Legionnaires' disease and the matter is dealt with in CIBSE TM13[8] and HSG 70[9]. One requirement is that the stored water should be heated to a minimum of 60°C . HWS are often a major source of energy wastage, particularly with large centralised systems. BRESCU GIR 40[4] gives recommendations for efficient design and operation.

5.4.5 Variable-volume systems

5.4.5.1 Pumping energy requirements

For a conventional heating system using pumped circulation of heating water, it has been estimated that energy used in the pumps represents some 2% of the total heating energy consumption and 7% of the energy cost[18]. Constant-volume systems operate the pumps at constant speed irrespective of system load requirement and use three-port valves to control the flow of fluid through the heat emitters. This has advantages in simplifying control and balancing problems at part load, but means that there is no reduction in pumping energy at part load. A variable-volume system, where the pump speed is reduced at low demand, offers the opportunity to save a substantial amount of pumping energy. With zero static delivery head, the power input to a pump varies as the cube of the flow. In practice, efficiency variations of pump and drive will affect this, but there are still substantial energy savings to be made.

Figure 5.27 illustrates the difference in controlling flow by throttling valve or by variable speed. Full flow of the system under consideration occurs at point A where the pump curve and system curve intersect. In a properly designed system, this occurs at a position where the pump is operating at near maximum efficiency. A reduction in system flow may be achieved by increasing the resistance in the circuit using a two-port throttling valve. The system curve becomes steeper and a new operating point B is achieved. The same reduction in flow can be achieved by reducing the speed of the pump. The pump curve is changed to the lower dotted line and now intersects the original system curve at C. The two operating positions B and C have the same system flow, but require different power inputs to the pump. The pump power is given by

$$\text{pump power} = (\text{pump head} \times \text{flow rate}) / (\text{pump efficiency} \times \text{drive efficiency})$$

Pump efficiency is a function of pump speed and flow. It shows little reduction over the upper portion of the speed range, but rapidly decreases as the speed falls below about 60% of the design value. The efficiency of the variable-speed drive also varies with speed. The overall relation between power consumption and flow is shown in Figure 5.28. With fixed-speed pumping and a throttling circuit there is some reduction in power consumption at reduced flow rates, while substantial reductions may be achieved with a variable-speed pump motor. With constant-flow three-port regulation there is no reduction in pumping power consumption.

The above considerations apply to system regulation as well as control. A variable-speed drive can be used when commissioning the system to provide design flow conditions at reduced pump speed, without the use of a regulating valve. This solution will use less pumping power than using a regulating valve with the pump running at full speed.

The selection of the optimum system for reducing energy costs is a function of several parameters. The extent of the savings will vary with:

- the degree of pump turn-down
- the gradient of the pump pressure/flow characteristic curve
- the gradient of the system curve
- reduction in static delivery head
- run time at different speeds
- energy costs
- variation in pump and drive efficiencies.

More information on estimation of savings in running costs is given by ETSU Good Practice Guide No.2[19] and Bower[20].

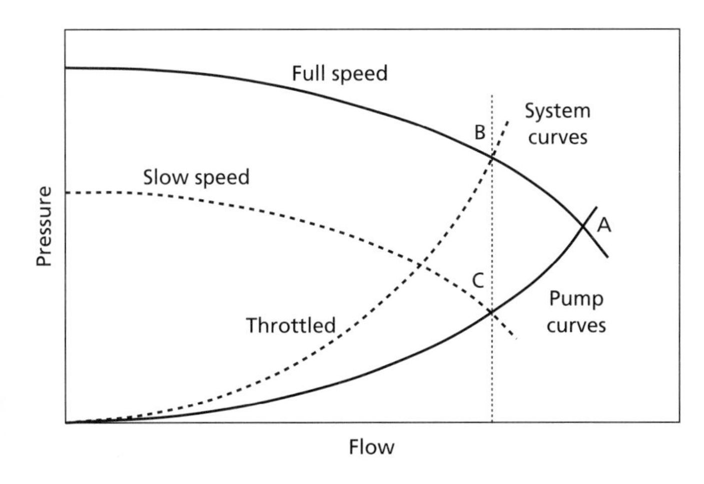

Figure 5.27 Flow control by throttling and speed reduction

5.4.5.2 Parallel pumping

Two or more pumps in parallel may be used where light load conditions could overload a single pump and as an alternative to fully variable-speed pumping. For any given pump head, two pumps in parallel provide twice the flow of a single pump. However, single-pump operation in a system produces more than half the flow of dual-pump operation; this is shown by the intersection of the system curve with the pump curves in Figure 5.29. Typically, a single pump

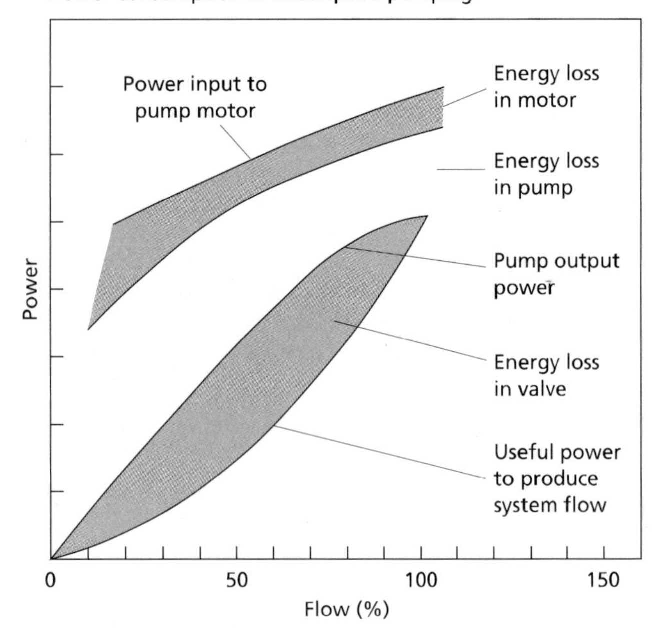

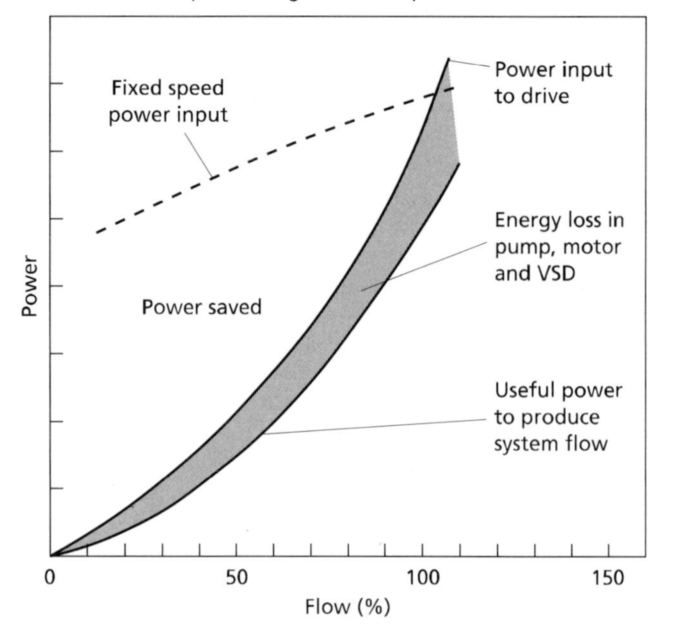

Figure 5.28 Pumping power consumption as a function of flow, showing saving from variable speed

produces about 80% of the flow with both pumps operating. Operation of the pumps is controlled by measuring the differential pressure across supply and return near the midpoint of the circuit. A control with large operating differential or built in time delay may be required to prevent rapid cycling between single and dual operation.

The pumps are connected so that each may be valved off for repair or replacement. The use of parallel pumps provides protection against failure of one pump. The remaining pump provides sufficient flow to maintain adequate operation until the failed pump is replaced. A non-return valve is fitted in the discharge of each pump to prevent backflow when one pump is shut down. Parallel pumps are sometimes used where the hydronic circuit is used for both heating and cooling. Both pumps operate during the cooling season to provide maximum design flow, but only one is required during the heating season, since the flow requirements are then lower.

To ensure satisfactory operation, the pumps should be identical. If for any reason it is decided to use different pumps, the composite pump curve should be constructed with care and all operating conditions examined to avoid possible undesirable interactions between the pumps[21]. Packaged twin pump sets are available which have built-in check valves; some have unequal-sized pumps to suit different operating modes.

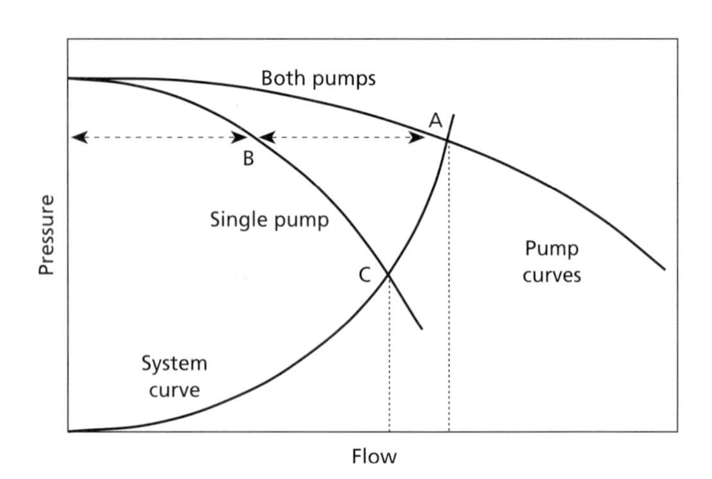

Figure 5.29 Parallel pump operation: identical pumps. System operating point with two pumps is at A; each pump operates at B with equal flow. With one pump in operation, flow reduces to point C

5.4.5.3 Fixed-speed circulating pump

Figure 5.30 shows a heating system consisting of boiler, pump, mains distribution pipes and a number of subcircuits. The subcircuits are controlled by two-port control valves which are modulated to satisfy local demand. The circuit is typical of those found in hospitals, schools and similar public buildings. The same circuit may be

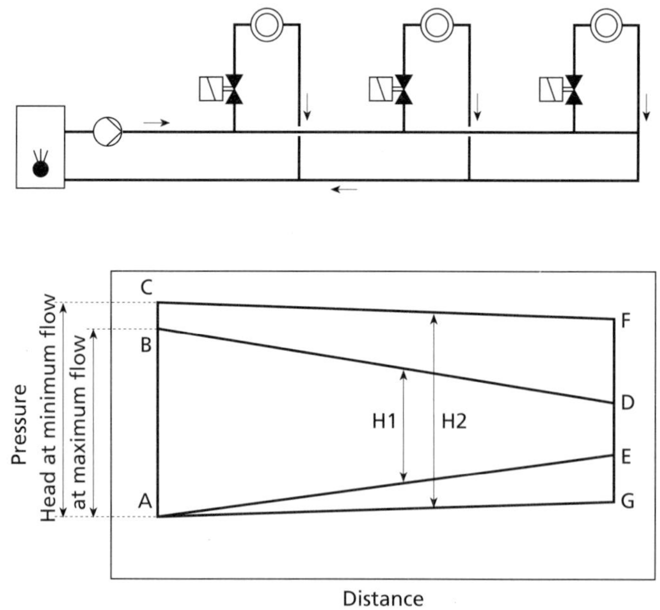

Figure 5.30 Fixed-speed circulating pump. Pressure distribution at maximum (ABDEA) and low demand (ACFGA). Balancing valves not shown.

applied for chilled water circulation. The subcircuits are connected in ladder configuration and a pressure–distance diagram can be drawn up to show the pressure distribution at maximum and minimum demand conditions. At maximum demand, AB represents the pump head at design flow rate, BD the frictional pressure loss in the flow pipe, DE the pressure loss in the furthest subcircuit and EA the frictional pressure loss in the return circuit. The circuit furthest from the boiler and pump is known as the index circuit. As heating demand decreases, the two-port valves in the subcircuits modulate towards the closed position and the system flow falls towards zero. At conditions of low flow the pipe pressure drops tend to zero, giving the pressure distribution ACFG. The pressure drop available across each subcircuit varies with the overall demand on the system. The variation is greatest for the index circuit, where the pressure differential varies from the full pump head to a small amount. This variation makes it impossible to balance the circuit at all loads.

5.4.5.4 Variable-speed circulating pump

The typical pump characteristics shown in Figure 5.27 results in the pressure head across the pump increasing as the flow though the circuit falls when the throttling valves modulate towards the closed position. Savings in pumping energy may be achieved if the speed of the pump is reduced at low loads. A common form of control is to reduce the pump speed so as to maintain a constant pressure differential across the pump as the flow in the circuit reduces. Circulating pumps are available which incorporate differential pressure measurement and variable-speed drive, providing constant pressure control in a single unit. Figure 5.31 shows the pressure distribution round the circuit at maximum and low demand. Savings in pumping energy are achieved compared with a constant speed pump, but the pressures across the branches are higher than necessary to achieve the required flows at low loads.

An alternative strategy is to control the pump speed to maintain a constant differential pressure across the index circuit. This produces the pressure distribution shown in Figure 5.32. At low loads, pressure across distant branches is maintained, but branches near to the pump risk starvation. In neither case is it possible to balance the branches in such a way that they remain in balance at all loads. The situation is worse when branches modulate independently. Figure 5.33 compares the two methods of pump speed control when a single branch modulates shut, for the case of a simple three-branch circuit[22]. With pump differential pressure control, a constant pressure is maintained at the pump. If a branch closes, flow in the circuit is reduced, giving lower pressure drops round the pipework. The result is higher pressures at the downstream branches, giving excess flow. The converse takes place when differential pressure control at the index circuit is used. Closure of the furthest branch results in reduced pressure and starvation of upstream branches.

Control by pump differential pressure control gives greater pump energy consumption, but is less likely to produce control problems. The recommended practice is to control the pump speed by differential pressure measured between flow and return, about two thirds of the way between pump and index circuit. The best position is difficult to predict and is best determined by trial and error. It is good practice to install additional pressure tappings along the pipework so that the effect of different measuring positions may be evaluated.

The difficulties in balancing the systems are caused by the pressure drop round the circuit; without this, all branches would operate at the same differential pressure without problem. It follows that the control problems will be reduced by sizing the pipework to give as low a pressure drop as possible. For systems of any significant size, designers would have to size pipes at less than 100 Pa/m pressure drop to achieve any real benefit[22]. This would significantly increase the pipe diameter and the increase in cost is unlikely to be acceptable.

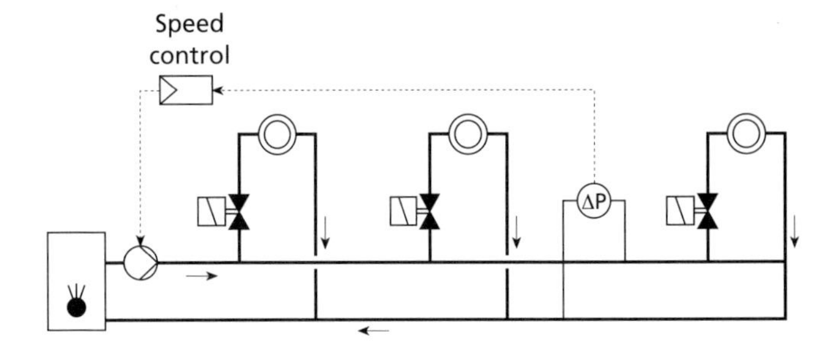

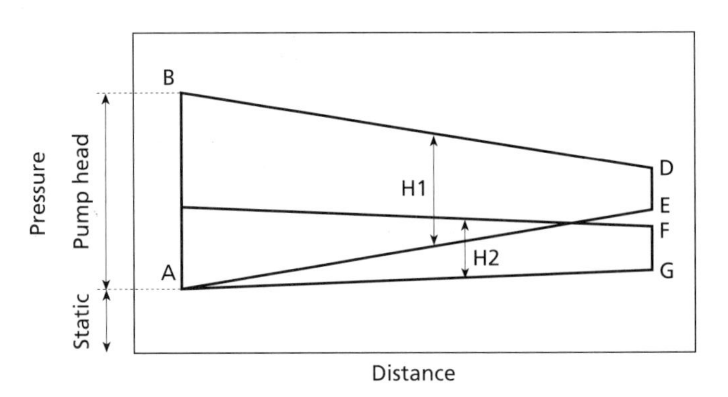

Figure 5.32 Control of differential pressure at index circuit. Pressure distribution at maximum and low demand

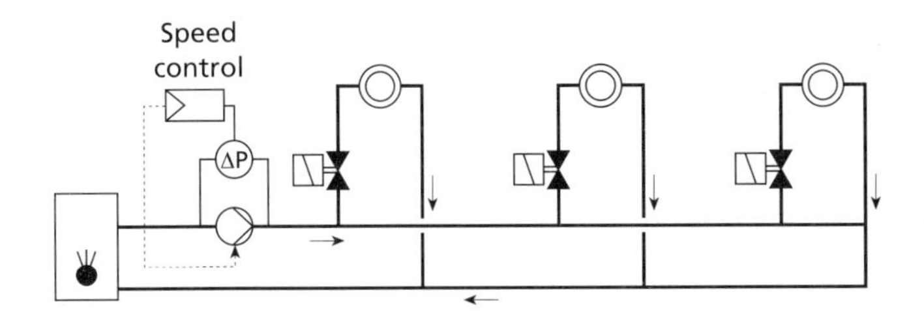

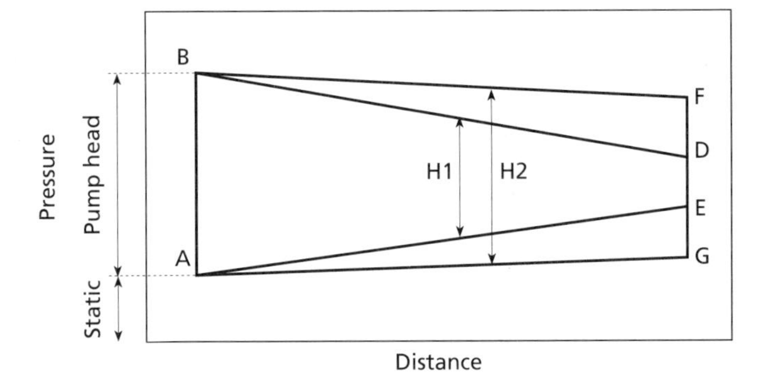

Figure 5.31 Pump differential pressure control. Pressure distribution at maximum and low demand when the branches throttle evenly

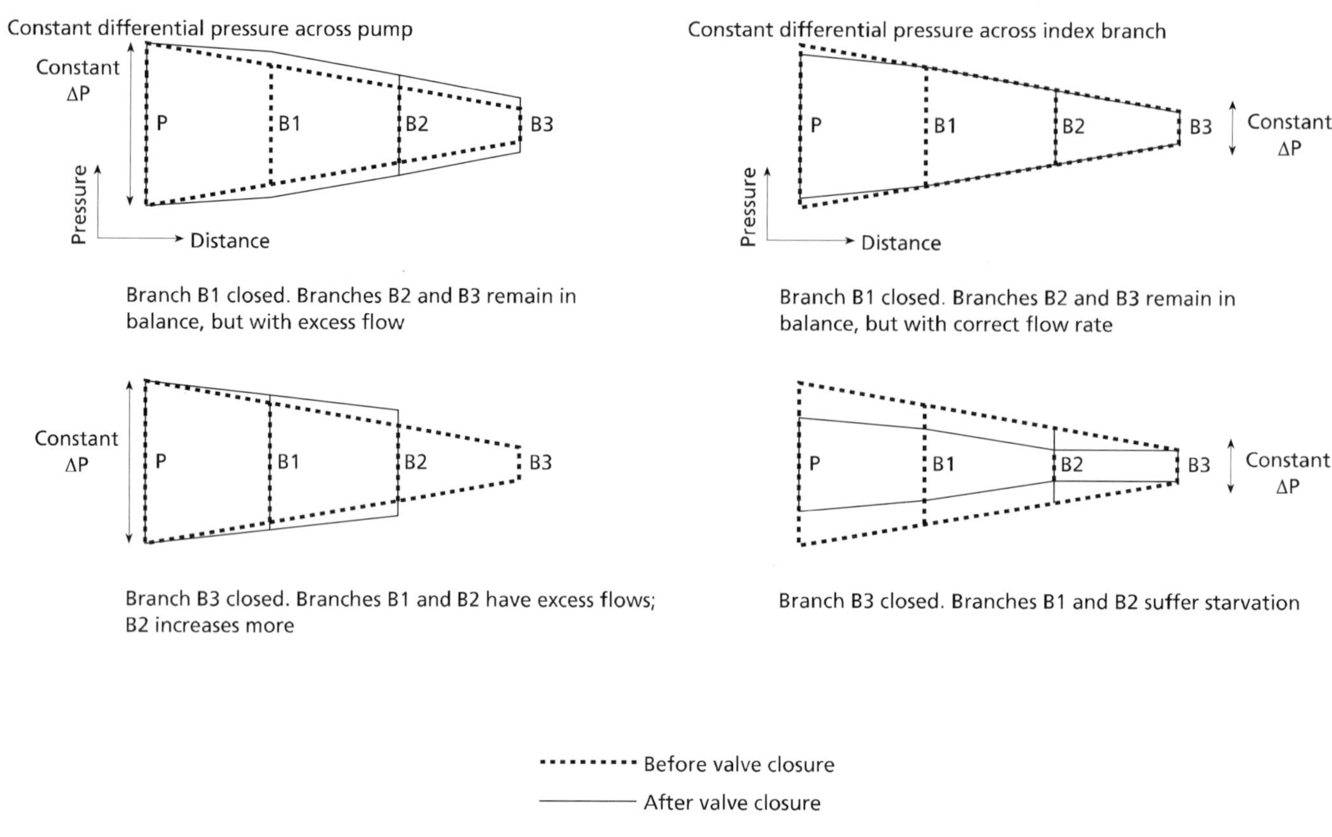

Figure 5.33 Control of three-branch circuit, showing effect of valve closures

It is inevitable that modulation of an individual two-port valve will produce flow variations elsewhere in the circuit and perfect balancing of the system is not possible. This topic is discussed by Rishel(23) who recommends that balancing valves are not used in such circuits; he states that balancing valves will interfere with proper control in some flow conditions and will cause additional consumption of pump energy. However, this technique produces its own problems of unbalanced flow during start-up conditions, as described in 5.4.1. The use of valves with high rangeability to cope with the variable conditions is required.

The main advantage in the use of variable-speed pumping is the saving in electrical energy, but it can introduce problems of control instability. The interaction of pump speed control and the two-port valves may produce instabilities. The response of the pump speed control should be faster than that of the two-port valves; this is not a problem in practice, since the two-port valves are typically in a control loop which involves a room temperature response and is therefore relatively slow. The use of automatic differential pressure control valves can avoid many of the problems of balancing or authority and is described below.

5.4.5.5 Differential pressure control valves

A differential pressure control valve is a mechanical device which can automatically control the flow of fluid in a circuit to maintain a constant differential pressure between two points in the circuit. Figure 5.34 shows a schematic drawing of such a valve. It is basically a two-port valve whose valve spindle movement is operated by a diaphragm. Pressure tappings from the circuit positions whose differential pressure is to be controlled are led to opposite sides of the diaphragm, and the out-of-balance force moves the spindle against the force of a spring until an equilibrium position is established.

Figure 5.35 shows a simple heating circuit with a number of secondary circuits connected to the primary circulation system in a ladder arrangement. As discussed above, the pressure drop round the primary circuit produces different pressure drops across each secondary circuit, which will vary with load. By using a differential pressure control

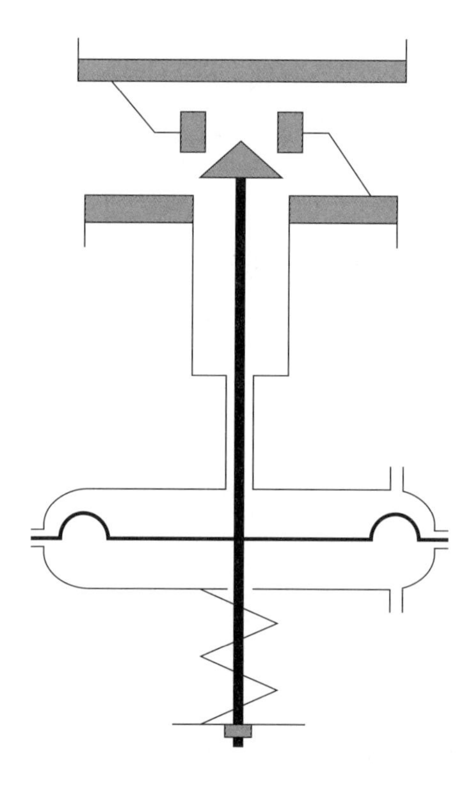

Figure 5.34 Differential pressure control valve

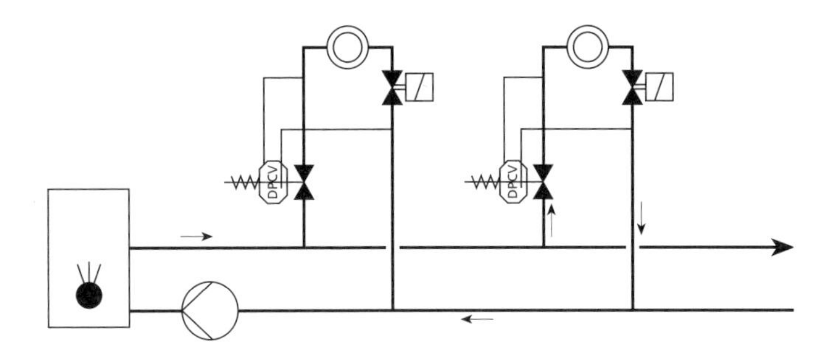

Figure 5.35 Differential pressure control valves installed in throttling circuits

valve in each circuit, it is possible to maintain the pressure drop across two-port valve and heat emitter at a constant value for each secondary circuit. Each two-port valve now operates under constant conditions at high authority. It may not be necessary to install a differential pressure control valve for each secondary circuit; it is often possible to install valves at strategic positions where they can provide suitable control for more than one circuit. This arrangement is to be found in large systems and district heating schemes.

5.4.5.6 Automatic flow-limiting valve

By incorporating a metering orifice into a differential pressure control valve, the device can be used as a flow limiter. The differential pressure across the orifice is communicated to the diaphragm of a control valve as shown in Figure 5.36. The diagram shows a separate metering orifice, but in a practical valve the metering orifice takes the form of an adjustable plug on the opposite side of the valve seat from the operating plug. If the flow in the valve rises above the preset limit, the increased pressure drop across the orifice causes the valve to close, controlling the flow at the limiting value. At flows below the set amount, the valve is fully open and does not exert a control function. Flow-limiting valves are used where it is required to ensure that no secondary circuit can deprive other circuits of their allocation of circulating water.

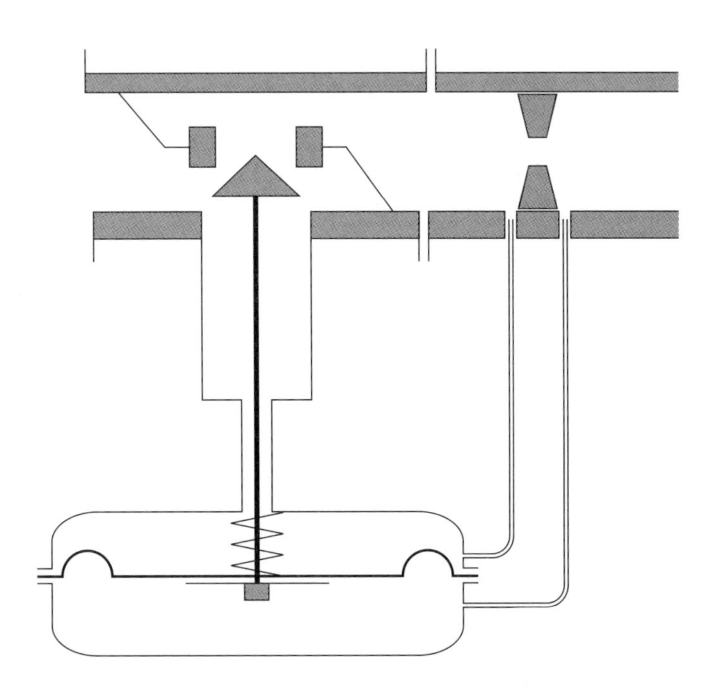

Figure 5.36 Automatic flow-limiting valve (principle)

In some circuits it is desirable or even necessary to control both the maximum flow rate and the differential pressure in each subcircuit. The reasons for this are:

- Limiting the maximum flow to each subcircuit ensures that no circuit can take more water from the distribution network than has been designed.
- Temperature control valves in subcircuits require to work below a maximum pressure differential. This is particularly important for thermostatic radiator valves.
- Subcircuit control valves should not be subjected to wildly fluctuating pressure differentials to maintain reasonable control authority.

The combination of a differential pressure control valve and an automatic flow limiting valve in each subcircuit will achieve the above requirements. However, this will be found too expensive for many applications. It is possible to obtain a valve which combines both functions in a single body. Another solution is to incorporate a flow-measuring orifice into the subcircuit together with an adjustable differential pressure control valve (Figure 5.37). During commissioning, the differential pressure control valve set point is set to a value which is just high enough to produce the design flow rate in the subcircuit, as measured at the flow measuring orifice. This adjustment is performed with any automatic control valves in the subcircuit fully open. The circuit thus operates with the minimum pressure drop across the load and control valve and the control valve operates under optimum conditions at all times.

5.4.5.7 Control of minimum flow rate

There is often a requirement to maintain a minimum flow rate through the boilers or chillers. This is usually achieved by the use of separate primary and secondary circuits or the provision of boiler protection circuits, which were discussed in 5.2 and 5.3. For the simple circuits described below, there is a single circulation loop. If all the throttling valves in the branches close down, the flow falls towards zero. Maintenance of the flow though the boiler may be obtained using a differential pressure control valve as a bypass. The DPCV valve operates in an opposite sense to that

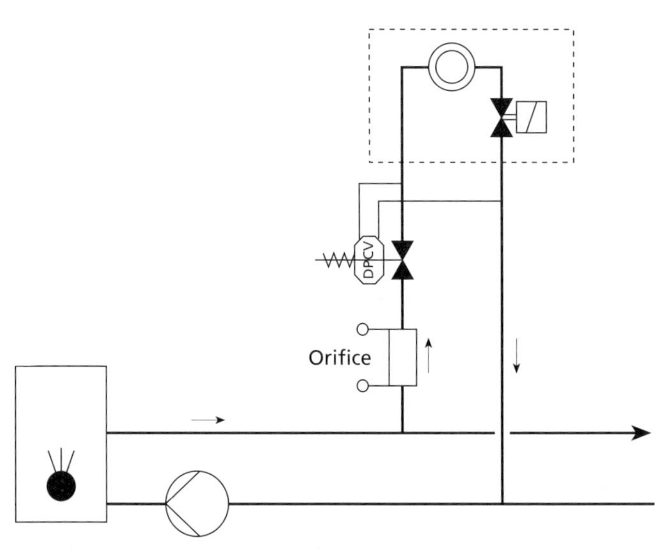

Figure 5.37 Use of a measuring orifice to set differential pressure control valve during commissioning

described in 5.4.5.5 and opens with increasing pressure differential. The valve is installed as a bypass, with the pressure tappings connected across the circulation pump as shown in Figure 5.38. As the throttling valves in the controlled circuits close down and reduce flow though the pump, the pump pressure rises and acts to open the DPCV. It is necessary to use a circulation pump with a relatively steep characteristic near the design operating point.

An alternative method is to use a three-port valve in diverting application at the furthest branch (Figure 5.39). This gives a constant flow through the furthest branch, which becomes the minimum circuit flow. As well as providing a minimum flow through the circuit and boiler, this solution has the advantage in ensuring that there are no dead legs in the system. Another important advantage is that it helps to reduce pressure variations in the rest of the circuit when the furthest branch closes down, since this is otherwise the worst case for producing underflows in other branches.

5.4.5.8 Design recommendations

The use of variable-speed pumping offers considerable economies, both in energy consumption and in the cost of valves and associated gear. However, the above discussion has shown that there are difficulties in achieving suitable valve authority and maintaining system balance at varying loads. Further information may be found in the literature[7,18,22]. Manufacturer's literature is also a useful source of practical advice.

- Control by differential pressure across the pump does not give minimum pumping power and may give excess flow in downstream branches.
- Control by differential pressure at the index circuit gives least pumping power but risks starvation of upstream branches.
- Control by differential pressure about two thirds of the way between pump and index circuit is a good compromise. The best position may be found by trial and error.
- Balance the system to ensure that the design flow is available at all terminals during start-up conditions.
- Valves of high rangeability are recommended.
- Size pipework for minimum pressure drop compatible with cost.
- Valves must be capable of closing against maximum pressure, which may be full pump pressure.
- Network analysis of flows at part load conditions is recommended.
- The use of differential pressure control valves eliminates balancing problems and gives good valve authority, but increases costs.
- Use a differential pressure control valve as a bypass or use a diverting circuit in the furthest branch to provide flow in the circuit when all branches shut.

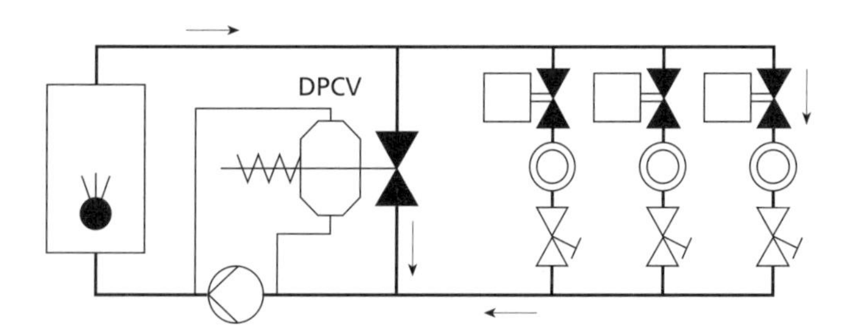

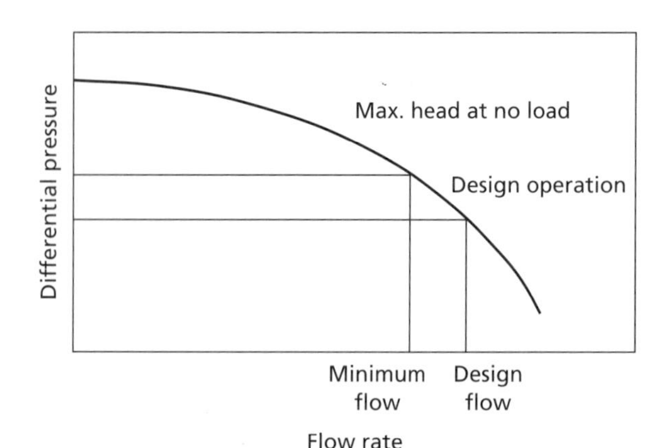

Figure 5.38 Automatic bypass. The DPCV opens to allow flow through the bypass leg as the differential pressure across the pump rises. This maintains flow through the boiler as the throttling valves close down

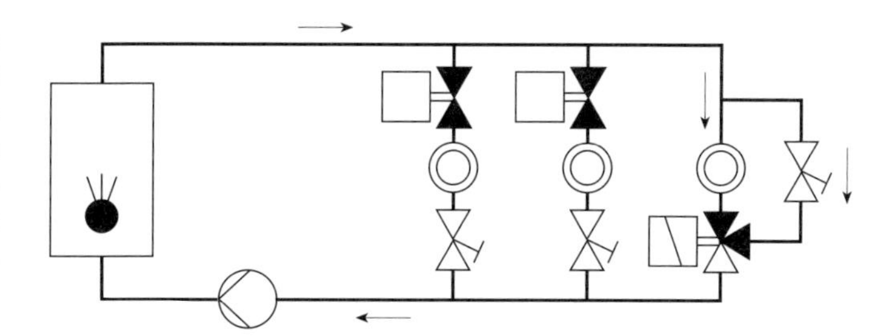

Figure 5.39 Constant flow at furthest branch. Replacing the two-port valve by a three-port diverting circuit at the furthest branch maintains circuit flow and reduces pressure fluctuations

5.4.6 Underfloor heating

Underfloor heating systems circulate hot water in coils embedded in the screed of a concrete floor. Systems employing heat spreader plates are available for use with timber floors, but are less common. Comfort considerations and restrictions on materials limit the flow temperature to a maximum of about 60°C and the floor surface temperature to about 30°C. The normal practice is to design for a maximum floor temperature of 26°C in the main body of the space; higher temperatures may be used in circulation spaces or zones of high heat loss, such as next to large glazed areas. The combination of low floor temperature and the high thermal mass of the floor produces a system that is slow to respond to changes in demand. This is compensated to some extent by the small temperature differential between floor surface and room temperature, which produces a degree of self-regulation.

An underfloor heating circuit suitable for general use is shown in Figure 5.40. This employs a primary heating circuit which circulates water from a conventional boiler through a low loss header. This primary circuit may be used to provide water for a variety of secondary circuits, such as other heating circuits or an HWS calorifier. The underfloor heating secondary circuit uses constant flow, variable temperature in the floor coils; this ensures an even temperature over the heated surface.

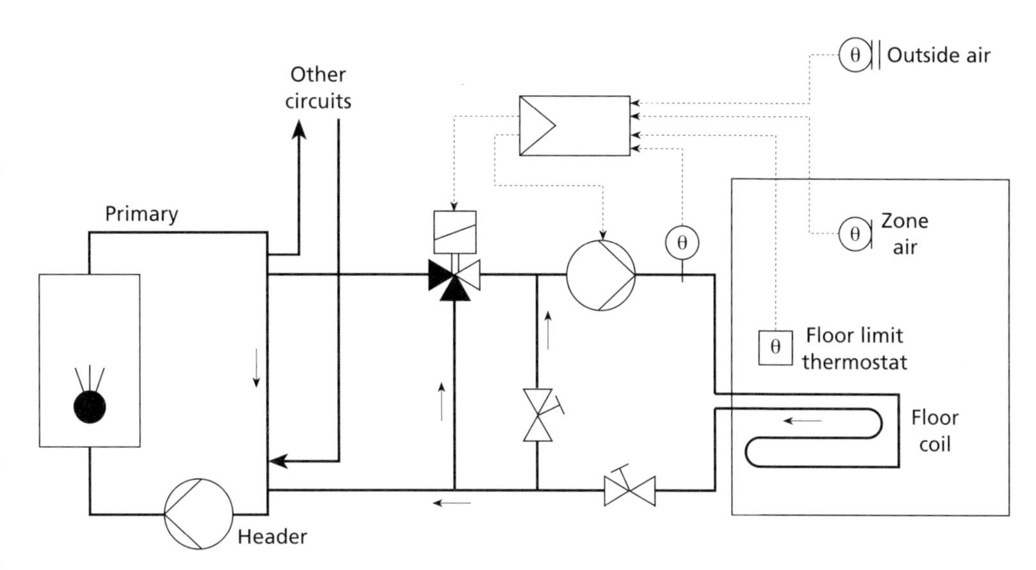

Figure 5.40 Underfloor heating. Water is circulated through the floor coil at constant flow rate, variable temperature. Flow temperature is controlled by a weather compensator; a floor limit thermostat disables the circulation pump

Weather-compensated flow control is used; the flow temperature is controlled over the range of typically 30 to 50°C as a function of the outside temperature. Control is by a standard weather compensator using PI control to modulate the three-port mixing valve. The bypass in the secondary circuit maintains some mixing flow when the three-port valve is fully open to the high temperature primary flow, limiting the secondary flow temperature to the required maximum value. A room temperature sensor in the occupied space acts as a high limit thermostat to disable the secondary circulating pump. If desired, additional protection can be provided by a high limit thermostat acting on either the floor surface temperature or the flow temperature. In practice, there may be difficulties in providing a suitable sensor for use as a floor limit thermostat. If a high limit temperature is exceeded, an alarm should be raised by the BMS, since the circuit balancing or weather compensation characteristic may need to be adjusted. Cascade control based on zone temperature may be used to reset the flow temperature as an alternative strategy to the use of weather compensation. This is a higher cost option. The long time constant of the underfloor heating systems makes them a suitable candidate for optimum start and stop control. Alternatively, night setback control may be used; for slow responding systems a setback of 5 K is adequate.

The low flow and return temperatures employed in underfloor heating makes it suitable for use with a low temperature heat source such as a condensing boiler. In this case, the heating circuit may form part of the boiler primary circuit. A simple control strategy can be used, with the circulating pump running continuously and the boiler switched on and off or modulated to control the zone temperature. Where several zones are to be controlled, it is common practice to use an outside compensator to provide control of flow temperature, plus separate room thermostats and valves to limit zone temperature.

5.4.7 Thermostatic radiator valves

Thermostatic radiator valves are widely used to provide local control of heating systems. The valves are self-acting and require no external source of power. The TRV is a two-port throttling valve which fits directly onto the flow or return connection of a conventional LPHW radiator. The head contains a liquid- or wax-filled capsule, which expands on a rise of ambient temperature and closes the valve against spring pressure. The manufacturer's mounting recommendations should be followed to ensure that the sensing head is properly exposed to room temperature. Incorrect positioning will result in the valve responding to heat from the radiator itself rather than to room temperature. The valve closes over a proportional band of about 3 K. However, the lack of valve characterisation, the highly non-linear relation between flow and radiator heat output and the difficulties of achieving suitable valve authority mean that the control characteristics are inferior to those obtainable with a motorised valve. The differential pressure against which the valve has to close and the flow temperature of the circulating water both have a small influence on the operating temperature. Most TRVs are self-contained and can be fitted as direct replacements for a conventional radiator valve; variations are available with remote sensing or adjustment. TRVs are also available which may be enabled or disabled by an electrical signal; the head contains a small heating element which acts to close the valve when energised. This system may be used for time scheduling or to provide a night setback facility. Heating circuits in most commercial buildings are required to use weather compensation under the present Building Regulations. In this situations, TRVs act as trimming devices, providing local temperature control. They also prevent overheating and make effective use of any incidental heat gains.

A heating system incorporating radiators fitted with TRVs falls within the category of throttling circuits described in 5.4.2.4. However, radiator circuits have characteristics of their own and the widespread application of TRV heating circuits fitted with TRVs demands special consideration. System balancing is important; without a properly balanced system it will be impossible to achieve proper control. During the warm-up period, all TRVs are likely to be fully open. If the system is not balanced, the flow though the radiators will be determined by pressure drops in the pipework, since the pressure drop in the radiators is low and virtually independent of their heat output rating. Favoured parts of the system will receive excessive flow and unfavoured parts insufficient flow. Since heat output from a radiator does not increase significantly at above design flow rates, the overall heat output from the system is below design output. The unfavoured parts of the system will only start delivering heat when the TRVs start closing in the favoured parts. It is therefore important to balance the system so that the design flow is available at all radiators when the TRVs are fully open.

TRV systems must be designed to operate successfully under part load conditions. This may be done by the use of variable-speed circulation pumps, or, if a constant-speed pump is used, by the use of differential pressure control valves. These methods avoid the production of noise in TRVs when subjected to a high differential pressure, and the difficulties in operation as the pump head increases at restricted flow rates.

Where a constant-speed circulation pump is used, a bypass is provided to provide a circulation loop as the TRVs close off. This has the advantage of providing nearly constant flow through the boiler, and is illustrated in Figure 5.41. Flow through the bypass leg is controlled by a differential pressure control valve, which opens as the pump head rises to maintain a nearly constant pressure differential between flow and return to the heating circuit. DPCVs were discussed in more detail in 5.4.5.5. This method of control requires a pump with a relatively steep characteristic near the design operating point.

Where the secondary heating circuit is fed from a primary circulation circuit with a differential pressure between flow and return of greater than 50 kPa, the circuit of Figure 5.42 can be used. The excess pressure is dropped across a DPCV which is controlled by the differential pressure across the secondary heating circuit. This circuit is an example of the use of automatic throttling valves in series; in general, this is not advisable because of the risk of instability. However, TRVs have a response time of 15 minutes or more, while a DPCV responds in a much shorter time, resulting in stable operation.

In both the circuits described, it will be generally be found suitable to set the DPCV to control between 10 and 20 kPa;

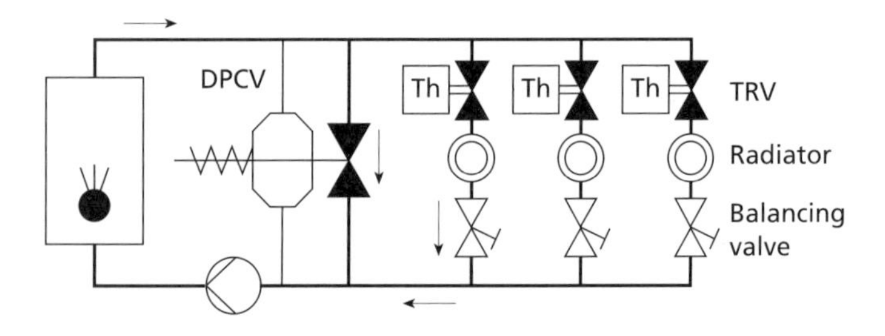

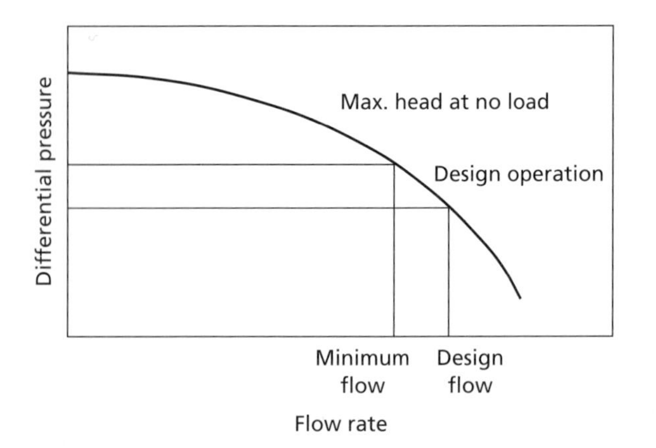

Figure 5.41 Control of radiators by TRVs. Constant-speed circulator. The differential pressure control valve acts as a bypass to control the pressure across the radiator circuit

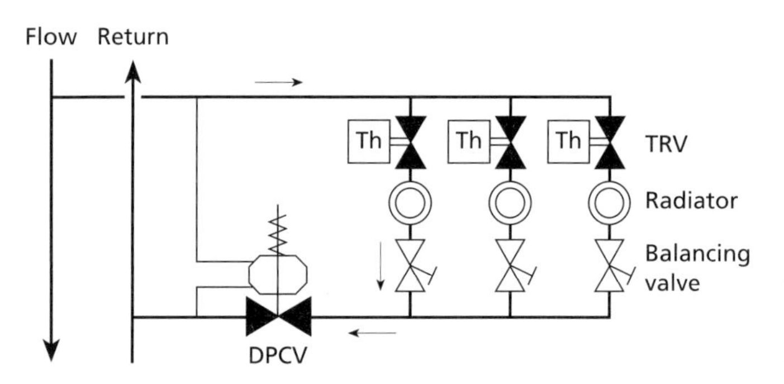

Figure 5.42 Control of radiators by TRVs. Pumped primary loop. This circuit is suitable for a mains differential of greater than 50 kPa

the setting should be consistent with the design flow though the radiators and the K_v value of the TRV to ensure that full flow is available at design conditions. There is a risk of noise being created in radiator valves at differential pressures of over 30 kPa.

The use of variable-speed pumps has several advantages, including reduction in pumping energy. A radiator system controlled by TRVs is a common example of a variable-volume system. The control of variable-volume systems was considered at length in 5.4.5. For a large or complex system involving TRVs and variable-speed pumping, this section should be consulted. However, for most simple TRV systems, it is sufficient to use a circulator pump with integral differential pressure variable-speed control. The motor speed is varied to maintain the pump head at the preset value. The circulators are designed to cope with the zero flow situation should all TRVs close simultaneously and there is therefore no need to provide a bypass valve. The effect on boiler operation of low flow rates must be considered.

5.5 Central air handling plant

Figure 5.43 illustrates a generalised air handling system with recirculation. Not all the components may be present in any actual system. The basic system is to be found with modifications in many HVAC systems.

Outside air is filtered, after which it passes to a preheater which serves to protect the rest of the system against low outside air temperatures. The air then goes to a mixing box, where it is mixed with a controlled quantity of recirculated air before being conditioned in the air handling unit. The AHU contains heating and cooling coils, humidification and the ability to dehumidify the air. It also contains the main supply fan which distributes air to the room terminal units. Supply air at the required temperature and humidity is distributed into the conditioned zone by room terminal units. Air is extracted from the occupied zone by the extract fan; the recirculation damper allows a proportion of the air to be recirculated; exhaust air is discharged to atmosphere. There may be supplementary direct extraction from the space where it is necessary to remove contaminated air directly to outside, such as from a kitchen or industrial process.

The air handling unit is the heart of a central air conditioning system. Its function is to supply conditioned air at the appropriate temperature and humidity, at the required flow rate and containing the correct proportion of

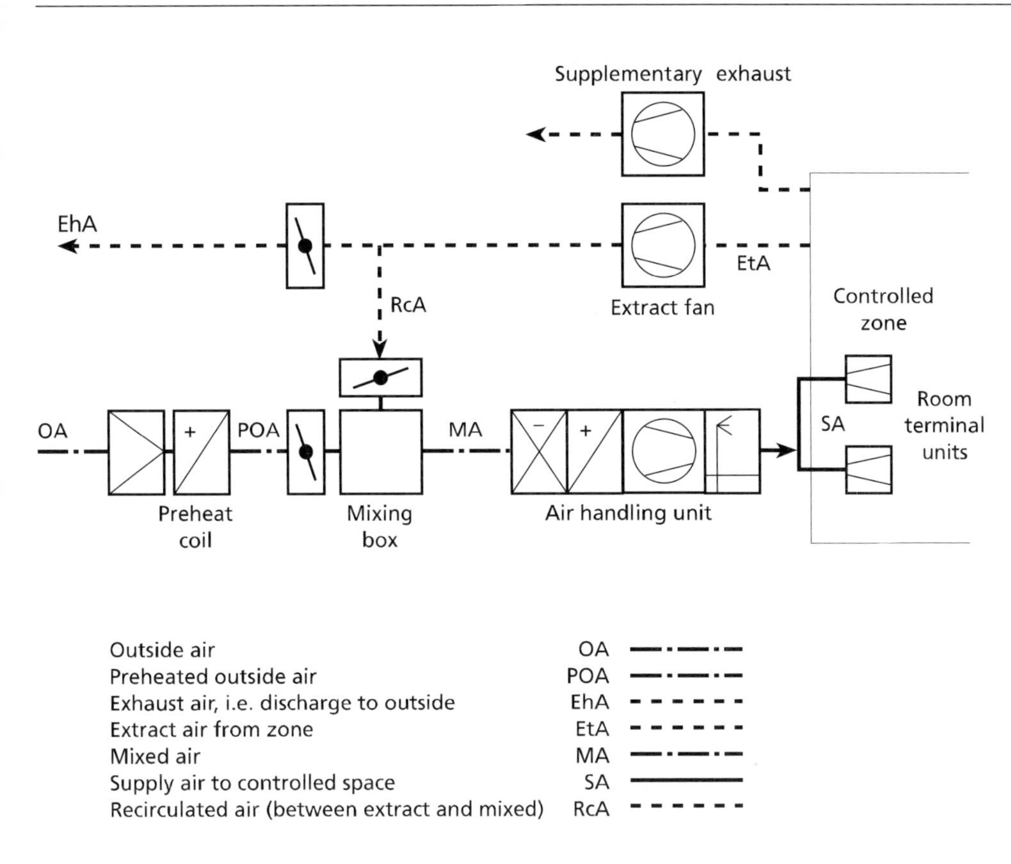

Figure 5.43 Major components of a central air conditioning system

outside air for ventilation. This section deals with the control of temperature and humidity plus the mixing of recirculated and outdoor air. The overall control strategy will depend on the type of air conditioning system, most importantly whether it is a constant volume or variable volume system. Control strategies for complete building systems are described in the next section.

5.5.1 Preheat coil

The preheat coil, when fitted, serves to preheat outside air to a minimum temperature before it passes to the rest of the plant. It also provides frost protection against cold air entering the air handling system.

— *Protection against freezing*. A frost thermostat downstream of the preheater prevents the supply fan from operating if it senses a temperature below its set point, typically 3°C. This overrides any other fan operation signal and raises an alarm.

— *Protection of the preheat coil*. During periods when the air handling plant is not in operation, the preheat coil is protected against freezing by circulating water through it. On receipt of a low outside air temperature signal or low primary water return temperature signal, the preheat coil control valve is opened to allow water to circulate.

— *Normal operation*. During normal operation, the control valve operates under PI control to maintain an off coil temperature at the set point, typically 8°C. In the event of a fan failure, the control valve moves to full bypass.

Control of preheat may alternatively be by using a face and bypass damper on the preheat coil. Where a preheat coil is not fitted, it will be necessary to transfer the protection functions to the first coil in the system.

5.5.2 Heating coil

Air heating is typically by an LTHW coil. Figure 5.45 shows the coil controlled by a three-way mixing valve in diverting application. In normal operation during the heating season, the valve modulates under PI control. The controlled variable depends on the air conditioning type, and is typically supply air temperature for a VAV system and zone air temperature for a constant volume system. Cascade control is common, where the deviation from the zone set point is used to reset the supply air temperature set point, which is controlled by the heating coil. If there is no preheat coil fitted to the intake air duct, the first heating coil in the AHU fulfils this function and freeze protection is installed, following the description in 5.5.1.

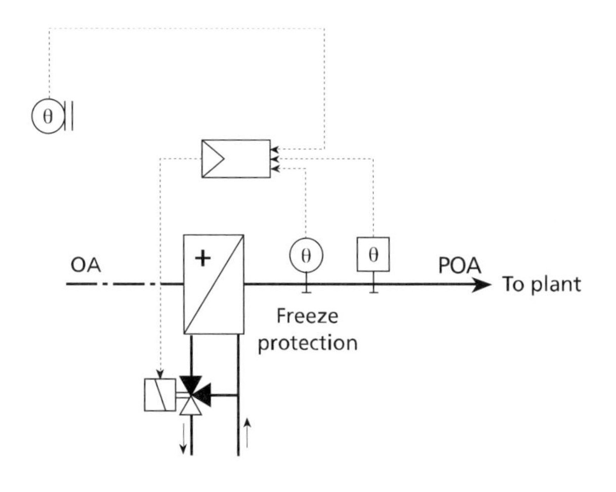

Figure 5.44 LTHW preheating coil. *Normal mode*: three-port LTHW valve is modulated under PI control to achieve the off-coil set point. *Valve closed (full bypass)*: plant shut-down, OSC or night cooling, cooling operation, economiser cycle. *Valve open*: low outside temperature, low primary return temperature, boost, OSH heating, freeze protection

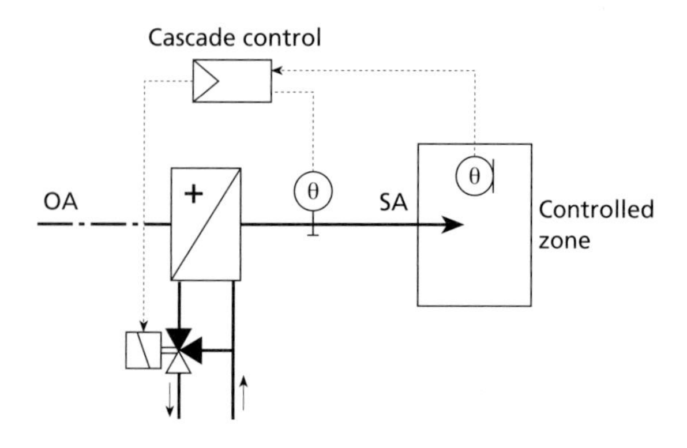

Figure 5.45 LTHW heating coil. Diagram shows cascade control of supply temperature reset from zone temperature. *Normal mode*: three-port LTHW valve is modulated under PI control to achieve the supply air temperature set point. *Valve closed (full bypass)*: plant shut-down, OSC cooling, night cooling, cooling operation, economiser cycle. *Variations*: the heating coil controls zone temperature directly. Freeze protection if no preheat coil

5.5.3 Cooling coil

Cooling of the supply air is achieved by a chilled water coil, whose operation is basically similar to the heating coil described above. Figure 5.46 shows a three-way valve modulating the flow of chilled water though the coil and also shows the cooling process on a psychrometric chart. The composition of the air approaching the coil is determined by the type of air conditioning system and is typically a mixture of exhaust air and outside air, controlled by mixing dampers described in 5.7.5. At high cooling loads, line 1, the cooling coil is required to cool the on-coil air to below its dewpoint, so that both sensible and latent cooling take place. The coil therefore acts as a dehumidifier. In practice, the air leaving the coil will be a little above the coil temperature and less than 100% saturation. At lighter cooling loads, line 2, sensible cooling only may be sufficient. The use of cooling coils for dehumidification and the implications of enthalpy control will be considered below. Here we treat temperature control only.

The flow of water in the coil is modulated using the control valve, controlled by a PI controller with either supply temperature (VAV system) or zone temperature (CV system) as the controlled variable. The valve remains in bypass position if free cooling from an economiser cycle or energy recovery device is sufficient to provide the cooling needs.

Direct expansion (DX) cooling coils may be used. The coil serves as the evaporator of a refrigeration circuit. Refrigerant flow to the coil is controlled by a two-position solenoid valve, which is opened when cooling is required. The thermostatic expansion valve is throttled to maintain a minimum refrigerant suction temperature; the expansion valve is normally supplied as part of the packaged control supplied with the coil. In normal operation the solenoid valve opens when the controlled air temperature, which may be supply or zone temperature, rises above the desired set point. The thermal expansion valve opens under self-acting control to maintain the refrigerant suction temperature set point (see Figure 5.47).

DX cooling differs from that provided by a chilled water coil in that the DX coil temperature is largely determined by the rate of heat transfer from the air flow to the coil. In the absence of sufficient air flow, the coil temperature may

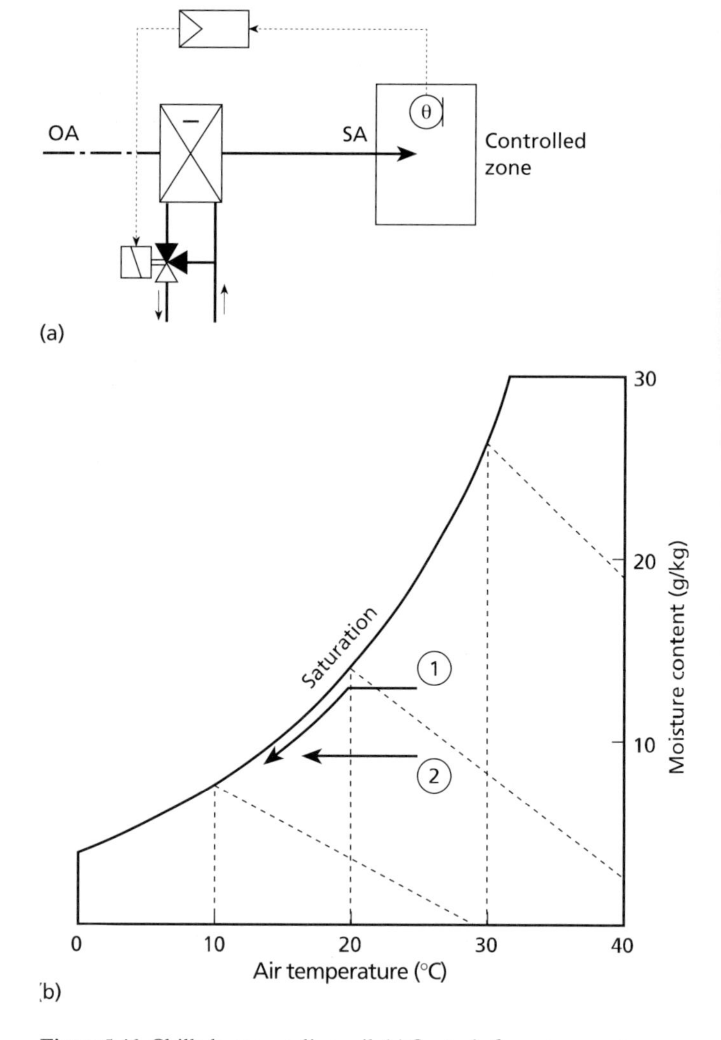

Figure 5.46 Chilled water cooling coil. (a) Control of zone temperature; there is no dehumidification control. *Normal operation*: three-port chilled water valve is modulated under PI control to achieve the desired set point temperature in the controlled zone. *Valve closed (full bypass)*: plant shut-down, OSH or boost heating, night cooling. *Valve open*: OSC cooling. *Variations*: control of supply air temperature is required for VAV systems. Existence of cooling demand from zone may be used to enable main chilled water plant. (b) Cooling coil psychrometrics. *Line 1*: high cooling load: both sensible and latent cooling. *Line 2*: low cooling load: sensible cooling only

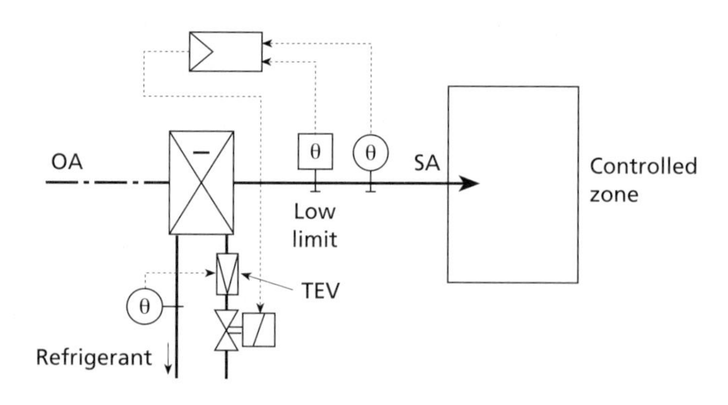

Figure 5.47 Direct expansion (DX) cooling coil. Refrigerant flow is controlled by a two-position refrigerant solenoid valve; the thermostatic expansion valve (TEV) controls the suction temperature T_{refrig} directly. The low limit thermostat prevents the off-coil temperature falling too low. *Normal operation*: refrigerant valve opens when the controlled air temperature rises above the set point. *Valve shut*: plant shut-down, OSH or boost heating, night cooling, low limit. *Valve open*: OSC cooling (low limit stat remains operative)

reach excessively low levels. The use of face and bypass dampers with a DX coil is not recommended because of the danger of ice formation. In general, a DX coil is capable of cooling to low temperatures and an off-coil air temperature low limit interlock prevents the supply air from becoming too cold.

5.5.4 Humidification

Indoor air humidities may fall to low levels during the heating season, when the moisture content of the intake air is low. If considered necessary, zone air humidity may be raised by:

— *Direct humidification*. Moisture is introduced directly into the controlled space. This method is most commonly found in industrial and horticultural applications. It may be used where a single zone requires control, such as a computer suite. Portable room humidifiers operating under their own humidistat are readily available. Independent humidifying units designed for ceiling installation can be used; these have a form similar to a fan coil unit and are designed for installation above a suspended ceiling.

— *Indirect humidification*. Moisture is introduced into the supply air at the AHU or elsewhere in the supply ductwork. This is the most common system for air conditioned buildings.

Some humidifiers, known as storage type, utilise a reservoir of water whose level is maintained using a ball valve or similar. The water is circulated in the supply air stream using a washer or sprayed coil to assist evaporation; unevaporated water is returned to the reservoir and re-used. Storage type humidifiers have been associated with health risks and their use is generally not recommended[24]. Non-storage humidifiers operate so that all the water is evaporated in the air stream and there is no reservoir which could harbour contamination. The major types of humidifier are:

— *Steam injection*. Steam from a local generator is injected into the supply air duct. It is possible to use steam from a central plant, but this is less common.

— *Atomised spray*. Fine droplets are generated from clean cold water and injected into the supply air, where they evaporate rapidly. Droplet generation is by an ultrasonic nebuliser or by using an air pressurised nozzle. Any dissolved solids present in the water supply will become airborne and water pretreatment may be advisable.

The two systems are compared on the psychrometric chart in Figure 5.48. Steam injection increases humidity with little or no warming effect (line 1). In contrast, the atomised spray humidifier is an adiabatic process and moves the supply air along a line of constant wet bulb temperature, thus producing a cooling effect. The direct operating cost of an atomising humidifier is much less than that of a steam humidifier, but adiabatic humidification reduces the temperature of the supply air and in most situations sensible heat must be added to restore the temperature (see line 2).

Figure 5.49 shows a steam humidifier controlling humidity in the occupied space. Steam is injected in the supply air

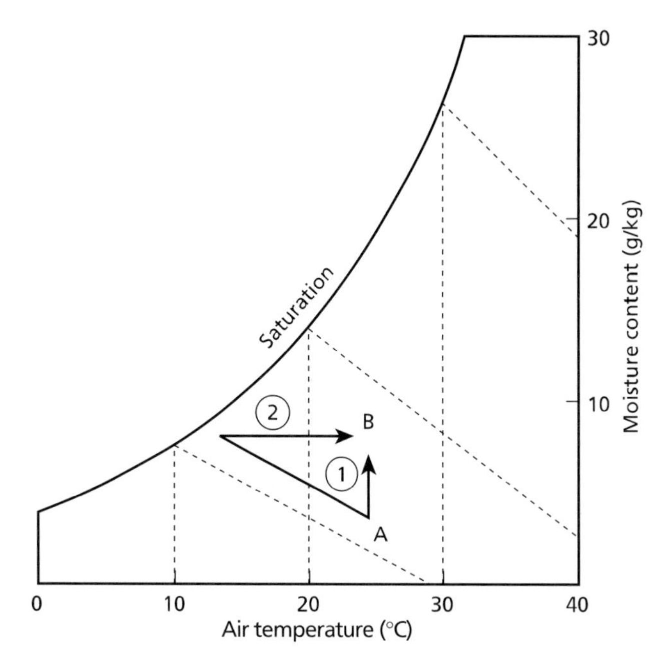

Figure 5.48 Humidification processes. It is required to humidify air from point A to B. Steam injection increases moisture content with little increase in dry bulb temperature, line 1. An atomising device humidifies along a line of constant wet bulb temperature, requiring the addition of sensible heat, line 2

downstream of the rest of the main AHU. The humidifier may be modulated continuously or in stages. Control of the humidity is achieved via a sensor which is positioned either in the controlled zone itself or in the extract air stream. PI control is used to modulate the humidifier. Response of the humidity in the zone to changes in output is slow; the absorption and release of moisture by the building fabric imposes a substantial inertia on the system. The settings of the PI controller must reflect this. A relative humidity sensor is used, with control on relative humidity. To prevent any risk of condensation in the supply ducting a high limit humidity sensor is positioned in the supply duct. This limits the humidifier if the supply RH rises above a high limit, typically 80% RH. The steam humidifier is enabled during normal operation provided that operation of the supply fan is proved. At all other times it is disabled.

The control strategy for an atomising humidifier is essentially the same as for the steam humidifier. The air is cooled by the humidifying process; this represents an additional heating load on the system and should be

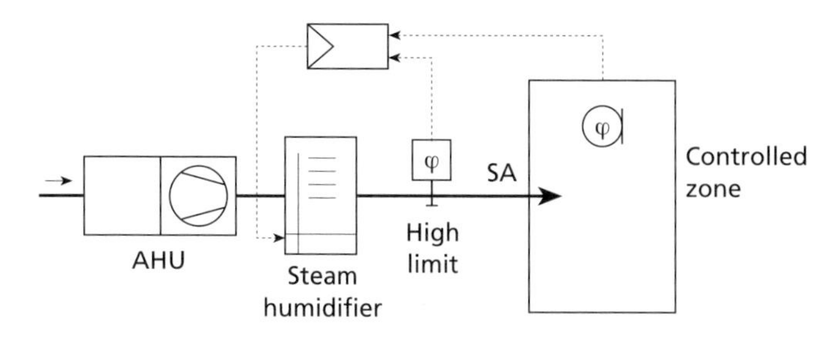

Figure 5.49 Steam humidification. Control of the humidifier is by PI control on the RH in the controlled zone. A high limit stat in the supply duct prevents condensation. *Normal operation*: supply of steam is modulated by a PI controller which controls RH in the occupied space. *Humidifier disabled*: at all times other than during normal operation with the supply fan proved. *Variations*: control may be on return air RH from the controlled zone

automatically dealt with by the temperature control system. A low limit thermostat may be installed in the duct downstream of the humidifier to bring in extra heating if the temperature falls too low; this is in addition to the high limit humidistat which is designed to prevent condensation in the ductwork.

5.6 Energy recovery

It is possible to recover energy from the exhaust air stream which would otherwise be discharged to atmosphere and transfer it to the intake air stream. Such devices may be used on full fresh air systems or between the exhaust and intake air streams of a recirculating system. Use of a heat recovery system makes it possible to use a full fresh air system without serious energy penalty, in situations where it is required to avoid recirculated air for hygienic or other reasons. Three major types of heat recovery system are in use and are listed in Table 5.4 with their characteristics. Energy recovery devices can be used during both heating and cooling modes of building operation. However, in the UK climate the periods during which exhaust air is discharged at a temperature appreciably below ambient are rare and so the application of energy recovery during cooling is limited. The table shows temperature efficiencies during heating operation; temperature efficiency is a measure of the effectiveness of the heat exchanger in recovering sensible heat. The energy recovery depends on the relative mass flows in the two air streams as well as on the temperature efficiency of the device. The efficiencies must be regarded as approximate and will vary with the make and model of the equipment. Efficiencies will generally be lower for cooling operation and lower again if enthalpy recovery is considered. When considering the overall cost benefit of a heat recovery installation, the running cost as well as capital cost must be considered. In addition to the operating cost of the pump or wheel motor, the devices introduce an extra resistance to flow in the air system, thus increasing fan costs. A comprehensive treatment of air to air heat recovery is given in an ASHRAE Handbook[25].

Another, the regenerative heat exchanger, operates by periodically reversing the two air flows over a heat absorbing mass. This type of device can achieve high operating efficiencies, but is more often found in process industries than in HVAC systems. Units are available which combine a plate heat exchanger with a heat pump; a small heat pump is used to transfer additional heat from the exhaust air, downstream of the heat exchanger, to the supply air. Such a unit is capable of discharging exhaust air at below outdoor ambient temperature and thus acts as a net supplier of heat to the supply air.

Table 5.4 Energy recovery devices

Type	Temperature efficiency	Contiguous air streams	Control	Cross-contamination
Thermal wheel	< 80%	Yes	Speed	Possible
Plate heat exchanger	< 70%	Yes	Face and bypass damper	No
Run-around coil	< 65%	No	Pump speed or valve	No

Table 5.5 Control strategy for heat recovery devices

ΔT between air streams	Control mode
Large	Energy recovery device operates to give maximum heat recovery. Control of supply temperature by HVAC plant
Medium	Control of supply air temperature achieved by modulating rate of heat recovery. No heating or cooling by HVAC plant
Small	Energy recovery not viable. Device remains off and control of supply air temperature is by HVAC plant as required

The thermal wheel and run around coil devices both require a power input for their operation (and supply and extract fans may use more power to maintain the design supply flow) and, for it to be worthwhile switching the unit on, the cost of operating the device must not exceed the worth of the energy transferred. This equation must be evaluated in financial terms, since the unit costs of the input power to the device and of the heat energy are unlikely to be the same. From this, a temperature differential, both positive and negative, between extract and intake air streams is calculated, below which it is not worthwhile operating the energy recovery device. When in operation, the rate of heat transfer between the two air streams may be controlled up to the maximum value of which the device is capable. There is a limited range of temperature difference between the air streams where the energy recovery can be varied to control the supply temperature without any additional heating or cooling. Outside this range, when the wheel has either reached maximum speed or stopped, the main plant is brought into play for temperature control. The control strategy for a variable-rate heat recovery device is simply summarised in Table 5.5.

5.6.1 Thermal wheel

The thermal wheel is a type of regenerative heat exchanger consisting of a shallow drum which contains a matrix capable of absorbing and releasing heat. The opposing air streams pass through the slowly rotating drum in an axial direction. Thus the packing absorbs heat as it passes through the exhaust air stream and subsequently releases it to the cooler intake air. Figure 5.50 shows a thermal wheel

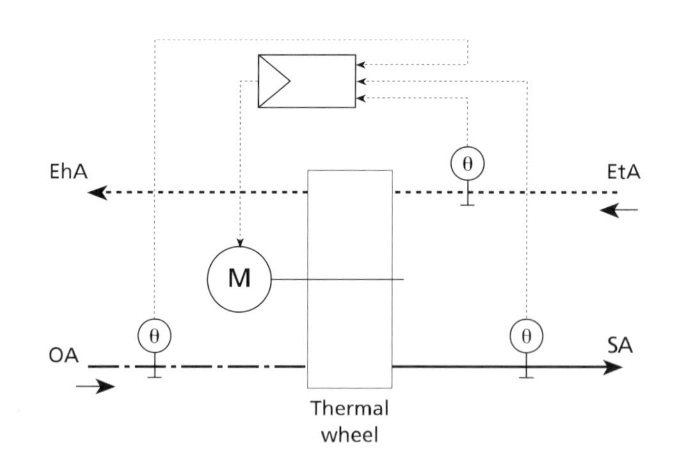

Figure 5.50 Thermal wheel heat recovery, variable speed. The wheel is enabled when the temperature differential ($T_{EtA} - T_{OA}$) is sufficient to justify operation. *Normal operation*: rotation of speed wheel is modulated to control supply or zone air temperature. Further temperature control in the AHU is enabled when wheel is at zero or maximum speed. Normal operation also during OSH heating or cooling or boost. *Wheel disabled*: plant shut-down or night cooling

used to transfer energy between extract and intake air streams. The wheel may be constructed so as to transfer latent as well as sensible heat, by coating the heat absorbing matrix with a hygroscopic substance. The wheel is driven by a variable-speed motor and to be effective, the cost of operating the wheel must not exceed the worth of the energy transferred. When in operation, the rate of heat transfer between the two air streams is an increasing function of the rotational speed of the wheel, up to a maximum speed above which there is only a negligible increase. When the wheel is in operation, there is a limited range of temperature difference between the air streams where the rotational speed of the wheel can be varied to control the supply temperature without any additional heating or cooling. This is realised using a PI controller to modulate the speed of the wheel, with either the supply or zone temperature as controlled variable. Outside this range, when the wheel has either reached maximum speed or stopped, the main plant is brought into play for temperature control.

Control interlocks follow the expected pattern:

— The wheel only operates when the main supply fan is running.

— The wheel is enabled when the BMS is signalling normal operation, optimum start heating or cooling, or boost operation.

— The wheel is disabled during night cooling or when the air handling system is shut down.

An alternative configuration for a thermal wheel uses a constant speed of operation of the wheel and varies the rate of heat transfer between air streams by use of a face and bypass damper. The control strategy is the same as for variable-speed operation.

5.6.2 Run-around coils

Run-around coils may be used when the intake and exhaust ducts are not adjacent to each other and when the potential energy recovery is sufficient to justify the extra initial and running cost. It has the additional advantage of preventing any possibility of cross-contamination. Figure 5.51 illustrates the control set-up of a pair of run-around coils. The two coils are connected with counterflow piping, with a pump circulating water, glycol or other thermal fluid around the circuit. There is a risk of icing on the exhaust coil if the solution entering falls to freezing point; the controller ensures that the solution entering the coil does not fall below a minimum temperature set point, typically 2°C. Control of the rate of energy recovery may be by a bypass valve, as shown, or by variable-speed pump. The control strategy follows the same lines as that described for the thermal wheel. The pump only operates if the temperature differential between intake and extract air is greater than a predetermined amount. The rate of heat transfer is modulated to control the supply air temperature up to the maximum possible, when the plant is brought in to provide the necessary heating or cooling.

5.6.3 Fixed plate heat exchanger

Fixed plate heat exchangers are available in a number of configurations. They require the two air streams to be brought together. Apart from any penalty due to the pressure drop though the heat exchanger, no external power is required for operation, so the minimum viable temperature difference described above is not relevant. If control of the rate of heat transfer is required, this may be obtained using a face and bypass damper system as shown in Figure 5.52. The control strategy for modulating the rate of heat transfer is exactly as described above for the thermal wheel with face and damper bypass. A pressure differential switch across the plates is used to provide an alarm in the case of a blocked or frozen heat exchanger.

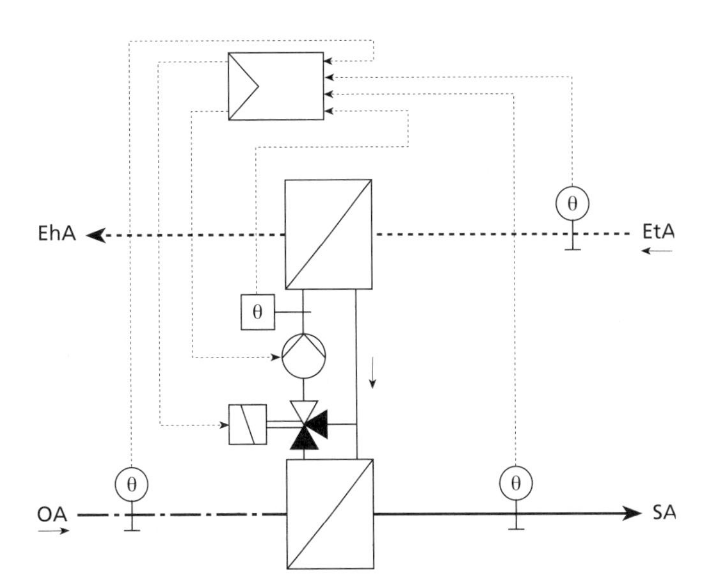

Figure 5.51 Run-around coil. Heat is transferred from the warmer to the cooler air stream. *Normal operation*: the pump is enabled when the temperature difference ($T_{EtA} - T_{OA}$) is sufficient to justify operation. The three-way valve is modulated to control the supply air temperature. Further temperature control by the AHU is enabled only if the pump is off or the valve is fully closed to the bypass. Normal operation also during OSH and boost heating, OSC cooling. *Pump off, valve to bypass*: plant shut-down or night cooling

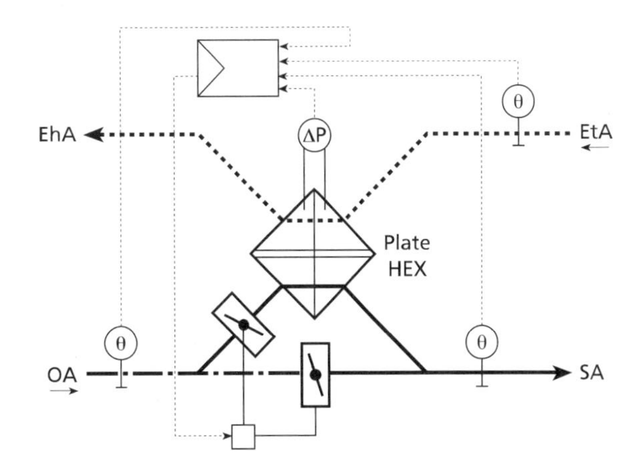

Figure 5.52 Fixed plate heat exchanger with face and bypass dampers. Heat recovery from warmer to cooler air stream via a fixed plate heat exchanger. Control is shown using face and bypass dampers in the intake air stream. A differential pressure sensor across the exhaust section warns of blockage due to icing. *Normal operation*: dampers modulate the flow of air through the HEX to obtain the supply air set point; the face and bypass dampers move in opposition. Further temperature control by the AHU is enabled when the face damper is fully open or closed. Dampers modulate during OSH or boost heating and OSC cooling. *Face damper open*: plant shut-down. *Face damper closed*: night cooling

5.7 Mechanical ventilation

All occupied buildings require a supply of outside air for ventilation purposes. A minimum supply of air is necessary for the removal of odour, carbon dioxide and any other contaminants produced by human occupation and other activities. The build-up of carbon dioxide and water vapour are normally the two most important determinants of the necessary ventilation rate. Other contaminants have received a great deal of attention in the context of sick building syndrome but as yet there is no reliable measurement which would allow automatic control of ventilation for indoor air quality. Minimum outdoor air requirements are set out in the CIBSE Guide Part A1[26]; international legislation is reviewed by Olesen[27]. In the absence of smoking, an outside air supply of 8 l/s per person is found sufficient to maintain acceptable air quality. In situations of low occupancy, emission of volatile organic compounds (VOCs) from building materials becomes the dominant air pollutant and it is good practice to set a minimum outside air supply rate of about 20% of the maximum value; this will provide acceptable air quality in conditions of low occupancy. Where ventilation with outside air is closed down during unoccupied hours to save energy, ventilation should be restarted before the beginning of the occupied period. The pattern of air movement in the ventilated space affects the ventilation efficiency, which may be defined as the ratio of contaminants in the extract air to that in the occupied region[28]. A fully mixing ventilation system has an efficiency of unity. A poor system, where the supply air short circuits the room and moves directly towards the extract, has an efficiency of less than one. A displacement ventilation system, where the incoming air moves over the occupants before picking up contaminants and removing them via the extract, can have an efficiency of greater than one.

Ventilation also has an important part to play in temperature control. In the absence of heating, indoor air temperatures are generally higher than outdoor ambient temperature as a result of solar gains and heat-generating activities in the building. Increasing the ventilation rate therefore dissipates heat gains and can reduce indoor air temperature towards a limiting value equal to the outdoor temperature.

Ventilation in domestic and small commercial buildings has traditionally been by natural methods, typically openable windows operated by the occupants. This does not guarantee adequate ventilation and the *Building Regulations Part F*[29] require the use of mechanical extract fans or passive stack ventilation. In recent years, the use of natural ventilation has been extended to large commercial buildings. Such systems must be integrated carefully with the design of the building itself to ensure that ventilation is achieved to extract heat gains in warm weather, thus achieving the goal of avoiding the requirement for air conditioning. The control of natural ventilation is discussed in 5.14.

The following deals with aspects of the control of mechanical ventilation. Where a central AHU is employed, it is normal to incorporate outside air ventilation. Figure 5.43 illustrates a generalised air handling system with recirculation. The basic system is to be found with modifications in many HVAC systems; a general description is given in BRECSU GPG 257[30]. A general description is given here, followed by a more detailed treatment of components.

Where the air supply is used to heat or cool the occupied zones, the volume flow of conditioned air is several times that of the outside air necessary for ventilation purposes. It is therefore normal practice to recirculate a proportion of the air which has been extracted from the conditioned space and mix it with the required outdoor air supply; this leads to economies in operation. A full fresh air system, with no recirculation, may be found in the following circumstances:

- The extract air is contaminated and should not be recirculated. In this situation, consider heat recovery.
- Non-air systems are used for heating or cooling, e.g. radiators, heated floors, chilled ceilings.
- Local air recirculation is used, e.g. fan coil units.

In these circumstances there is no recirculation at the central plant. Outside air supply rate may be fixed at the required minimum ventilation rate or controlled as a function of ventilation demand or cooling requirement.

5.7.1 Demand-controlled ventilation

Subsequent sections deal with the control of ventilation systems which, in general, provide the required minimum amount of outside air for ventilation purposes, plus a variable amount of recirculated air and further outside air as required for temperature and humidity control. For most purposes, it is sufficient to calculate the required minimum outside air supply from a knowledge of typical building occupancy, plus any further requirements to remove additional contaminants. Where the building has a large and unpredictable occupancy, there are potential energy savings in matching the outside air supply to the ventilation demand; this is termed demand-controlled ventilation (DCV)[31,32].

Carbon dioxide level is the usual controlled variable for occupied spaces. Carbon monoxide monitors can be used for controlling the ventilation of enclosed car parks. Other air quality sensors are available which respond to a mixture of pollutants, but have not found widespread applications; see 3.1.3.5. As an alternative to direct measurement of CO_2 it may be possible to estimate the number of occupants in the controlled space and adjust the ventilation rate accordingly. This has the advantage of providing the required outdoor air supply immediately, without waiting for CO_2 levels to build up. Occupancy may be obtained from the access control system in an integrated BMS.

The control system for DCV should always ensure a minimum outdoor air supply, regardless of CO_2 level, to deal with outgassing from building materials. Set point control, where outdoor ventilation rate is switched between maximum and minimum rates, is suitable for situations where the occupancy is either high or low, but at unpredictable times; examples are meeting rooms, lecture theatres. A set point of between 600–1000 ppm CO_2 will be found suitable, with a differential band of about 100 ppm. Proportional control is more suitable where the occupant density is variable. Where outdoor ambient CO_2 levels are variable, which is often found in urban locations, it is better to control on the difference between outdoor and indoor concentration. The lower limit of the proportional band should be set at a level 100 ppm above outdoor, with a proportional band of about 500 ppm.

5.7.2 Poor outdoor air quality

Effective ventilation requires that outdoor air is cleaner than the indoor air which it replaces. This is not always the case and it may be necessary to take precautions to ensure that polluted air does not enter the building. The first line of defence is to ensure that air intakes are placed in clean air and away from any likely sources of pollution, such as car parks or loading bays. Outdoor air ducts incorporate filters to remove particulate matter; the increased fan energy cost associated with better filtration must be evaluated at the design stage. However, filters will not remove gaseous contaminants, such as carbon monoxide or oxides of nitrogen.

Where demand-controlled ventilation is used, there is a danger that the indoor air quality (IAQ) sensor which controls ventilation rate may react to any ingress of pollutants by increasing the ventilation rate and so making the situation worse. Outdoor air quality (OAQ) is often variable over the day, at its worst during peak traffic or at some other time associated with the generation of pollution. This provides an opportunity for control. Air handling systems with recirculation can vary the balance between recirculated and outdoor air, and it is advantageous to reduce the outdoor air contribution during times of high pollution, increasing it to compensate at other times. Such a solution can only operate satisfactorily where the pollution episodes are of relatively short duration. Two strategies may be used[33]:

- Where the OAQ is known to be poor at the same time each day, a timeswitch can be used to set outdoor air supply to its minimum level during these times.
- Where OAQ is variable, an OAQ sensor can be used to reduce outdoor air supply during pollution episodes.

A carbon monoxide sensor is recommended for as a sensor. CO is found to show a high correlation with other urban pollutants such as oxides of nitrogen and other compounds and reliable sensors are available. The recommended strategy is

- Use a set point of 5 ppm CO for ventilation control.
- Use CO values averaged over 5 minutes to avoid responding to short term spikes.
- If the set point is exceeded, switch the ventilation system to full recirculation for a short period, say 15 minutes. Maintain the required average minimum outdoor ventilation rate over periods of 1 h. Do not allow reduced ventilation rates to compromise thermal comfort.
- When the building is unoccupied and OAQ is satisfactory, purge with outside air for a time equivalent to four full air changes in the building.

The above strategy will reduce intake of polluted air, while giving priority to sufficient ventilation to remove indoor pollutants.

5.7.3 Full outside air damper system

Figure 5.53 shows a simple full outside air system. The exhaust and intake dampers move together. Both move

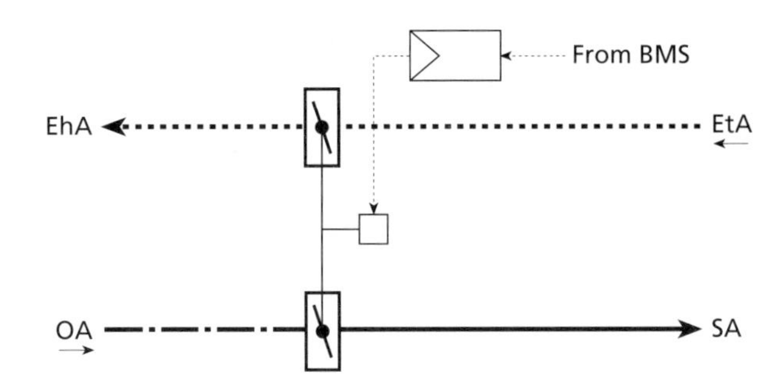

Figure 5.53 Full outside air damper system. Exhaust and intake dampers move together to maintain system balance. *Normal operation*: both dampers are fully open to admit outside air and discharge exhaust air. Dampers are proven open by end switches to enable supply fan. *Dampers open*: normal operation, OSC heating and cooling, boost heating, night cooling. *Dampers closed*: plant shut-down, fan failure

fully open when the plant is operating and close during periods of non-operation or when a fan-failure signal is received. An interlock prevents fan operation if the respective damper fails to prove open.

5.7.4 Variable-volume extract fan

It is normal practice to run a building at a slight overpressure to prevent ingress of outside air through leaks in the fabric. This is done by ensuring that the supply air flow is greater than the extract flow; the balance is made up by exfiltration through the building envelope. The volume flow rates of the supply fan and extract fan are normally linked to prevent excessive over- or under-pressurisation of the building. For constant volume systems this is simply done by linking the operation of the two fans, but for a variable-volume system it is necessary to control the speed of the extract fan to follow the supply air volume. Control of supply and extract flows is linked with the control of outdoor air supply and 5.7.6 should be read in conjunction with this discussion. There are four techniques which may be used to control the operation of the extract fan: open loop, building static pressure, duct static pressure and air flow tracking.

Open loop control

Open loop controls the speed of the extract fan in step with that of the supply fan, but without any feedback from the result (Figure 5.54). Open loop control requires supply and

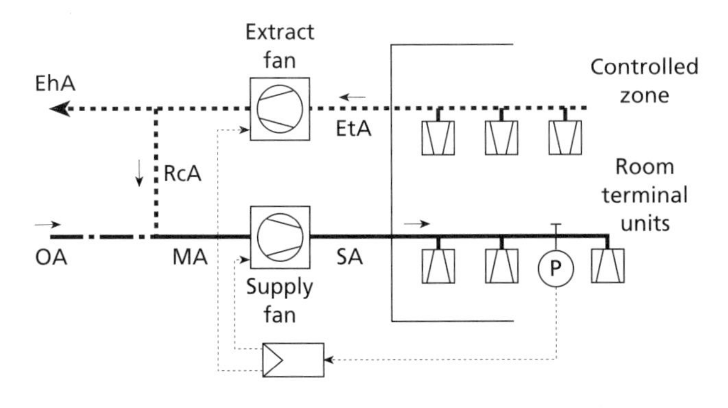

Figure 5.54 Open loop control of extract fan: a VAV system with supply fan controlled by duct pressure. Extract fan is controlled in tandem with supply fan

extract fans with similar operating characteristics. Balancing the two air flows is done during commissioning, but the balance will be affected by changes in building operation in an unpredictable way. This form of control may be acceptable in small systems which are tolerant of variations in building pressurisation.

Building static pressure control

The speed of the extract fan is varied to control the building static pressure (Figure 5.55). The pressure is compared with the static pressure outside the building and may be set to a slight over- or under-pressure. Careful location of the pressure sensors is important; the indoor sensor should be distant from opening external doors or liftshaft, while the outdoor sensor should be mounted at least 5 m above the building and shielded from wind. A hallway on an upper floor is usually suitable for the indoor location.

Duct static pressure control

Where individual returns from the space are damper controlled, it is preferable to control the extract fan by the return duct static pressure (Figure 5.56). This is similar in principle to the control of a VAV supply fan by duct pressure; in the case of the extract fan the pressure is negative.

Air flow tracking

Velocity sensors, or other flow-measuring devices, are positioned after the supply fan and before the extract fan and the measures converted to volumetric supply rates (Figure 5.57). The supply and extract volume flow rates are compared and the extract fan speed is modulated accordingly. The fan speed may be controlled so that the extract flow rate is a fixed proportion of the supply flow rate. In a typical VAV system, the supply fan speed is controlled using a measure of the static duct pressure and the extract fan using a measure of the difference in supply and extract flow rates. The sampling rate for the latter must be slower than the former to ensure stability. The extract fan is disabled in the event of plant shut-down or supply fan failure; in all other situations the extract fan is modulated to follow the supply flow rate. Where there is a supplementary exhaust system, it may be necessary to

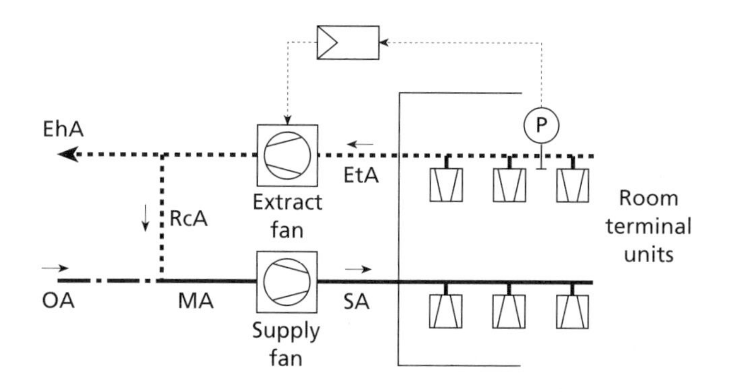

Figure 5.56 Control of extract fan by extract duct static pressure. The extract fan speed is controlled to ensure that there is sufficient negative pressure in the extract duct. Applicable for systems with multiple damper controlled extract terminals

measure the supplementary exhaust flow and subtract this from the supply flow to find the desired extract rate. Air flow tracking presents some problems in control in practice and improved techniques are discussed in 5.7.6.

5.7.5 Mixing damper systems

Many air handling systems employ some form of mixing damper system, in which a proportion of the return air from the occupied space is recirculated and mixed with outside air. The three sets of dampers work in unison with each other so that the proportion of recirculated air and outside air may be varied while keeping the air flows in balance.

The principles governing the control of mixing damper systems are:

- At least the minimum outside air requirement is supplied during occupied hours.
- Dampers are controlled to give minimum energy demand from the HVAC system.

Details of control strategy vary between systems. Where humidity control is provided, control is more energy efficient using enthalpy control rather than simple dry bulb temperature. VAV systems require a minimum outside air supply to be maintained against variations in supply air volume flow. Figure 5.58 shows the basic mixing damper system. During part of the year it may be possible to satisfy

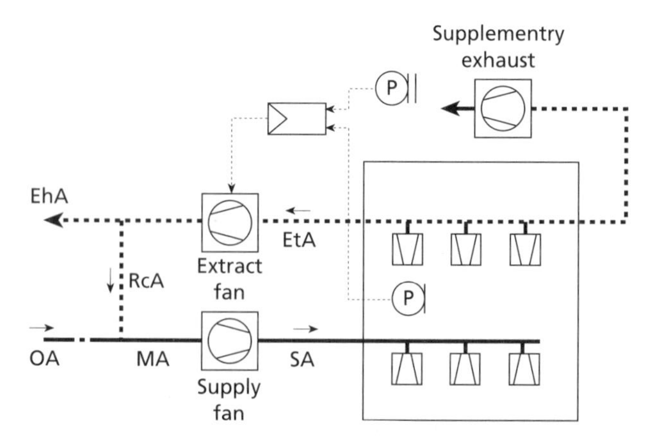

Figure 5.55 Control of extract fan by building static pressure. The extract fan is controlled to maintain building pressure at a fixed set point. Supply fan control not shown

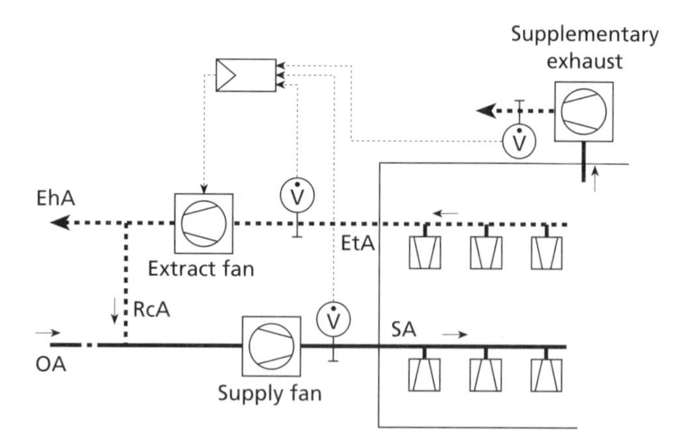

Figure 5.57 Control of extract fan by air flow tracking. Extract fan speed is controlled so that the exact flow rate balances the supply flow, less any supplementary exhaust flows

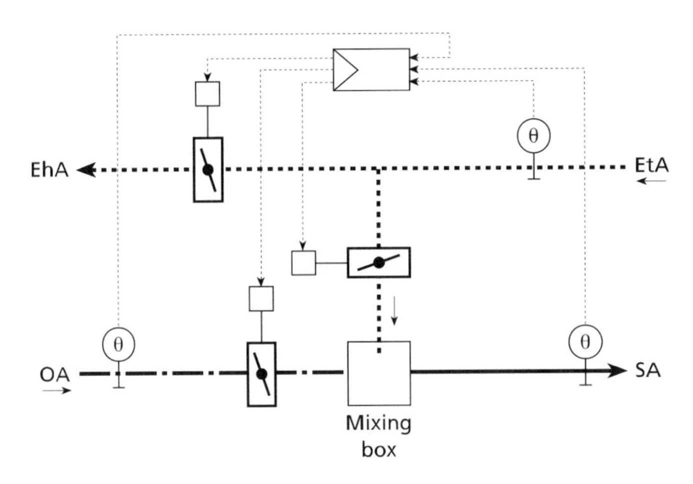

Figure 5.58 Mixing damper control. The dampers are moved together to provide a balanced supply. The proportion of outside air is varied to give economic operation, subject to minimum ventilation requirements

the requirements of temperature control by mixing recirculated air and outdoor air in suitable proportions. When this is not possible, the outside air supply rate is set to either minimum or maximum value, depending on the conditions.

The minimum outside air damper position is set during commissioning; it may be set physically at the damper or, more commonly, within the controller software. For constant volume systems, where the supply fan runs at constant speed during normal operation, the minimum damper position is fixed. For variable-volume systems, the outside air damper position needs to be controlled as a function of the supply air flow rate. The outside air damper opening is increased at low supply flow rates to maintain the required minimum volume of outside air. This may be achieved in two ways:

— linking the outside air damper position to the fan speed control signal

— controlling the outside air damper position using a local control loop which utilises a flow sensor in the outside air duct. More details are given in 5.7.6.

The operation of the mixing dampers to provide the most energy efficient recirculated air supply is termed the economiser cycle. The principle is illustrated on the psychrometric chart in Figure 5.59; this shows control on dry bulb temperature only. The diagram shows a situation where extract air from the controlled space is at a temperature T_{EtA} and air is to be supplied at a temperature T_{SA}. For near steady state, four operating modes may be distinguished:

— At low outside temperatures the air is cold enough to bring the mixed air temperature to below the required supply temperature even at the minimum outside air supply rate; at these temperatures, below T_{OAmin}, the outside air damper is set to its minimum position and heating is required to bring the supply air temperature up to T_{SA}.

— When the outside air is within the range $T_{OAmin} < T_{OA} < T_{SA}$ the mixed air can be controlled at the required supply temperature by modulating the dampers, up to 100% outside air when $T_{OA} = T_{SA}$. In this range the system is operating in free cooling mode.

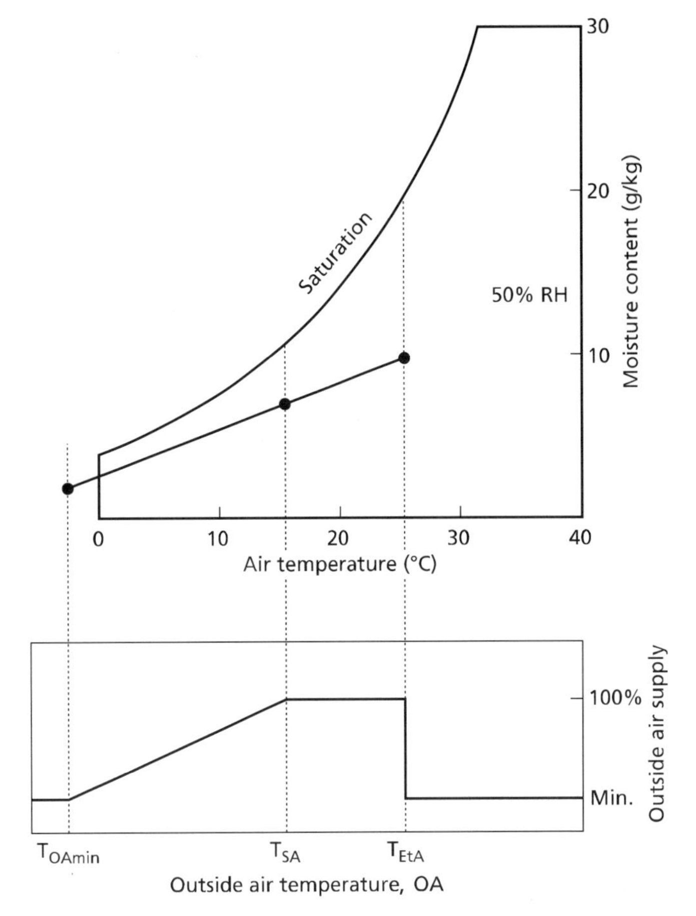

Figure 5.59 Principle of economiser control

— Within the range $T_{SA} < T_{OA} < T_{EtA}$ the outside air is maintained at 100% to give the lowest temperature of the recirculated air entering the air handling unit for cooling.

— As the outdoor air temperature rises above the return air temperature, the outside air damper moves to its minimum position to give the lowest possible mixed air temperature.

The figure shows control of outdoor air flow as a function of supply and extract temperatures; in practice these will vary with cooling load.

Thus in the free cooling mode, temperature control can be achieved without the need for heating or cooling from the AHU. The simple example assumes a fixed supply air temperature. For a constant volume system, the mixing dampers may be operated by a PI controller with input from the controlled zone air temperature; the zone set point may be changed according to season. For a VAV system, it is the supply air temperature that is controlled, but there is additional control of the outside air damper to ensure that the required minimum outside air flow is maintained. The control of the dampers may be generalised:

The dampers modulate towards the open position under the following circumstances:

(1) The return air temperature is greater than the outside air temperature and cooling is required ($T_{SA} < T_{EtA}$) (free cooling mode).

(2) The return air temperature is less than outside air temperature and heating is required (non-steady state situation).

The dampers modulate to the minimum outside air position under the following circumstances:

(3) The return air temperature is greater than the outside air temperature and heating is required ($T_{SA} > T_{EtA}$).

(4) The return air temperature is less than the outside air temperature and cooling is required. (non-steady-state mode).

Situations 2 and 4 would only be met during changing temperatures and are not usual.

Where the humidity as well as the temperature of the supply air is controlled, account must be taken of the total energy content of the air streams if maximum efficiency is to be obtained. This is known as enthalpy control. This is illustrated on a psychrometric diagram in Figure 5.60, which shows an all air system where the extract air from the occupied space has a temperature and humidity defined by the point A on the diagram. The normal operating situation is considered where the supply air to the space requires dry and wet bulb temperatures which are both below that of the return air, i.e. the supply air condition lies somewhere in region 1 of the diagram. This applies in virtually all situations, whether heating or cooling, since the sensible and latent loads in the occupied space raise the dry and wet bulb temperature of the extract air above that of the supply air, whether the building is in heating or cooling mode. Two lines are shown on the diagram: a line of constant air temperature a–a and one of constant enthalpy e–e; this is practically indistinguishable from a line of constant wet bulb temperature. In region 4 of the diagram, the enthalpy of the outside air is less than that of the return air and economies will be obtained by opening the outside air damper, even though the dry bulb temperature is greater than that of the extract air. Conversely, in region 2 of the diagram, the high humidity of the outside air would create an energy penalty on dehumidification and the outside air supply should be set to minimum. The principle of enthalpy control, during normal operating conditions where the return air wet and dry bulb temperature are above those of the required supply air condition, is as follows:

- When the outside air has a lower enthalpy than the extract air but a higher temperature (region 4), use 100% outdoor air with no recirculation.
- When the outside air has a lower enthalpy and temperature than the extract air (region 1), use the optimum mix of outdoor and recirculated air.
- When the outside air enthalpy is more than the extract air enthalpy (regions 2 and 3), the dampers provide the minimum outdoor air supply required for ventilation.

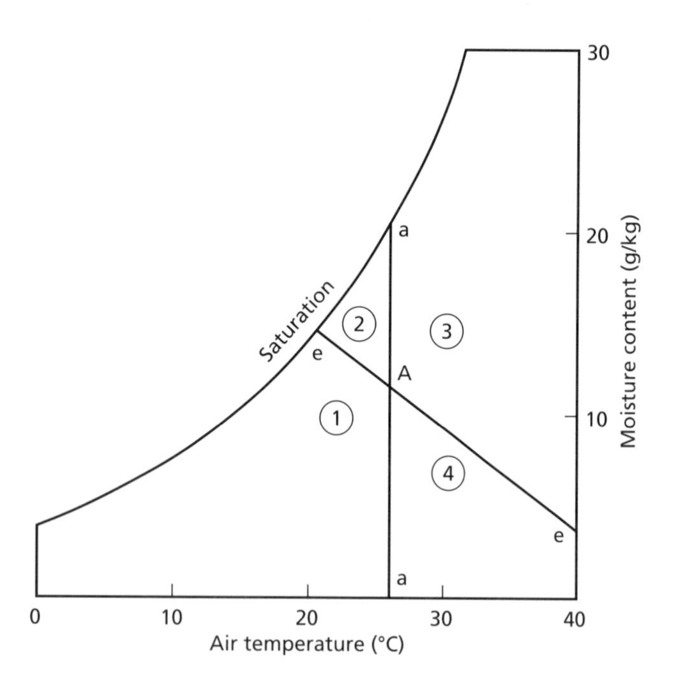

Figure 5.60 Principle of enthalpy control. Point A is the extract air condition and the required supply air condition is somewhere in region 1. Minimum outside air is used when the outdoor conditions are in regions 2 and 3; enthalpy economiser control mixes outdoor and extract air when outdoor conditions are in regions 1 and 4

Note that the temperature and humidity of the supply air do not appear in Figure 5.60, but simply the relationship between the return air and outside air. Examples of the different situations are shown in Figure 5.61. There is no single sensor for the measurement of enthalpy, which has to be calculated from separate measurements of temperature and humidity. Sensors are available which combine both sensors and carry out the necessary computation.

5.7.6 Outdoor air supply

The general air handling system shown in Figure 5.43 supplies air to the occupied space; the supply air is a mixture of outdoor air and recirculated air. The proportion of outdoor air varies with the quantity of extract air that is recirculated. Since the volume flow of supply air is itself variable in a VAV system, there can be problems in ensuring that the required minimum volume of outdoor air for ventilation purposes is always provided.

In a constant volume system, the supply fan provides a constant volume of air. Where economiser control is employed, the proportions of outdoor and recirculated air are varied for optimum energy efficiency. The outdoor air damper has its minimum opening position set during commissioning, so that when the economiser control is in the maximum recirculation position, the required amount of outdoor air is supplied. After the initial balancing, no further control is required. The system will always provide at least the minimum outdoor air ventilation requirement, rising to 100% outdoor air during maximum free cooling.

In a VAV system, the supply fan is modulated according to the load requirement of the building and a control strategy is required to ensure that the minimum ventilation requirement is met. The difficulties of ensuring reliable control of outdoor air supply are discussed by Kettler[34], who surveys the available strategies:

(1) *Limiting the outside air damper to a minimum opening position*. This must be set during commissioning to give the required outdoor air flow with the supply fan at minimum speed and with maximum recirculation if an economiser control is fitted. This solution is the simplest, but does not give minimum energy consumption, since outdoor air flow may be excessive at high supply flows.

(2) *Limiting the outside air damper to a minimum opening position which is a function of fan speed*. The

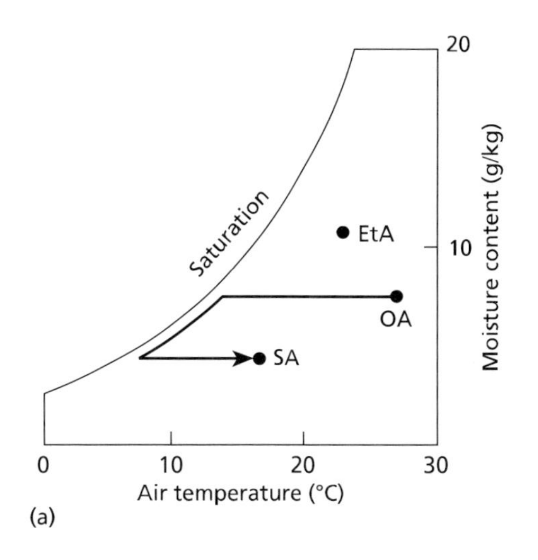

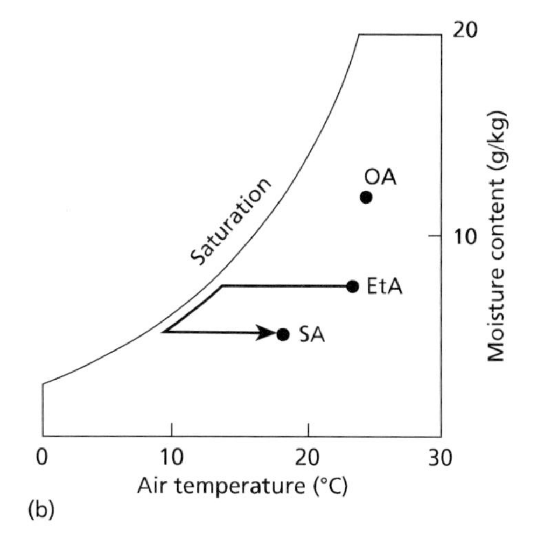

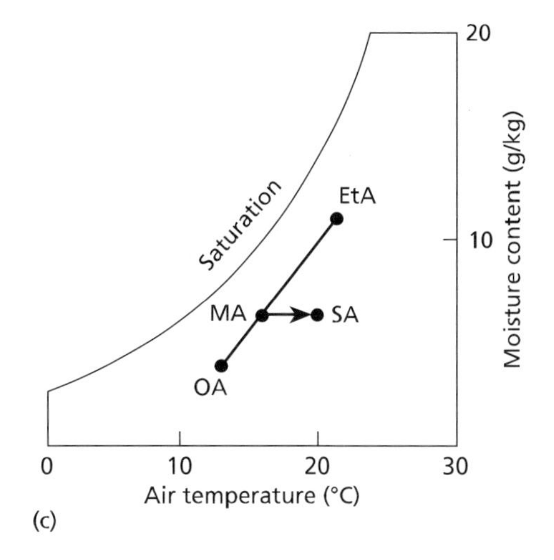

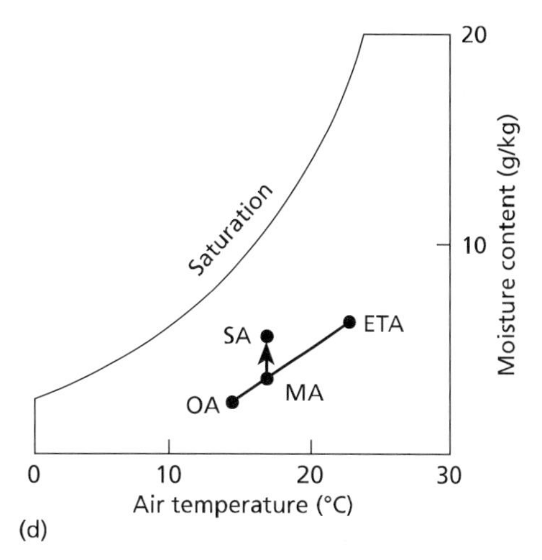

Figure 5.61 Examples of enthalpy control. (a) 100% outdoor air (region 4); (b) minimum outdoor air (region 3); (c) economiser control, heating required (region 1); (d) economiser control, humidification required (region 1). EtA, extract air; SA, supply air; OA, outside air; MA, mixed air

relationship is established during commissioning and must provide the required ventilation supply at all combinations of supply fan speed and recirculated air proportion.

(3) *Modulating the outside air damper to provide the desired air flow as measured in the outdoor air intake.* An air flow sensor in the intake duct is used with a PI controller in a separate control loop.

(4) *Modulating the extract fan speed in a technique known as volume matching or air flow tracking.* 5.7.4 describes the air flow tracking technique, whereby the extract fan speed in a VAV system may be controlled to follow the supply fan. If the extract fan is controlled to maintain constant difference between extract and supply air flows, the difference is made up by the outdoor air intake flow; the difference is set to be equal to the ventilation requirement.

(5) *Injection fan method.* An extra fan is fitted to run in parallel with the main intake duct. This fan runs continuously and injects the required minimum outdoor air flow downstream of the mixing box; see Figure 5.62. The injection fan controls the air flow in the duct with its own PI control loop, which ensures that the required volume flow of air is delivered irrespective of wind effects or back pressures. It is especially suitable where the outdoor air supply has to be reset, as in demand-controlled ventilation. It has the disadvantage of requiring additional ductwork.

(6) *Plenum-pressure method.* The basis of the plenum-pressure method is shown in Figure 5.63. A differential pressure sensor is fitted across the outdoor air louvre and intake damper. With the recirculation damper open and the supply fan giving maximum air flow, the outdoor air damper is set during commissioning to give the required minimum outdoor air supply. The combination of louvre and minimum damper setting is now treated as a fixed orifice for air flow control. The plenum-

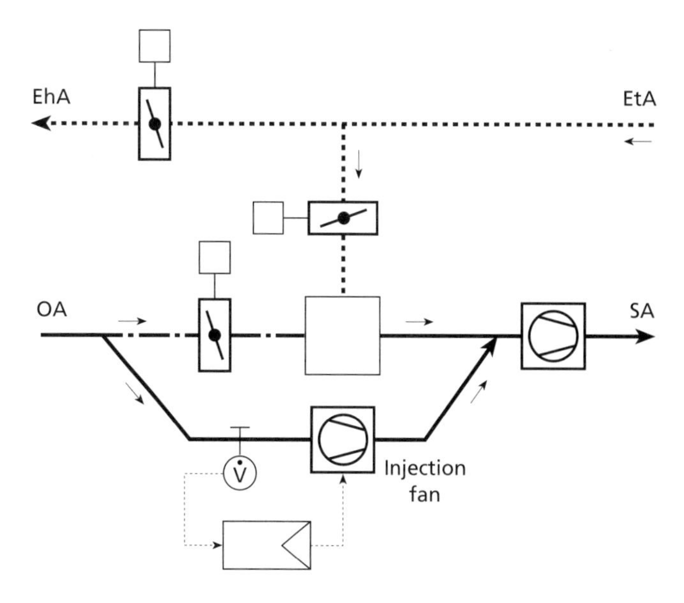

Figure 5.62 Injection fan system. The required outdoor air volume is injected into the supply duct by a separately controlled fan. Control of dampers is not shown

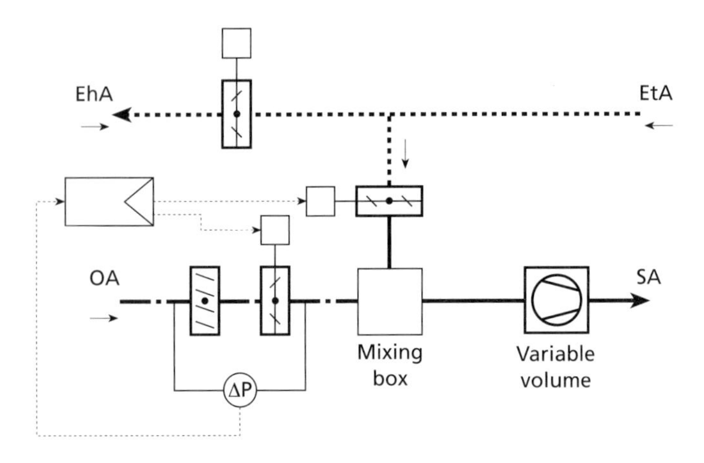

Figure 5.63 Plenum-pressure control of outside air. When the economiser circuit calls for minimum outside air, the intake damper moves to minimum open position. The recirculation damper modulates to maintain pressure differential across louvre and damper. Other controls not shown

pressure strategy operates in conditions where minimum outdoor air is required. As the supply fan reduces air flow in response to reduced demand from the VAV boxes, the controller modulates the recirculation damper towards the closed position to maintain the pressure drop and hence required flow through the outdoor air damper. This ensures that the required ventilation flow is maintained in situations of maximum recirculation. When the mixed air control requires more than minimum outside air, the outdoor air damper is allowed to open further and the plenum pressure control loop is deactivated, since the dampers are now positioning to bring in more than the minimum flow required for ventilation. This system is claimed to give reliable and simple control with a small energy penalty[35].

The first two strategies have the advantage of simplicity. The first will provide excess outdoor air in all but conditions of minimum supply fan speed. The second is a simple open loop control and should be recommissioned at suitable intervals. While apparently offering direct control of the outdoor air supply, the third and fourth strategies can present problems in practice. Measurement of air flow in the outdoor air intake duct presents practical problems arising from turbulence and the low face velocities demanded by filters. Inaccuracies in flow measurement can also produce problems with the volume matching method. Normally acceptable errors in flow measurement may translate into unacceptable error in the difference between extract and supply. It can be shown that a 5% error in flow measurement may produce an unacceptable 50% error in outdoor air supply[35].

Where volume-matching control is used in conjunction with recirculation, there can be situations where outdoor air is drawn in via the exhaust air outlet. This tends to occur in situations of high recirculation, when the intake and exhaust dampers are modulated towards closed. Air entering the system via this route bypasses any intake filters and risks introducing poor quality air to the system, especially if the exhaust is situated in a dirty area, such as a vehicle loading bay. The problem may be avoided by leaving the outdoor intake damper wide open during normal occupancy; the speed of the extract fan is controlled using the volume matching technique and the exhaust and recirculation dampers are linked conventionally[36].

5.7.7 Dehumidification with mixing damper control

A typical air handling plant will use a combination of mixing damper control and a chilled water dehumidification coil to provide air at controlled temperature and humidity; humidification is considered separately. The control strategy ensures that cooling and dehumidification are provided by the most economical method; this is a practical application of the principle of enthalpy control described above. The arrangement of the system is shown in Figure 5.64. A heating coil is shown in addition to the cooling coil, since reheat is required to provide independent control of both air temperature and humidity.

The use of mixed recirculated and outside air offers economies in the dehumidification process in a way analogous to the economiser cycle. The process is illustrated on a psychrometric chart in Figure 5.61. It is desired to condition the supply air to the temperature and maximum moisture content indicated by the point SA on the chart; humidification is considered elsewhere. If the desired moisture content of the supply air lies between that of the extract air EtA and the outside air OA, then the mixing dampers are modulated to produce mixed air of the desired moisture content.

Proper dehumidification requires a heating coil to be in operation downstream of the cooling coil. In some installations it is the practice to place the only heating coil upstream of the cooling coil. By this means it is possible to dispense with a separate preheating coil by combining the functions of pre and main heating coils. In this case, during operating conditions when dehumidification is required, air leaves the cooling coil just above saturation. This creates a risk of condensation in the ductwork and also creates problems with control. Fully independent control of temperature and humidity in the controlled space is not possible with this arrangement. However, for situations where dehumidification is seldom required during a year, it may provide acceptable conditions.

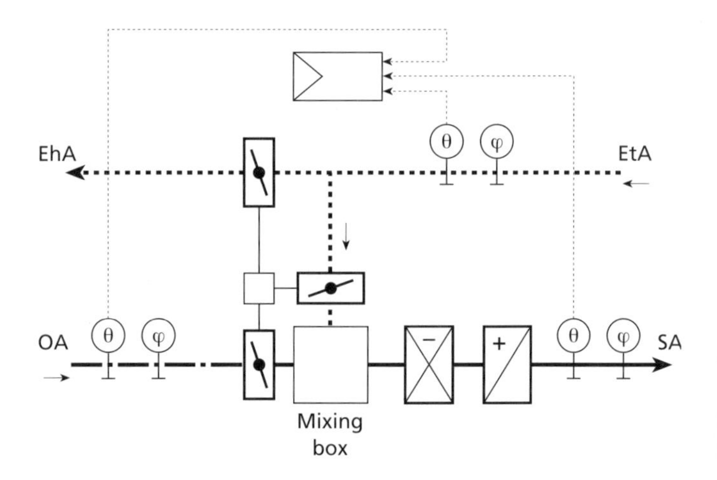

Figure 5.64 Chilled water dehumidification with mixed air control. The dampers are moved together to provide economic operation. Further dehumidification and reheat provided by the heating and cooling coils

5.7.8 Displacement ventilation

Conventional mixing ventilation aims to mix supply air with room air; both temperature and the concentration of contaminants are approximately uniform over the controlled space. The extract air also has a similar temperature and composition to the air in the occupied zone. This is not the most efficient way of ventilating a space and ignores the general rule that heat and pollutants should be extracted at source, before they mix with the general air. Displacement ventilation aims to supply clean air directly into the occupied zone (Figure 5.65). As the air makes its way towards the extract, it picks up heat and contaminants, so that the extract air is significantly hotter and dirtier than the air experienced by the occupants. In this way, it is possible to provide effective ventilation at lower supply volumes compared with a mixing system, with resulting economies. Various geometries are possible and are applied in special situations. The use of downflow systems is general in clean rooms and operating theatres. For open office spaces, an upward air flow is employed. Air is introduced through floor diffusers and is extracted at ceiling level. Sources of heat, such as people or office machinery, produce a rising plume of warm air. Many pollutant sources are warm, so the dirty air will rise towards the ceiling without entering the breathing zone.

The result is a vertical temperature gradient in the occupied space. There may be a shallow layer of cool air where the supply air is produced at floor level, termed the pool zone. The occupied zone extends to a height of 1.5 m or so, above which the temperature rises towards the ceiling. The cooling power of a displacement system is limited. It is not possible to reduce supply temperature or increase supply volumes because of the risk of local discomfort. If additional heating or cooling is required it must be provided by a separate system. The common combination of chilled ceiling and displacement ventilation is discussed in 5.12.

It is normal to run a displacement system at constant flow rate with the supply temperature controlled as a function of outdoor temperature. Supply flow rates are typically 2.5 air changes per hour. These flow rates are low compared with a mixing system and the fresh air requirement represents a large part, or even all, of the supply flow. The system therefore lends itself to a full fresh air system, with a heat exchanger for energy efficiency. Figure 5.65 shows a displacement ventilation system. The system shown offers full control of temperature and humidity of the supply air. Dehumidification is required if the ventilation system is used with any form of chilled panel cooling system, to prevent the formation of condensation on cold surfaces. This is discussed in 5.12. A full specification and points list of the displacement system shown in the figure is given in Section 3.16.4 of the BSRIA *Library of System Control Strategies*[37]. The main points of the strategy are:

- Fan and extract fan run at constant speed during normal occupancy. The volume flow must be sufficient to meet outdoor air requirements at maximum occupancy; it is typically 2.5 air changes per hour.
- Supply air temperature is linked to outdoor temperature. The supply air set point is at its maximum of 22°C when outside temperature is 12°C or below; the set point is at a minimum of 18°C when the outside temperature is 21°C or above, with linear interpolation between these points.

Control of the thermal wheel, humidification and dehumidification follows the strategies detailed elsewhere in this Guide.

The minimum supply air temperature is primarily governed by comfort considerations, since the air is introduced at floor level over the occupied area. The minimum acceptable temperature depends on the design and location of the air inlets and in some cases may be lower than 18°C. Some installations link the supply air temperature to zone temperature, rather than outside temperature as described above. A master–submaster controller is used, which resets the supply air temperature set point according to the zone temperature. Careful consideration must be given to the placement of the zone air temperature sensor, bearing in mind the vertical temperature gradient produced by displacement systems.

5.8 Variable air volume

Variable air volume (VAV) was probably the most widely installed type of air conditioning system during the last two decades. Conditioned air is supplied by a central plant to several zones; the air supply is controlled independently to each zone by one or more room terminal units (RTU) which control zone temperature by modulating the volume flow of air to the zone. The main air handling unit supplies

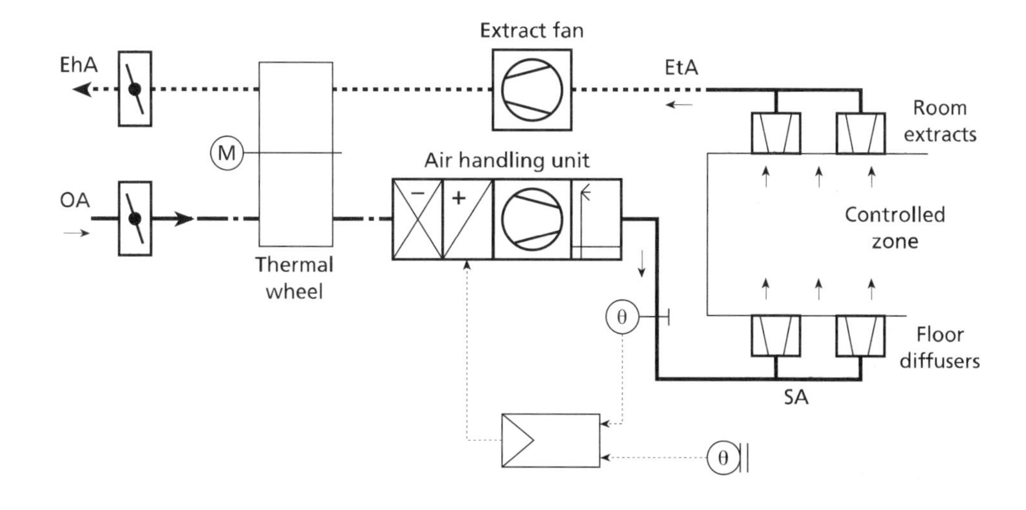

Figure 5.65 Displacement ventilation. Diagram shows reset of supply air temperature as a function of outside temperature

conditioned cool air to all room terminal units; this supply contains the necessary proportion of make-up outside air for ventilation. The air is supplied at constant temperature, though the set point may be varied with season. The room units vary the air flow to each controlled space to provide the necessary temperature control. The units may be provided with reheat coils or heater batteries, in which case they provide all heating and cooling requirements. It is, however, common to use VAV systems for cooling only and to provide any necessary heating by the use of a separate heating system such as perimeter radiators. The diversity in demand between different zones means that part load economies are achieved in both fan power and cooling power when using a VAV system. The air supply from the terminals to the conditioned space varies with cooling or heating demand and a variable-volume extract fan may be used to ensure a constant building pressure. In this case the total measured flow from the VAV units is summed and made available to the extract fan controller. The controller can scale the signal, so that the space operates under positive or negative pressure as desired. As an alternative, the building pressure may be measured directly and used to control the extract fan.

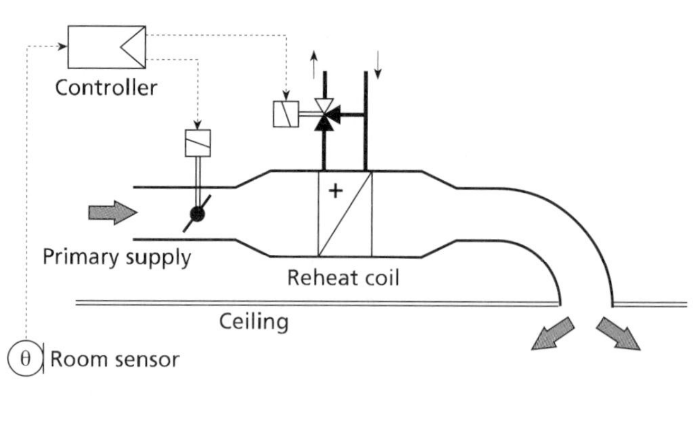

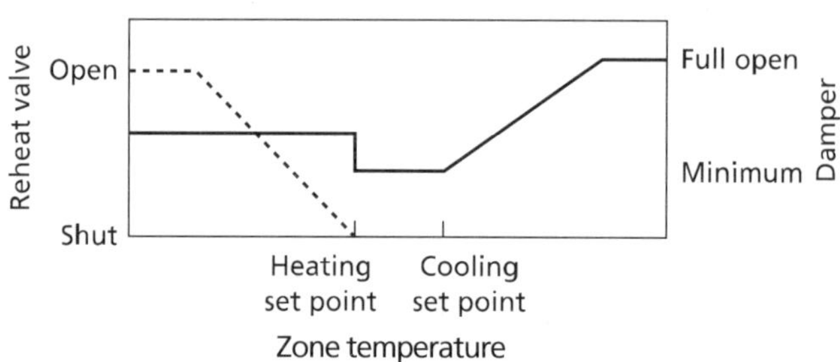

Figure 5.66 Pressure dependent VAV terminal unit. Unit shown with optional reheat coil fitted. Normal operation is shown in the diagram. Additional modes are: *primary supply damper and reheat valve open* — boost and OSH heating; *supply damper open, reheat valve to bypass* — OSC and night cooling, plant shut-down

5.8.1 Room terminal units for VAV systems

VAV terminal units are fitted with their own local controller; the control functions for the various types of VAV terminal are described below. The term intelligent VAV terminal has no precise definition. At the very least, it implies that the controller is connected to the main BMS network, allowing common reset of temperature set point for groups of terminals and the linking of the terminal damper control to the operation of the central AHU. However, the term intelligent is better reserved for controllers which have full exchange of information with the BMS, including feedback of damper position.

5.8.1.1 Pressure dependent VAV unit

A simple VAV unit is shown in Figure 5.66. Flow through the unit is controlled by a throttling damper, whose position is controlled by a proportional controller as a function of room temperature. If a reheat coil is fitted, it is brought into action when the room temperature falls below the heating set point, using a PI controller to modulate the valve controlling the hot water supply to the coil. An electric heater battery may be fitted in place of the reheat coil. The control characteristic is shown in Figure 5.66. When the room temperature is below the cooling set point, the damper is held at a minimum opening position to ensure sufficient air supply for ventilation and room air distribution. The air supply to the space at a given damper position is a function of the static air pressure in the supply duct, which varies with the state of the system. A signal may be provided to the BMS when the damper is open.

5.8.1.2 Pressure independent VAV terminal

The static air pressure in the supply duct to the VAV terminal will in general vary with the operation of other terminal units in the system and any consequent control of the main supply fan. A pressure independent VAV terminal, also known as variable constant volume, delivers a supply of air to the conditioned space which is a function of space temperature only and not of the supply air pressure. The control characteristic is shown in Figure 5.67. Proportional control of air supply as a function of space temperature is achieved using a cascade controller. The first stage of the controller resets the air flow set point as a function of the measured space temperature using a P or PI controller. A

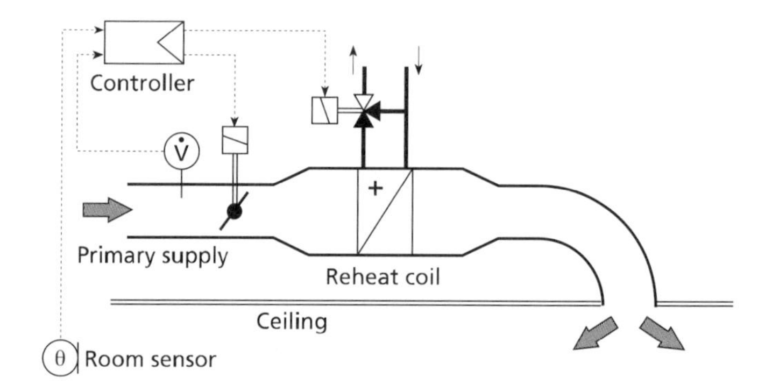

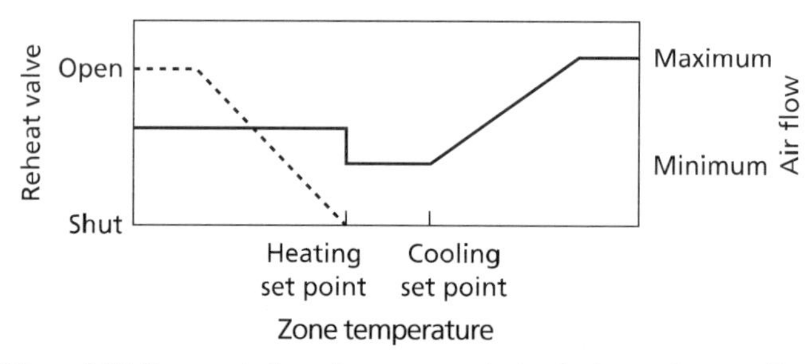

Figure 5.67 Pressure independent VAV terminal unit. A cascade controller is used to reset the air flow as a function of zone temperature. Normal operation shown in diagram. *Damper and reheat valve open*: boost and OSH heating. *Damper open, reheat valve to bypass*: plant shut-down, OSC or night cooling

second PI controller is then used to operate the damper to produce the required air flow. The VAV box incorporates a differential pressure sensor for the measurement of air flow. Intelligent damper actuators are available which incorporate a differential pressure sensor, controller and actuator in a single unit.

5.8.1.3 Induction VAV terminal unit

The induction unit combines cool air supply from the central AHU (primary air) with warm return air from the conditioned space (secondary air). The VAV box is located in the ceiling plenum of the zone, where an adequate supply of return air is available for recirculation. A diagram of an induction unit is shown in Figure 5.68. A pressure independent type controller is shown. The flow rate of the primary air is controlled as a function of zone temperature by operating the primary inlet damper, as described for the standard pressure independent VAV unit. The induction dampers are moved simultaneously in the opposite direction, so that as the primary air supply is reduced the amount of recirculated air is increased. The total volume of air discharged to the room is roughly constant with load, giving improved room circulation compared with a simple VAV terminal. Induction systems require increased main duct pressure to accomplish effective entrainment and so result in higher operating costs compared with the basic VAV unit.

5.8.1.4 Fan-assisted VAV unit: series fan

The fan-assisted VAV box also mixes cool supply air with recirculated air from the conditioned space. A small continuously operating fan in the VAV box is used to mix the primary and secondary air and to ensure good air distribution in the zone. The improved circulation is of value at low cooling loads and during heating duty if a reheat coil is fitted. The air supply to the space is at constant volume and at a temperature which varies with load. A diagram of a unit is shown in Figure 5.69; the fan serves to draw air in from the return plenum, mix the two air supplies and overcome any downstream pressure drop. The control characteristic is shown for the example of a pressure independent unit, where the primary inlet damper is modulated to maintain the primary air flow set point, which is reset as a function of room temperature by a cascade controller.

5.8.1.5 Fan-assisted VAV unit: parallel fan

A parallel fan box is sometimes used where heating performance is of importance. A diagram of the unit is shown in Figure 5.70. Parallel boxes rely on the primary cooling air supply for effective distribution during the cooling season, when the fan is switched off. When heating is required, or when cooling needs are low, the fan operates to recirculate secondary air from the room and to provide better air distribution.

5.8.2 Supply fan control for VAV systems

The central fan in a VAV system supplies conditioned air to a number of VAV room terminal units. Each room terminal

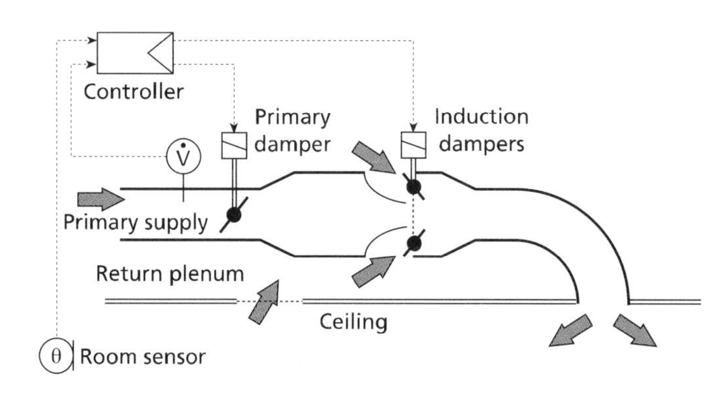

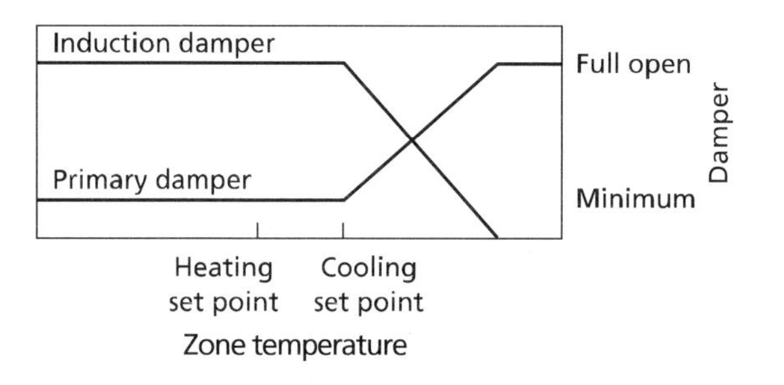

Figure 5.68 Induction VAV terminal unit, pressure independent. No optional reheat coil shown. Normal operation shown in the diagram. The primary damper is modulated to control the primary supply at the set point, which is reset from room temperature. Where a reheat coil is fitted, control is as for simple pressure independent VAV box. Other modes: *Primary damper open, heating valve open*: OSH and boost heating. *Primary damper open, heating valve closed*: plant shut-down, fan failure, OSC and night cooling

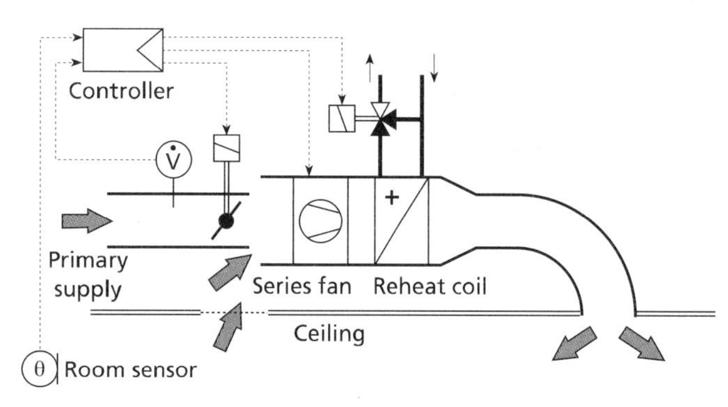

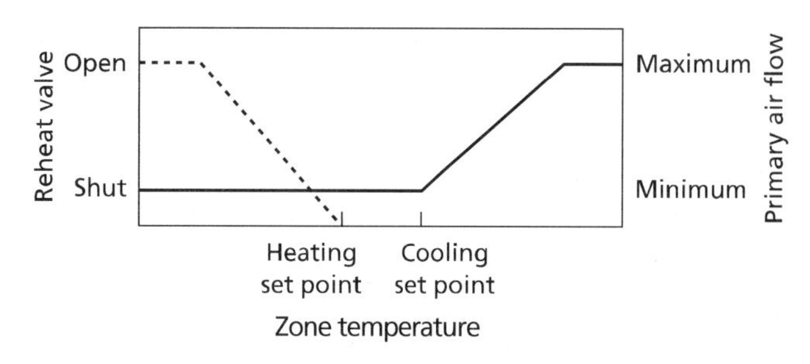

Damper	Fan	Heating valve	Operating mode
Open	On	Open	Boost, OSC heating
Open	On	Closed	OSC or night cooling
Open	Off	Closed	Plant shutdown, fan failure

Figure 5.69 Fan-assisted VAV terminal unit — series fan, pressure independent. Fan runs continuously. Normal operation shown in diagram, other modes in table

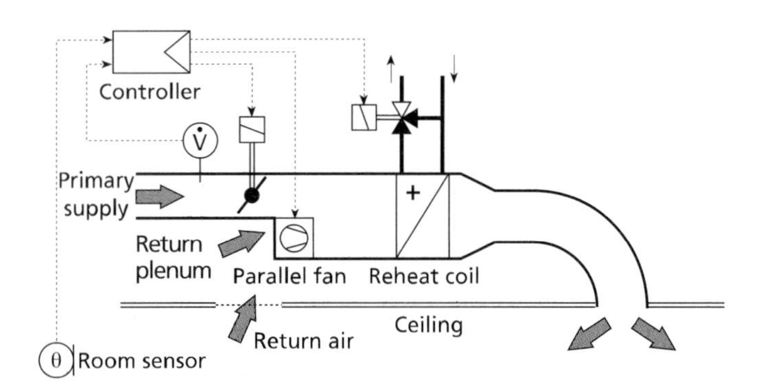

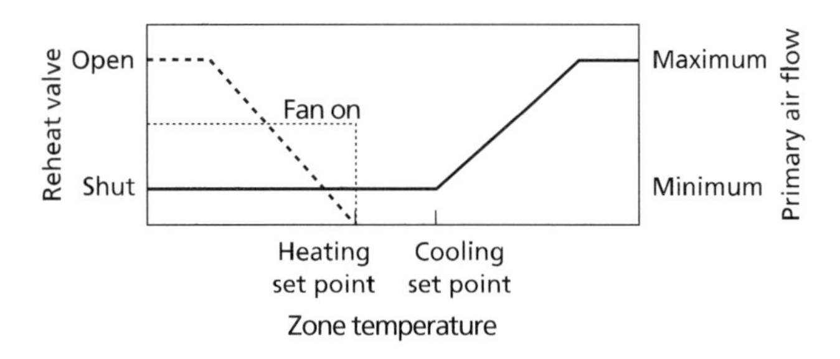

Damper	Fan	Heating valve	Operating mode
Open	On	Open	Boost, OSC heating
Open	Off	Closed	OSC or night cooling
Open	Off	Closed	Plant shutdown, fan failure

Figure 5.70 Fan-assisted VAV terminal unit — parallel fan, pressure independent. Normal operation is shown in the diagram. Fan operates during heating only. Other modes shown in table

unit (RTU) supplies air to a room or zone and controls the temperature in the zone by modulating the supply of air that it delivers. The central supply fan is therefore required to supply a varying volume of air, whose quantity is determined by the independent actions of a number of room terminal units. The use of a constant-speed supply fan would produce an increase in duct pressure as room units modulated towards closed, resulting in an increase in energy consumption and a risk of damage. It is therefore universal practice to modulate the supply fan in a VAV system to match the demand of the terminal units. The supply fan may be controlled by a number of methods. Variable-speed drive is the most efficient and its use is becoming widespread.

Several control strategies for fan speed control are in use. For simple layouts, a single measure of duct static pressure may be used to control the supply fan. As the complexity of the duct layout increases, it is necessary to use several sensors to ensure that all branches receive adequate air supply and do not suffer starvation in some part load conditions. Where intelligent VAV boxes are used, it is possible to monitor the condition of all room terminal units and modulate the fan for optimum operation.

Single pressure sensor

The supply fan speed is controlled using a static pressure sensor positioned in the supply duct (Figure 5.71). Fan speed is controlled using a PI controller to maintain duct static pressure at the set point. The room terminal units modulate towards closed as the cooling load lessens; this causes the duct pressure to rise and the fan speed is then

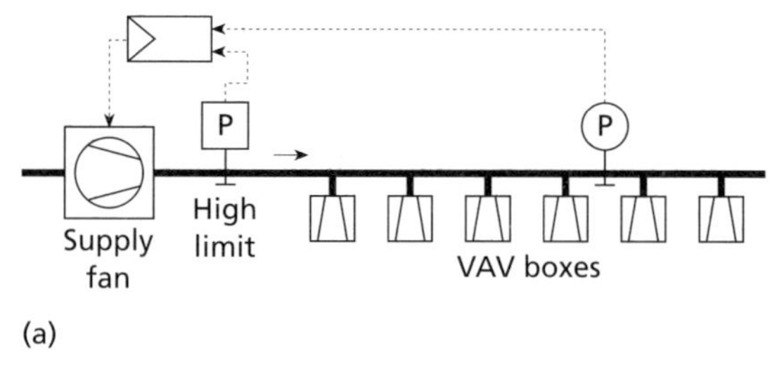

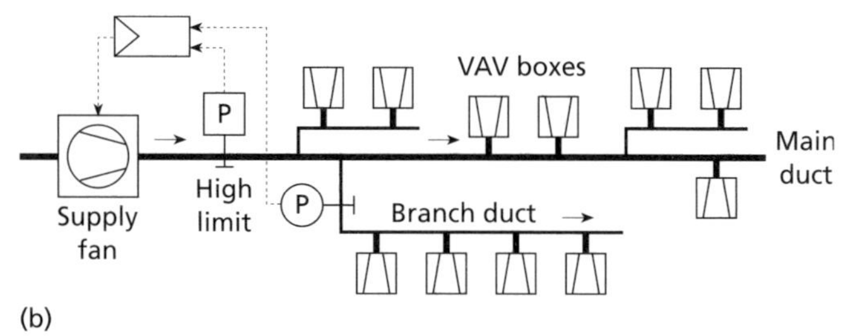

Figure 5.71 Control of VAV supply fan using single duct static pressure sensor. (a) Single main duct. Supply fan speed is controlled to maintain a constant static duct pressure at a position two thirds of the way down the duct. (b) Several branch ducts. The single sensor is positioned at the head of the branch with the largest pressure drop along it

reduced to maintain the pressure. Choosing the best position for the pressure sensor presents some difficulties. Where the terminal units are positioned along a single duct, it is possible to use a single sensor. Positioning the sensor towards the end of the duct remote from the supply fan gives the most efficient control. The pressure set point is chosen to give sufficient pressure at the RTU in design conditions to overcome the pressure drop in the terminal and air diffuser; because of the pressure drop down the duct due to duct loss, terminal units upstream will always operate at a higher pressure. However, placing the sensor at the far end of the duct may give rise to excessive variations in pressure following disturbance by events such as opening and closing doors, and a position about two-thirds of the way along the main duct is conventionally adopted as a satisfactory compromise. Where the main duct feeds branches and a single pressure sensor is to be used, the sensor should be placed just before the branch which has the highest pressure drop in design conditions.

Multiple pressure sensors

In general, however, a single pressure sensor is inadequate for a branched duct layout. If the independent operation of the RTUs causes the relative pressure drop in the branches to differ notably from design conditions, a branch upstream of the sensor may risk starvation. The problem may be overcome by using a pressure sensor for each branch (Figure 5.72). The pressure sensors are placed about two-thirds of the way down each duct; placing the sensors at the head of the branches will often produce satisfactory results. The controller is programmed to adjust the fan speed so that all pressure set points are reached or exceeded; this ensures that all branches receive an adequate air supply.

Intelligent VAV boxes

Where intelligent pressure independent VAV room terminal units are used, it is possible for the BMS to obtain

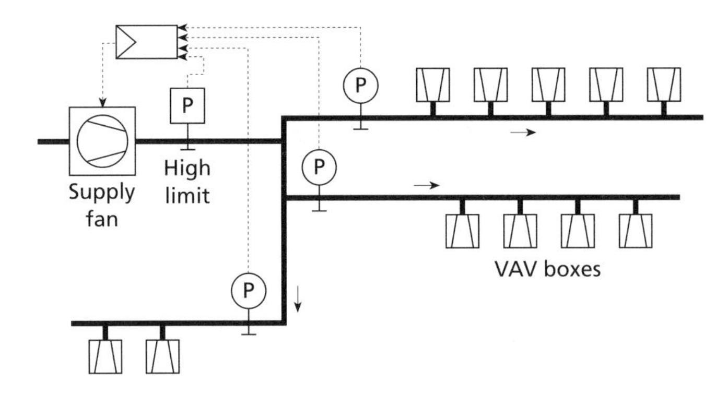

Figure 5.72 VAV supply fan control: multiple branches and sensors. Supply fan speed is controlled to ensure that all pressure sensors achieve at least the set point pressure. The signal to the PI controller is the largest negative deviation from set point

information on the setting of each individual RTU and to use this information to calculate the required fan speed setting (Figure 5.73). The pressure sensors are positioned at the head of each branch. The controller uses a pre-programmed knowledge of the duct layout and the position of the pressure sensors. The RTU setting gives the required air flow at each terminal, from which it is possible to calculate duct pressure drops and find the static pressure at the sensor positions which are necessary to feed each branch. To avoid overloading the communications net, the pressure set points are reset every 15 minutes. The PI controller then controls the fan speed to satisfy these requirements.

An alternative approach is possible where intelligent VAV boxes are used, whereby the fan speed is controlled so that the terminal units operate towards the middle of their range. The controller employs a form of floating control, which increases fan speed if the controller finds too many boxes open and reduces fan speed if it finds too many closed. More precisely, the strategy is[18]:

— The controller reviews the positions of the dampers in all the VAV boxes.

— If any box has a damper position of more than 90% open, the fan speed is gradually increased until all dampers are below 80% open.

— Else if every box damper is below 60% open and at least one VAV damper is at or below its minimum setting, then gradually reduce the fan speed until at least one VAV damper opens to 70%.

— Else leave the fan speed unchanged.

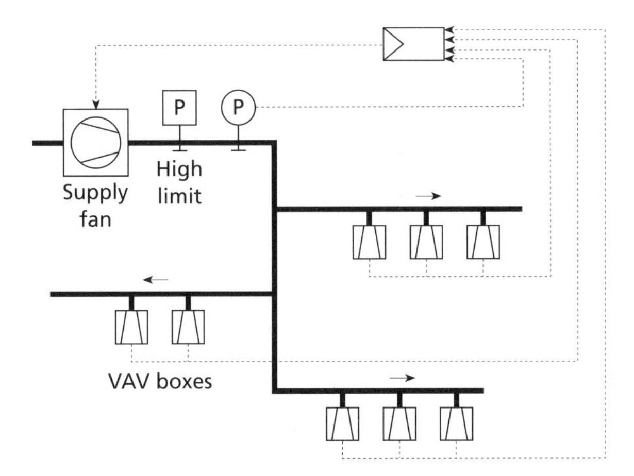

Figure 5.73 VAV supply fan control: intelligent VAV boxes. The controller sums the demand from all individual room units and resets the duct static pressure set point

Whichever sensor mounting position is used to control fan speed, it is necessary to mount a high limit pressure sensor in the supply fan discharge. This provides protection against the development of high pressures in the duct, which could result from a duct blockage. On receipt of a high pressure signal, the supply fan is disabled and a warning sent to the BMS. For all the above control systems, the supply fan is disabled on plant shut-down. The dampers in the RTUs should be left open when the system shuts down, so that they are open when the supply fan is started; intake and exhaust dampers should be open on start-up.

5.8.3 Supply air temperature control for VAV systems

Several levels of control are available for VAV systems. The most straightforward is the provision of cooling only; the room terminal units modulate to provide room cooling and any necessary heating is provided by a separate system, typically water radiators. In all but the simplest of systems, the speed of the main supply fan is linked to demand; this provides economic fan operation at times of low demand. Additional energy savings can be achieved by resetting the temperature of the supply air according to load, e.g. by raising the supply air temperature at times of low cooling load. Two major control strategies are available. The first requires the use of intelligent room terminal units which can signal the position of their dampers to the BMS; the second links supply air temperature to fan speed.

5.8.3.1 VAV supply air temperature reset with intelligent terminal units (BSRIA 3.14.2)

The control strategy supplies air from the central air handling plant at an optimum supply temperature, aimed at minimising the supply air flow rate and the amount of terminal reheating; the strategy is described in Martin and Banyard[37], Section 3.14.2. The room terminal units act to control room temperature in the normal way, employing a local control loop to modulate the damper positions and to use reheat where room heating is required. The intelligent terminal units signal damper positions to the BMS; in particular, they signal when any dampers have reached predefined maximum or minimum opening positions. Control of the supply fan speed is by one of the standard methods, such as duct static pressure. The supply air temperature is to be controlled between predefined maximum and minimum temperatures, typically 22°C and 12°C. The basis of the control strategy is as follows:

— All, or nearly all, zones require heating: the supply air temperature set point is gradually raised until one damper has closed to its minimum opening position.

— All, or nearly all, zones require cooling: the supply air temperature set point is gradually reduced until one damper has closed to its minimum opening position. The set point is reduced further if any dampers are at maximum opening position.

— Zones require a mixture of heating and cooling: in this situation, control of the cooling set point takes precedence over that of heating. The air temperature set point is adjusted so that the terminal units supplying the greatest cooling load are at the maximum open position.

The interaction of the RTU dampers, supply fan speed control and supply air temperature reset presents a risk of instability. The control of supply air temperature should be slow acting and should follow changes in the other two control subsystems.

5.8.3.2 VAV supply air temperature reset scheduled to fan speed (BSRIA 3.14.3)

This strategy can be used where information on the RTU damper positions is not available. It is generally applicable and can be used where the RTUs incorporate reheat coils. However, in this situation the control strategy of the RTU should maintain the dampers at the minimum open position when the unit is supplying heating; some strategies increase the air flow though the RTU during reheating to ensure adequate air circulation. The object of the strategy is to adjust the primary air supply temperature so that the main supply fan operates within its efficient speed range, between predetermined high and low set points.

During normal operation, the supply fan speed is compared with the upper and lower speed limits and adjusted if outside the desired operating range.

— If the fan speed is above the upper limit, the system is having difficulty meeting the cooling load. The supply air temperature is gradually reduced. This has the effect of reducing damper opening in the RTUS with a consequent reduction in supply fan speed. The reduction in supply air temperature continues until the fan speed falls below the high limit or the minimum air temperature set point is reached.

— If the fan speed is below the lower speed limit, the supply air temperature is gradually increased until either the fan speed is above the lower limit or the supply air temperature set point reaches its maximum value.

The interaction of the RTU dampers, supply fan speed control and supply air temperature reset presents a risk of instability. The control of supply air temperature should be slow acting and should follow changes in the other two control subsystems. The following rules will assist stability:

— On starting the main supply fan, or changing to normal operation following optimum start, night cooling or boost operation, a period of at least 15 minutes should be allowed for the system to settle before initiating supply temperature set point control.

— Before initiating a change in set point, the fan speed should have been above or below the limits for a predetermined time delay period.

— The change in set point should be gradual. Steps of 0.5 K every 10 minutes is typical.

— Control of the high and low speed limits should incorporate a differential to avoid undue cycling.

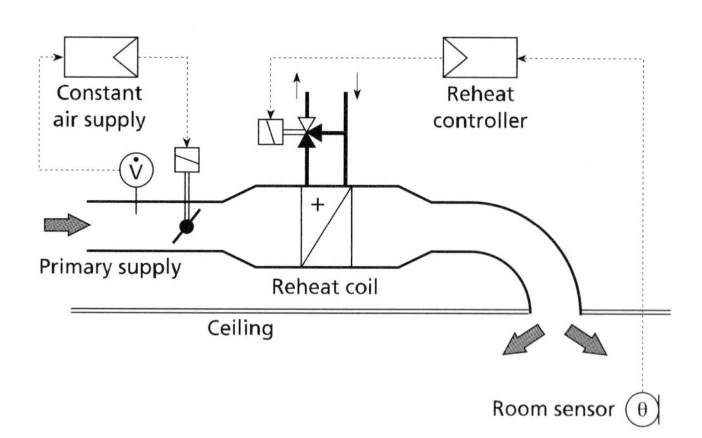

Damper	Heating valve	Operating mode
Open	Open	OSC and boost heating
Open	Closed	Plant shut-down, OSC and night cooling

Figure 5.74 Constant-volume air terminal unit, pressure independent. *Normal operation*: damper modulates to provide constant design air flow. Heating valve modulated to maintain space temperature. Other modes in table

5.9 Constant-volume room terminal units

The central plant of a constant-volume air conditioning system supplies conditioned air to all terminal units. The constant-volume RTUs supply a constant volume of air to individual zones. Correct balancing of the air supply of the RTUs during commissioning ensures that the zones remain in balance. Control of conditions as the overall load changes is performed at the central plant. The RTUs may be equipped with reheat coils for the adjustment of individual zone temperatures.

Individual RTUs therefore require adjustment to balance the system and to ensure that they deliver the required constant air supply. Self-adjusting mechanical volume regulators may be used but is generally better to use positive control using an air flow sensor and damper. A schematic diagram of a constant-volume RTU is shown in Figure 5.74. Air supply is maintained constant by a PI controller operating the damper, irrespective of pressure variations in the main supply duct. If fitted, a reheat coil comes into action when the room temperature falls below the heating set point. There is no local control of cooling or humidity, which is carried out at the central plant. Separate controllers are shown in the diagram to emphasise that there is no input of zone temperature to the air supply controller, in contrast to a VAV unit. The controllers are connected to the main BMS for such functions as optimum start, boost heating and shut-down.

5.10 Fan coil units

The use of water as a distribution medium for heating and cooling offers considerable economies of space compared with the use of air. Fan coil units provide local temperature control, using a fan to recirculate room air over water/air heat exchange coils. A wide range of units is available, typically fitted under windows or in a suspended ceiling; other variations are possible. Fan coil units (FCUs) control space temperature by recirculating room air and provision must be made to ensure a supply of outside air for ventilation. This may be done independently of the FCUs. For more sophisticated systems, a central unit conditions outside air to suitable temperature and humidity and supplies it to the FCUs, where it is mixed with the recirculated air. The necessary distribution ducts are much smaller than those required for an all air system. Fan coil units are not suitable for accurate humidity control. Some control of latent load can be achieved by dehumidifying the supply of outside ventilation air and the cooling coil in the unit can handle some latent load, in which case provision must be made to drain the condensate.

Control of the output of a fan coil unit may be by:

- *Fan speed control.* Water is circulated at a constant temperature in the coil of the heat exchangers and the fan speed modulated according to demand. This method may cause disturbance to occupants, since changes in noise level are more noticeable than the noise from a fan running at constant speed; there may also be noticeable changes in local air movement. If variable fan speed control is used, the controlling temperature sensor should be mounted in the controlled space, since a sensor mounted in the return air path in the unit will give an unsatisfactory indication at low or zero fan speeds. A manual boost fan speed may be provided for rapid temperature change.
- *Waterside control.* The rate of flow of hot or chilled water is controlled using a two- or three-way valve.
- *Airside control.* Water is circulated through the heating and cooling coils in the unit. The air flow path through the unit is controlled by internal flaps which direct the air over the coils according to demand.

The water circuits may be classified into two groups:

- *Two-pipe water systems.* A single supply of water is available at the FCU. This may be either hot or cold depending on season, but it is not possible for different units within the same zone to heat and cool simultaneously.
- *Four-pipe systems.* Both hot and chilled water are available to the FCU. The unit may employ separate heat exchangers for heating and cooling, or a single heat exchanger coupled with the use of a changeover valve. The type of heat exchanger will have implications for the selection of water supply temperature.

5.10.1 Fan coil units. Waterside control

Figure 5.75 shows a four-pipe unit which incorporates ventilation air supply from a central AHU. The fan runs continuously during normal operation. Zone temperature is controlled by modulating the supply of hot or chilled water to the coils using three-port valves; heating and cooling are never provided at the same time. Multi-speed fans are commonly fitted to fan coil units, with the appropriate speed for each unit selected during commissioning. A higher speed may be selected by the controller during boost operation or to supply a high demand.

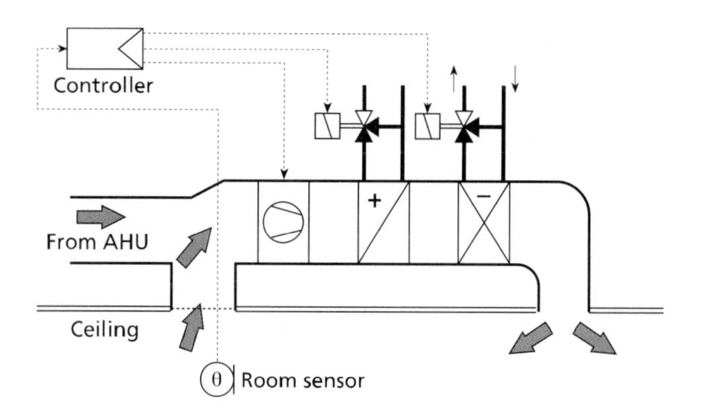

Operating mode	Fan	Cooling valve	Heating valve
Normal (heating)	Low/high	Off	Modulate
Normal (cooling)	Low/high	Modulate	Off
Boost, optimum start heating	High	Closed	Open
Optimum start cooling	High	Open	Closed
Shutdown	Off	Closed	Closed
Low outside temperature, low return temperature	Off	Open	Open
Fan overrun	High	Closed	Closed

Figure 5.75 Fan coil unit, four-pipe waterside control. Fan runs continuously and coils modulate to control room temperature. Optional ventilation air from central AHU. Other modes in table

There are several variations in the type and control of FCUs. Control of a four-pipe unit may be by a simple heating/cooling thermostat with central deadband, which operates an on/off solenoid valve to the coils. Fan coil units are in widespread use and can operate successfully under local control. Locally controlled fan coils have found widespread use in such applications as hotels and multi-cellular offices, where the units are accessible and local control by occupants is satisfactory. The effective control of fan coil units in large buildings and spaces is greatly facilitated by the use of intelligent fan coil units, fitted with controllers that communicate via the BMS network. This has additional practical advantages where units are not readily accessible, e.g. when mounted behind a suspended ceiling, since adjustments to set points and balancing may be carried out from the head end without requiring access to the unit. The use of fan coil units controlled via the network allows:

- incorporation of the FCUs in optimum start programs
- remote adjustment of set points, individually or in groups
- flexible time scheduling of operation
- linking to occupancy detection
- simple adjustment of control parameters.

Since heating and cooling are never employed simultaneously within a single FCU, it is possible to use a single coil which is used for both heating and cooling duty. Such four-pipe single-coil fan coil units require the use of special three-way valves on both the supply water side and return water side of the single coil. When the thermostat calls for heat, the valves to the chilled water supply and return circuits remain closed, while the supply valve hot water port is throttled to provide heating control. There is a deadband between heating and cooling operation.

Two-pipe fan control units are supplied with either LTHW or chilled water; the selection is made centrally and it is not possible for different units to provide simultaneous heating and cooling. Operation of the two-pipe FCU is controlled by an air thermostat, mounted in the return air path of the unit or else mounted remotely in the room. The thermostat is a heating/cooling type and requires a changeover signal. This may be provided remotely or from a pipe-mounted thermostat on the supply water pipe, which is used to change the operational mode from heating to cooling. The thermostat controls a motorised two- or three-port valve admitting water to the coil. The fan runs continuously during operation; the fan may be switched by central time switch or locally.

5.10.2 Fan coil units. Airside control

The operation of a four-pipe FCU with air side control is shown in Figure 5.76. Outside air for ventilation can be supplied directly to the room or, less commonly, for distribution via the FCU. Water supply to the coils is not modulated and the coils use two-port valves with open/close control. Output from the FCU is modulated by varying the air path over the coils by the use of dampers, which adjust the proportion of air flowing across each coil. The dampers are interlinked so that the air does not flow across both coils at the same time. Airside FCUs may be chosen in preference to waterside units where it is desired to avoid the large number of small water control valves required for waterside units. Where waterside units are installed, it is important to take care over the flushing procedures to ensure that no debris is left which could lodge in the valves; isolation of the valves during flushing is recommended.

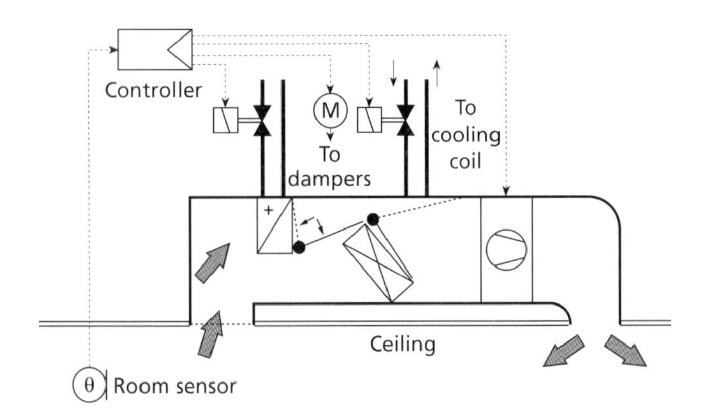

Figure 5.76 Fan coil unit, four-pipe airside control. Fan runs continuously. Dampers are connected to provide air path for heating, bypass or cooling; heating mode shown. Room temperature control is by modulating dampers and on/off control of coils

5.11 Dual duct systems

Both heated and cooled air are circulated throughout the building in separate ducts. The air streams are combined at the room terminal to provide the required temperature control. Dual duct systems require a substantial amount of space within the building to accommodate the two ducts and in their simple forms have poor energy economy, since heated and cooled air streams are mixed to provide control. They have fallen out of favour and new systems are no longer specified. However, many examples of dual duct systems are still to be found and some possible control strategies are described below. By using variable-volume control of the supply fans and temperature reset of the two supply temperatures to match demand, substantial economies in energy consumption may be made over a simple constant-volume system. The major types of dual duct system are described below; many variations are possible. Control strategies for the extract fan and air recirculation follow normal practice.

5.11.1 Constant volume, dual duct, single supply fan

The central air handling unit uses a single supply fan which runs at constant speed, supplying air to the two supply ducts. The hot duct, often known as the hot deck, has its own heating coil and the cold deck an independent cooling coil. Each coil is controlled by its own temperature controller. Each room or zone to be controlled has a terminal unit consisting of a mixing box and coupled dampers, which control the proportion of hot and cold air admitted to the box from the supply ducts. The dampers are modulated by a zone controller and are coupled to provide a constant mixed air flow into the room, at the temperature required to maintain the zone set point. The supply fan runs at constant speed. This system has high energy consumption; some improvement may be made by resetting the hot and cold duct temperatures to match the required demand.

5.11.2 Variable volume, dual duct, single supply fan

With this type of dual duct system, the room terminal units have independent dampers on the heated and cooled air supplies. The RTUs provide a variable air flow to the controlled zones and at conditions of low load, both air flows are reduced to the minimum level required for ventilation supply. The main supply fan speed is controlled via pressure sensors in the two supply ducts. This operates in a manner similar to that used for VAV systems. A comparator control is used, giving priority to the lower pressure in the two ducts, to ensure that each duct has the sufficient pressure to deliver air to all zones.

5.11.3 Variable volume, dual duct, dual supply fans

This variation of the dual duct system offers the most control and greatest potential economy. Separate fans are used for the two ducts together with room terminal units that give independent control of the heated and cooled air supply to the room. The room terminal units control the zone temperature by providing the requisite amount of hot or cold air. The most energy efficient units do not mix the air supplies, but supply either hot or cold air as required. A diagram of the unit is shown in Figure 5.77. The units are termed variable constant volume. This implies that the air supply rate is varied under proportional control to meet the room temperature control requirements, but that the air flow is independent of variation in supply duct pressure.

The twin supply fans are controlled to give a fixed static pressure in their respective supply ducts, in exactly the same way as a VAV system (Figure 5.78). If all the room terminal units call for heating, the cold air dampers in the units close down and the cold deck supply fan reduces in speed to maintain the supply duct pressure at the set point. This gives savings in fan energy compared with a simple constant volume system. Further economies are made by resetting the hot and cold supply temperatures. The aim is to heat and cool the two air streams just enough to provide the heating and cooling required by the controlled zones. For this method of control intelligent room units are required which signal their damper positions back to the supply air controller via the BMS. The hot air supply temperature is reset so that just one hot air damper in the RTUs is fully open; if more than one damper is open the supply temperature is increased to provide the required heat. The equivalent strategy is used for the cold deck temperature. This control strategy is capable of providing the greatest economy in operation. However, in practice it is sensitive to sensor malfunctions; a single sensor error or a faulty thermostat will confuse the whole system. An alternative is to predict the required supply duct temperatures from parameters such as outside air and mean indoor air temperatures.

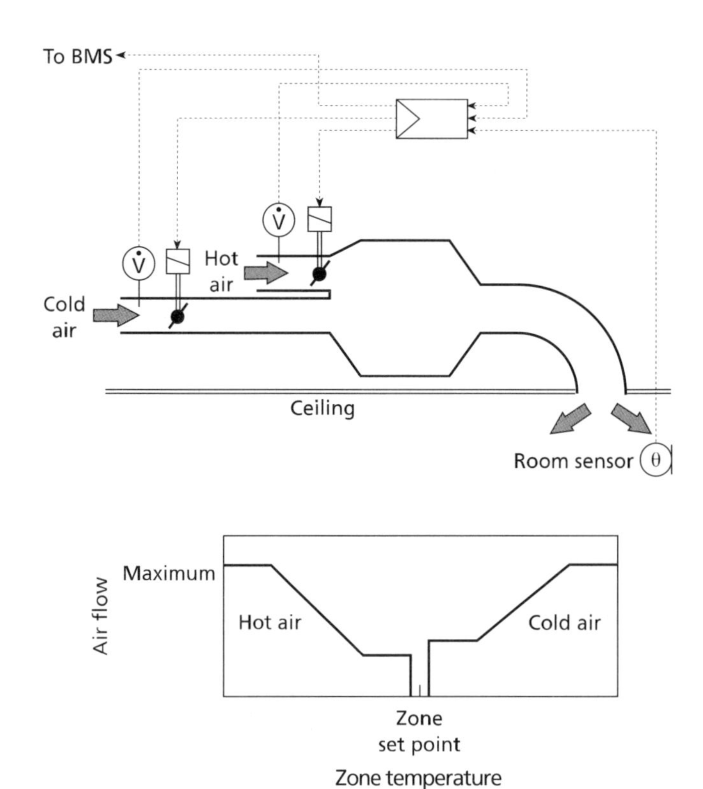

Figure 5.77 Dual duct variable constant volume room terminal unit. The controller modulates the hot and cold air flows to maintain zone temperature. No mixing of the air flows is allowed. Damper positions are made available to supply fan controller

5.12 Chilled ceilings

Chilled ceilings provide cooling by a mixture of radiation and convection. Panels present a large surface area to the room and have a substantial radiant component. Chilled beams employ natural convection to cool room air, which slowly descends into the occupied space. While some beams are available which combine cooling, heating and ventilation air supply, generally beams and ceiling panels provide cooling only and ventilation must be provided separately. The control strategy for chilled ceilings is governed by their limited cooling capacity and the necessity to avoid condensation occurring on the cold surface[38]. The ceiling panel and associated pipework run

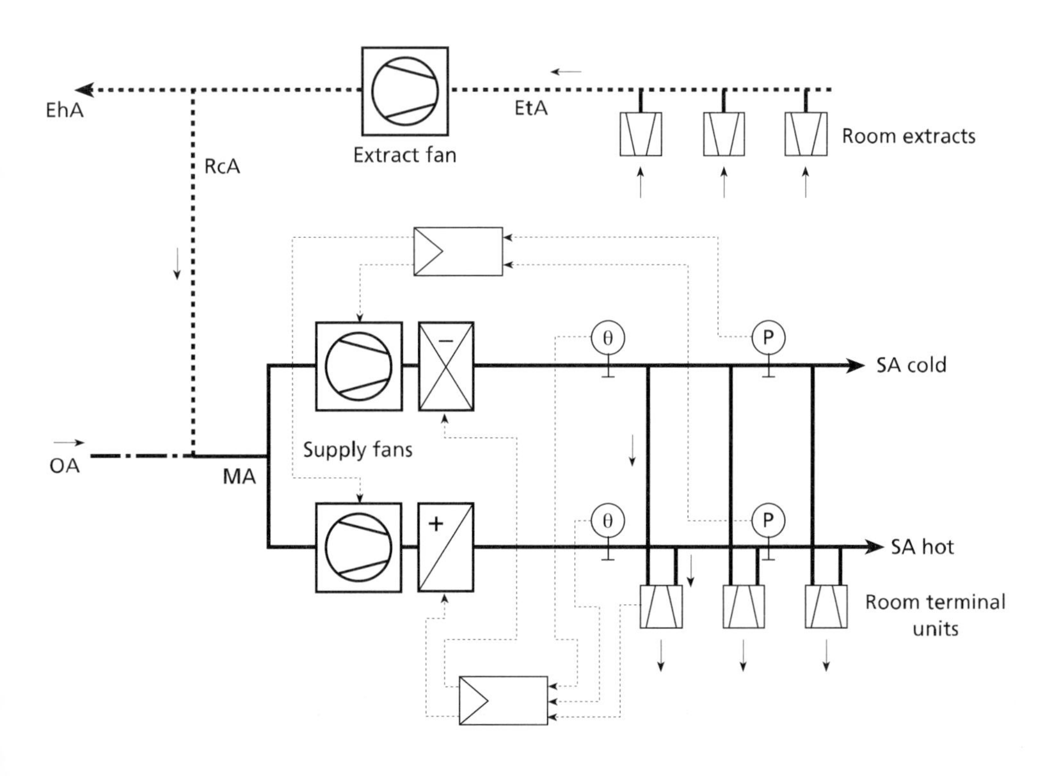

Figure 5.78 Variable volume, dual duct, dual supply fan. Fan speeds controlled to give constant duct static pressure. Duct heating and cooling controlled to satisfy room demands. Control of extract fan and recirculation follows conventional practice

several degrees below room temperature and condensation will form if the coldest surface, typically the incoming chilled water flow pipe, falls below the local air dewpoint. Unlike fan coil units, chilled ceilings have no provision for the collection or removal of condensate and the control system must ensure that no condensation is formed which could drip into the occupied space. Condensation is prevented either by:

— control of the zone air dewpoint by dehumidification of ventilation air

— reset of the chilled water flow temperature to maintain it above room air dewpoint by a small temperature differential, typically 2 K.

Dewpoint is measured using a RH and temperature sensor in the same enclosure and calculating the dewpoint with the appropriate algorithm. Whichever method of condensation control is used, there should be an additional sensor to detect the initiation of any condensation and stop the chilled water flow.

For systems using mechanical ventilation that incorporate dehumidification of the supply air, the dewpoint of the supply air is controlled so that the dewpoint of the zone air remains below the ceiling temperature. The minimum chilled water flow temperature for a ceiling is typically 14°C and the dewpoint of the zone air needs to be limited to 12°C. The energy implications of dehumidification should be considered at the design stage by estimating the frequency of high outdoor dewpoints during hot weather. Consider whether comfort criteria may be relaxed by allowing the chilled water flow temperature to rise on those occasions when the dehumidification plant is at maximum capacity and unable to meet the 12°C target. Use of a single internal dewpoint measurement position is normally suitable for mechanically ventilated buildings where latent loads are similar throughout the building. The sensor is normally situated in the extract air duct. If one part of the building is known to have a high latent load, that area should be used as the reference location. Where sensible and latent loads are non-uniform over the building and maximum cooling power is needed from the ceiling, zoning is required. The dewpoint should be measured in each independently controlled zone and the chilled water temperature adjusted accordingly. This will ensure that each zone receives the maximum potential cooling.

Where no dehumidification of ventilation air is available, the ceiling flow temperature must be controlled to be at least 2 K above the zone dewpoint. Where natural ventilation is employed, it is sufficient to control on external dewpoint, since the internal dewpoint will be close to the external dewpoint, unless there are significant sources of moisture vapour in the building. The chilled water flow temperature is controlled typically at 14°C, or reset to 2 K above the external dewpoint, whichever is higher. This will give relatively high ceiling temperatures in summer, with correspondingly reduced cooling power. The same strategy for controlling flow temperature is used for mechanical ventilation without dehumidification; dewpoint may be measured in the zone air or in the extract duct.

The small temperature differential of 2 K between dewpoint and water temperature is sensitive to measurement error; three sensors are involved in its computation. Good quality sensors must be used and be recalibrated at regular intervals. It is strongly advisable that a condensation detector be used which will disable the secondary circulation pump in the event of condensation being detected. This should be regarded as a safety mechanism and not used as a control method. The condensation detector is mounted near the flow connection of the ceiling panel, since this presents the greatest condensation risk.

The flow of chilled water in the ceiling is of course primarily used to control the temperature of the occupied space; the above paragraphs deal with the necessity to avoid condensation and emphasise the limited cooling power available from a chilled ceiling. Chilled ceilings are not suitable for rooms with high sensible or latent loads. Night cooling or optimum start cooling may be used to support the limited cooling capacity. Chilled ceilings are to some extent self-regulating, since the cooling power increases as the room temperature rises. The generally recommended method for zone temperature control is by on/off control of the chilled water flow to the ceiling panel, either by control of the circulation pump or by a two-port on/off valve. Better comfort control is achieved by modulating the flow temperature of the chilled water, subject to the above provisions of condensation prevention. Where chilled ceilings are used to cool individual rooms which have variable occupancy, a presence detector may be used to enable the circulation pump for each room; this is combined with separate dewpoint measurement for each controlled space.

Figure 5.79 illustrates the control strategy for a chilled ceiling system that does not employ dehumidification of ventilation air. It shows individual zone control, with control of chilled water flow temperature using a three-port valve mixing circuit. The cooling circuit is enabled when the space temperature rises above the cooling set point, typically 23°C. It is normally sufficient to use simple on/off control of the secondary pump for cooling control, since the system is partially self-regulating. Flow temperature is controlled at a fixed interval of 2 K above the zone dewpoint and the circulation pump is disabled if condensation is detected. Where both mechanical and natural ventilation are available, natural ventilation should be restricted or prevented while the chilled ceiling is enabled.

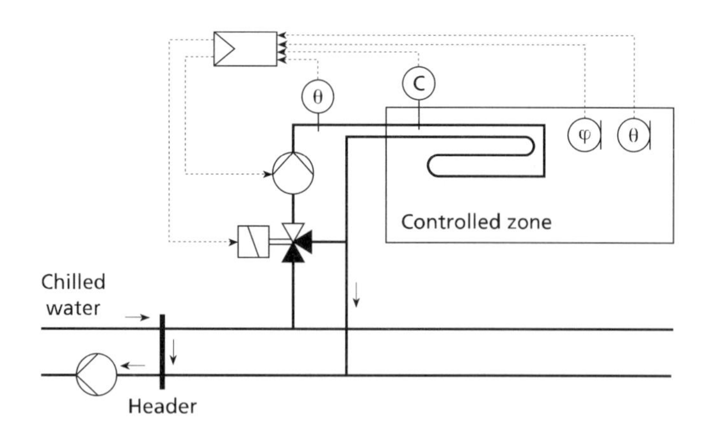

Figure 5.79 Chilled ceiling, with control of flow temperature on zone dew point. Secondary circulation pump is enabled when zone air temperature rises above cooling set point. Flow temperature is controlled at 2 K above zone dew point. If condensation is detected, pump is switched off

5.13 Heat pumps

5.13.1 VRF split systems

This type of system is characterised by a type of fan coil unit mounted indoors, containing a direct expansion cooling coil. The unit is connected by refrigerant lines to an outdoor unit containing the compressor, condenser and cooling fans. Many systems operate as a reversible heat pump and so can provide heating or cooling from the indoor unit as required. No air treatment other than heating or cooling is involved, so the system cannot strictly be described as air conditioning. Split systems come in a range of sizes and may be classified as:

— *Packaged split units*. These comprise a single room unit and outdoor unit. The simpler units provide cooling only; heating and cooling are available if the unit can operate as a reversible heat pump.

— *Multi-split unit*. Several room units are connected to a single outdoor unit. Individual room temperature control is possible, but not simultaneous heating and cooling from the same outdoor unit.

— *Variable refrigerant flow (VRF) systems*. Several room units are connected to a single outdoor unit, which has a variable-speed compressor to give variable load. In the more advanced systems, it is possible for different room units connected to the same outdoor unit to heat and cool simultaneously.

The distinction between multi-split and VRF systems is not clear cut and is one of size and sophistication. Large VRF systems may contain several hundred indoor units and be connected to the BMS. VRF systems have the advantage of requiring no central plant room and needing little space for installation; the room units are compact and are connected by refrigerant lines to the outdoor unit. A single outdoor unit can typically serve up to 16 indoor units; extended runs of refrigerant piping are possible and large buildings may be served by the appropriate number of outdoor units. Consideration must be given to the safety implications of the use of long runs of refrigerant pipework in occupied spaces. Refrigerant vapours are neither toxic nor inflammable, but permissible exposure limits are laid down. The design of the installation and use of the building must pose minimum risk of damage to the pipework. Each group of units served by one outdoor unit operates as an independently controlled system, but groups can be linked together as described below. Energy efficient operation occurs when indoor units within a group provide heating and cooling at the same time. This depends on the units being served by the same outdoor unit, so the design of a large system should take this into account, e.g. by grouping core and periphery, or north and south zones within the same group. Due regard must be paid to ventilation requirements, which have to be provided independently.

Controls for split units are provided by the manufacturer and can provide a hierarchy of control levels, from occupant control of individual indoor units up to full BMS capability (Figure 5.80). The outdoor unit of a multi-split unit is required to operate over a wide range of loads and normally incorporates a variable-speed drive compressor. The outdoor unit contains its own controller to adjust the refrigerant flow to the indoor units in response to demand. The modular nature of the system means that that an outdoor unit and its associated group of indoor units can continue operating under local control in the event of a high level control failure. While the details vary from manufacturer to manufacturer, the general principles are the same and are identified by the different controllers available.

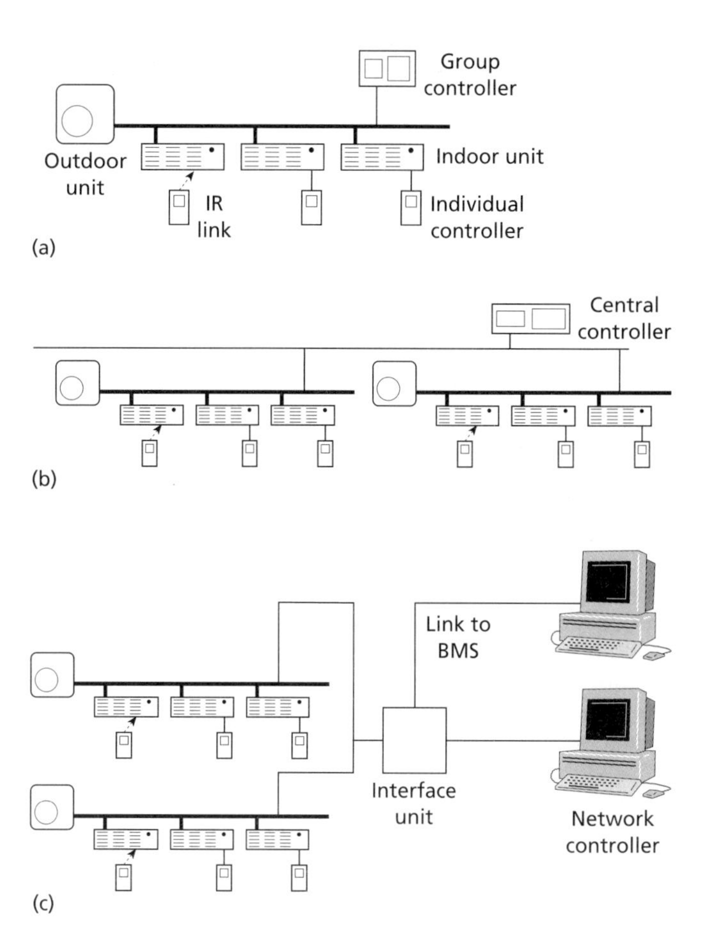

Figure 5.80 Control hierarchy of a VRF multi-split system. Large systems may be built up while retaining occupant control of individual units. (a) Group of VRF room units, showing individual room controllers and group remote controller. (b) Several groups of units under the control of a central controller. (c) Large system with PC-based control, interfaced with BMS

Individual room unit controller

This is a simple unit designed for use by the room occupant and may be wall mounted, or hand held operating via an IR link. It gives control of room temperature and fan speed only, plus on/off. It may be overridden by a higher-level controller.

Group remote controller

This provides control for a group of units connected to a single outdoor unit. If required, it can override individual room controller settings. The room units can be controlled together as a group or individually addressed. The group controller provides the following control functions:

— mode: on/off, heating, cooling

— time scheduling

— room temperature set point

— fan speed

— air direction.

In addition, the controller provides various checking functions, information displays and fault diagnoses. A group remote controller can operate as a stand-alone unit or be incorporated into a higher level system.

Central controller

This provides central control of up to a few hundred units. The units may be organised into groups and be operated from the central controller, which allows group settings of set points and time schedules. The individual room units retain their own local controllers, which may be used by the occupants unless overridden by the central controller. The central controller provides monitoring and fault diagnosis for all the units under its control.

PC-based controller

For large systems, the central controller is replaced by a PC-based operating system which can interface with thousands of individual indoor units. This provides the ability to organise the units into groups, set flexible time scheduling and set points and monitor performance and fault reporting. Functions such as optimum start may be incorporated. The PC-based system connects via interface units to the indoor units, or can operate via a layer of central controllers if it is desired to retain some redundancy. This supervisor may be linked to the BMS via a suitable interface. This provides the BMS with the information to monitor faults and temperatures and to carry out limited control of set points and on/off switching. Alternatively, for some systems it is now possible to obtain an interface which will provide full integration with the BMS.

5.13.2 Reverse cycle heat pump units

This system consists of a number of room units which are interconnected by an uninsulated two-pipe water loop. Each room unit contains a reversible heat pump, which can accept heat from or reject heat to the water loop. An integral fan recirculates room air to provide local heating or cooling as required.

Each room unit is supplied by the manufacturer complete with a control unit, which has a thermostat and fan speed selector. The thermostat selects the heating or cooling mode of the heat pump; there is a deadband between heating and cooling set points of about 3 K, within which the compressor does not operate. The units may be individually controlled by the occupants or arranged in a master–slave configuration where several units serve the same space. Central control of temperature setting and time scheduling is possible via the BMS. Damage to the units may result if they are operated with water loop temperatures outside prescribed limits and interlock switches are fitted to prevent this. Ventilation must be provided independently of the heat pump units. This can be by a central AHU, which filters and tempers the air before distributing it to the occupied space. Where possible, the ventilation air may be supplied near the inlets of the heat pump units, so that it will mix with the recirculated air; this is practical in the case of units mounted in the ceiling void.

The units must be provided with a circulating loop of water of controlled temperature. Figure 5.81 shows the basic control of the loop. The control system is required to ensure

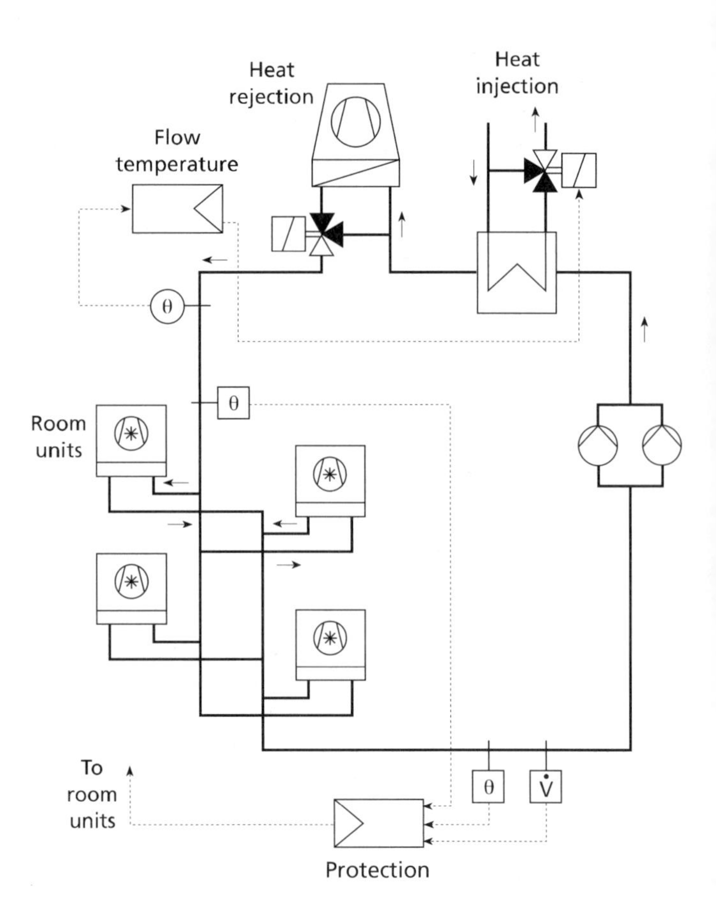

Figure 5.81 Unitary water to air reverse cycle heat pump system. Circulating water flow temperature is controlled at about 33°C; room heat pump units are switched off if water goes outside range 20°–43°C

that there is a flow of water maintained between limits of typically 20 and 43°C at all times that the room units are operating. Heat rejection from the loop is performed by a dry cooler or evaporative closed circuit cooler; an open cooling tower is unsuitable, since it is important to maintain high water quality in the circulating loop. For the same reason, heat injection to the loop is preferably via a heat exchanger, though direct injection from a boiler primary circuit is permissible. Heat injection and rejection never occur at the same time. The water loop is fitted with conventional pressurisation and make-up control; these are not shown in the figure. During normal operation, the heating and cooling valves are modulated to maintain a flow set point of 33 ± 2°C; this temperature gives efficient operation of the heat pump units, particularly in buildings with year round heat gains. In conditions of high heating or cooling demand, the flow temperature may move outside this control band. High and low limit thermostats are installed which switch off the heat pump units if the flow temperature falls below 18°C or rises above 45°C. Operation is restored at 22 and 41°C respectively.

The control sequence on start-up is as follows:

(1) Start water pumps.

(2) Energise heat injection circuit.

(3) Energise reverse cycle heat pump units.

Step 3 may only take place when the water loop is up to temperature. It is recommended that a limit thermostat be placed in the return of the water loop circuit. When this temperature reaches the required limit, of at least 20°C, the heat pump units are enabled and the limit thermostat is switched out of circuit. Alternatively, the control system may impose a time delay between energising the heat injection circuit and enabling the heat pump units. When operating the system under a time schedule control, this delay must be allowed for to ensure that the room units are enabled by the start of the occupied period. Where a central time switch is used, each unit should have a random time delay to prevent high start currents resulting from all units being switched simultaneously. When individual time-switches are used, there will normally be sufficient variation.

The control sequence on shut-down is:

(1) De-energise the heat pump units.

(2) De-energise the heat injection circuit.

(3) Turn off the water pumps.

5.14 Natural ventilation

This section deals with buildings where natural ventilation is an integral part of the environmental control system. One of the goals of natural ventilation design is avoidance of the need for air conditioning and such buildings are likely to feature:

- solar shading and fenestration design to reduce solar gains
- use of daylighting and high efficiency lamps to reduce internal gains from lighting
- high thermal mass construction with exposed internal mass to give a thermal flywheel effect
- design features to enhance natural ventilation flow, such as atria and ventilation stacks
- user involvement, often by openable windows.

The forces driving natural ventilation are low and variable, and the design of successful buildings presents some problems. The CIBSE Applications Manual AM10[39] sets out the principles of successful design; a summary is given in BRECSU GPG 237[40]. Complete control of natural ventilation by manual window opening under control of the occupants is possible only in small buildings. For larger buildings some form of automatic control is essential, with at least some of the ventilation control devices being operated according to the appropriate control strategy via the BMS. It is common for the occupants to have some manually operated vents at their disposal. Experience shows that occupants in naturally ventilated buildings are more tolerant of temperature variations than their counterparts in air conditioned buildings, particularly if there is some form of user control in the form of openable windows. Variation of internal temperature with external climatic conditions must be expected. Further information on natural ventilation and passive cooling may be found in Kolokotroni[41], Limb[42] and Allard[43].

5.14.1 Controlled devices

Controlled devices for natural ventilation consist primarily of modulating inlet and outlet ventilation openings[39]. Natural fluctuations in wind speed and direction give rise to variations in ventilation rate and make precise control difficult. When the ventilation devices have a large area relative to the space being ventilated, some form of modulating control will be required. For smaller openings it may be possible to use open/shut control, particularly if a number of openings may be operated in sequence. Consideration must be given to the performance of the openings in rain and wind. If adequate protection is not provided by some form of baffle or cowl, it will be necessary to restrict the opening position in bad weather, to avoid damage or rain ingress. Where natural ventilation is used during unoccupied hours for night cooling, consideration must be given to any compromise of building security presented by the open air inlets. It is common to provide opening windows under occupant control in addition to the vents operated by the BMS.

5.14.2 Measurement devices

Technical aspects of sensors are covered in section 3. Natural ventilation control has its own measurement requirements and some or all of the following may be required:

- *External air temperature*. This forms an essential input for natural ventilation control systems. The sensor should be mounted on the north face of the building and be provided with a radiation shield against solar and sky radiation.
- *Rain intensity sensors*. These are fitted so that windows and vents may be shut automatically to prevent rain ingress. For a large building it may be advisable to mount sensors at opposite ends to detect approaching rain.
- *Wind speed*. Windows and vents may be shut or partially closed at high wind speeds to avoid damage or to control over-ventilation. If combined with a rain sensor, wind speed sensors can be used to give protection against driving rain ingress.
- *Wind direction*. This can be used to select windward or leeward vents as appropriate.
- *Solar radiation*. A solar radiation sensor may be used to increase ventilation rates at times of high solar gain or to automatically control the operation of solar screens or window blinds.
- *Slab temperature*. The building fabric plays an essential part in passive temperature control and some natural ventilation control strategies use a measure of fabric temperature rather than internal air temperature. The sensor should be inserted to a depth of approximately 25 to 50 mm into the slab, packed with thermal paste and cemented in. As a less satisfactory alternative, a surface mounted sensor may be used which is covered with insulation. A position should be chosen on an internal wall or ceiling remote from temperature disturbances.
- *CO_2 concentration*. This can be used as a set point for ventilation control in addition to internal air temperature and is of value in buildings with a large but variable occupancy.
- *Occupation*. The number of people in the controlled space may be used to set the target ventilation rate. This is feasible where the building has controlled entry and exit; in an integrated system it may be possible to obtain information from the security system.

The low driving forces for natural ventilation preclude the use of mechanical filtration. It is therefore important that the air intakes are positioned in clean areas. In some situations, there may be times of day when external pollution is unavoidable, and the inlets should be controlled during these periods.

5.14.3 Control strategies

The aim of ventilation control is to provide sufficient outdoor air for ventilation purposes and to remove any excess heat gains. For any given ventilation opening, the rate of supply of outside air is greatly affected by external ambient conditions of wind and temperature and so the provision of a minimum ventilation opening does not guarantee adequate ventilation; on the other hand, in windy conditions, there may be excessive ventilation. Many naturally ventilated buildings have a large but variable occupancy and hence a changing need for ventilation. It is therefore common to use some form of demand controlled ventilation (DCV), normally employing CO_2 as the measured variable. Other air quality sensors have not found widespread application. The level of occupation may be used to estimate the rate of CO_2 production and hence the required ventilation.

In naturally ventilated buildings, heating is provided by an independent system, typically LTHW radiators. Natural ventilation is used to maintain satisfactory levels of CO_2 at all times and to provide cooling when required. Permanent trickle ventilation is provided so that the legal minimum openings are maintained when the ventilation control system has closed all controlled ventilation openings. An Approved Document (AD Part F1) of the *Building Regulations* sets the minimum size for the opening area in office rooms as 4000 mm^2 in rooms of up to 10 m^2 floor area, and 400 mm^2 per m^2 of floor area in larger rooms. The air flow through the ventilators is subject to wind pressure and some means of control is desirable. Hit and miss ventilators under occupant control may be provided, or automatic ventilators which self-regulate under different pressure conditions. Alternatively, background mechanical ventilation may be used to provide the minimum ventilation air supply. The use of night ventilation to dissipate heat gains acquired during the heat of the day and to precool the building fabric is an important strategy in avoiding the use of mechanical cooling and it is possible to keep the internal temperature below daytime external ambient temperature by this means.

Ventilation rates will inevitably fluctuate under the influence of changes in wind speed. Where modulating control of natural ventilation openings is provided, it will not be possible to use fast acting control loops to try and achieve more accurate control. Large differential bands should be used and it is not normally necessary to use PID or PI control loops[44]. Control of natural ventilation system can be described in terms of two strategies[45,46], one which operates during normal occupation periods the other which provides night-time cooling of the structure.

5.14.3.1 Occupation period

During warm weather when no heating is required, the ventilation openings are modulated by the control system to control the internal air temperature. During colder weather, when the internal zone temperature falls below the heating set point and the heating system is activated, the vents are set to their minimum position, providing sufficient air for ventilation only. When demand control ventilation is present, ventilation to maintain air quality takes precedence. The target air quality is typically a CO_2 concentration below a set point of 1000 ppm. The decision tree is shown in Figure 5.82. Depending on the complexity of the installed system, vent opening may be determined by wind direction to select the optimum choice of windward and leeward openings.

The decision tree does not show any special action to be taken when the outside temperature rises above the internal set point, other than to open all vents to the maximum. Bringing in outside air at a higher temperature than the internal zone temperature will result in a heat gain, and there is an argument that vents should be closed to minimise the ingress of warm outside air. However, this does not allow for the effect of internal heat gains which will tend to increase the inside temperature to a level above that outside. In general it will be more effective to use warm outside air for ventilation than to allow the gains to build up inside without ventilation. Where the building has openable windows under occupant control, users will usually open windows to the maximum in hot weather, nullifying any attempt to reduce ventilation. This will aid comfort by increasing local air movement. In practice it will be generally found adequate to maintain maximum ventilation in hot weather; this may have to be reconsidered in climates other than the UK[45].

5.14.3.2 Night cooling

The aim of night cooling is to use ventilation during the unoccupied period to dissipate heat gains absorbed by the building during the day. Buildings designed for natural ventilation typically have high thermal mass with exposed concrete slabs. These have a flywheel effect on temperatures, absorbing heat during the day and so preventing air temperature from rising too high. It is therefore advantageous to ensure that the slabs are cooled during the night. The design of a natural ventilation system is integral with the architectural design of the building. This extends to the control system and there are few off-the-shelf systems for general application. The recommended control strategy simply uses ventilation to bring the zone temperature down to the lower limit of the comfort band, i.e. as low as possible without bringing the heating system into action. More complex control strategies which bring the zone temperature below the comfort band in an attempt to maximise cooling have not proved worthwhile in practice[46]; see also Kolokotroni[41].

The night cooling system is enabled if:

— the mean outside air temperature over the period 1200 to 1700 > the night cooling (mean ambient) set point (typically 20°C)

or the average daytime internal temperature over the period 0900 to 1700 > the night cooling (mean zone) set point (typically 22°C)

or the slab temperature > the slab cooling set point (typically 23°C)

or peak zone temperature > peak set point (typically 23°C).

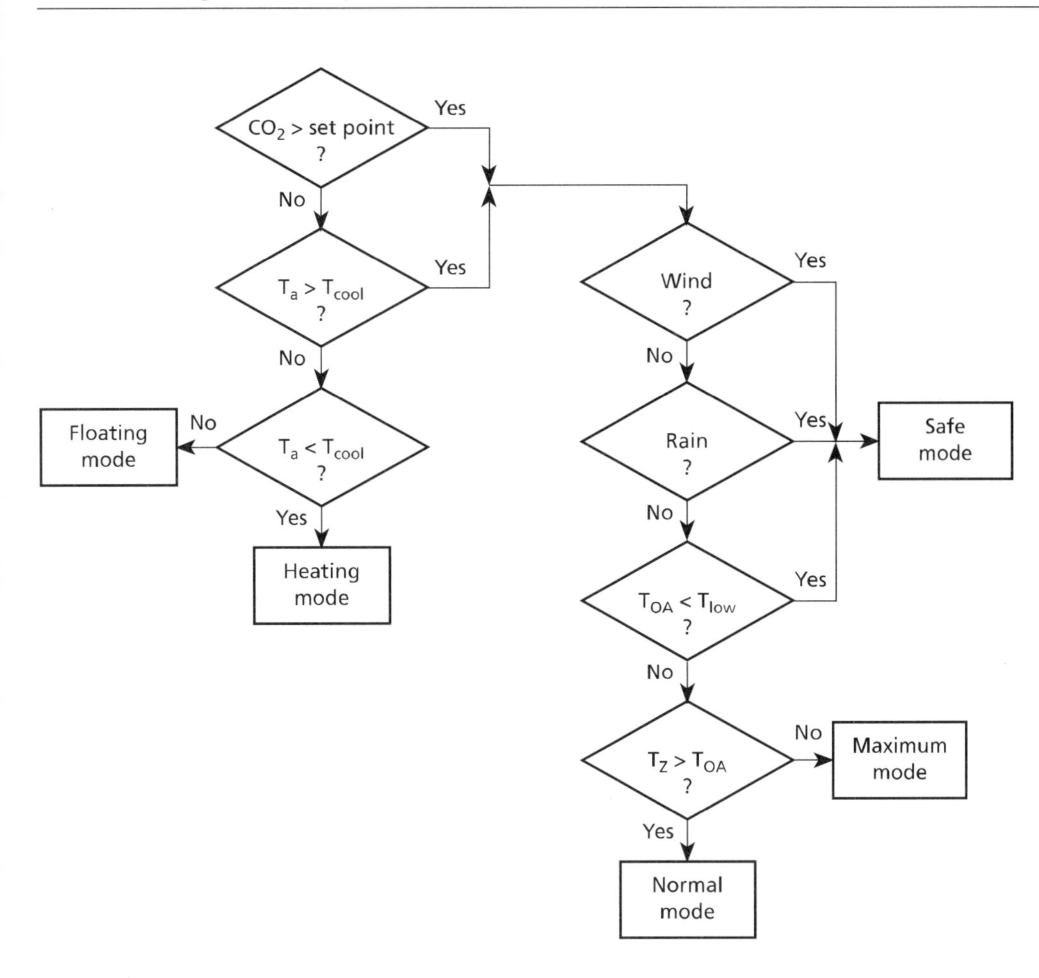

Figure 5.82 Decision tree for natural ventilation during occupied period

Mode	Description
Heating	Ventilation openings closed
Floating	Vents maintain position
Safe	Vents moved to safe position in wind etc.
Maximum	Vents fully open
Normal	Vents modulate to control T_Z and CO_2

Symbol	Meaning	Typical value
T_{OA}	Outside air	
T_Z	Zone temperature	
T_{cool}	Cooling set point	22°C
T_{heat}	Heating set point	19°C
T_{low}	Cold outside air	12°C
CO_2	Carbon dioxide	1000 ppm

If night-time cooling is enabled, during the non-occupied period the vents are opened for cooling if:

— zone temperature > outside air temperature

and zone temperature > heating set point (typically 19°C)

and minimum outside air temperature > 12°C

and no bad weather interlock.

The heating system should not be operated when night cooling is enabled. Night cooling should be available throughout the whole non-occupied period and for seven days per week. Because of the long time constant of the building structure, when the night cooling has been applied for a period of five days or more, it should be continued for a further two nights after the afternoon outside temperature has fallen below the night cooling set point.

5.14.4 Mixed mode

The term mixed mode implies a combination of natural ventilation with mechanical ventilation and possibly mechanical cooling. The general philosophy is that natural ventilation should be used to the greatest possible extent, with mechanical services employed only when necessary. By these means it should be possible to minimise energy consumption. Mixed mode has no precise definition and different types have been classified[45,47]:

— *Contingency*. This describes a building that has been designed primarily for natural ventilation, with the provision for the addition of mechanical services later if found necessary.

— *Zoned*. Some zones of the building are provided with mechanical services where there is a specific requirement. An example might be a building with naturally ventilated office spaces, a mechanically ventilated kitchen area and an air conditioned computer room.

— *Complementary*. Both natural ventilation and mechanical services serve the same zone. The systems may be operated in either changeover or concurrent mode. Concurrent systems often provide a basic level of mechanical ventilation to ensure a permanent draught free ventilation air supply, while allowing occupants to open windows as they choose. Changeover systems operate in distinct modes to satisfy different requirements. For example, windows may be openable in mild weather, but locked shut when mechanical plant is used for either heating or cooling. Changeover systems present practical problems of producing effective control strategies and ensuring that occupants are fully conversant with the different requirements of the various modes of operation.

The use of mechanical services to complement natural ventilation has two main functions:

— *Winter ventilation*. During the heating season, it may be difficult to control a natural ventilation system to provide a supply of ventilation air which is sufficient for ventilation purposes, but not so large that it causes excessive heat loss or uncomfortable draughts. It may be preferable to close the natural ventilation system and employ mechanical ventilation during the heating season.

— *Summer cooling*. The use of natural ventilation to control summer temperatures was described above. In some conditions of warm still weather, the natural driving forces may be insufficient to provide the required level of natural ventilation and this is supplemented by the use of mechanical ventilation. The same ventilation paths my be utilised, e.g. by positioning a fan in an extract stack.

The control strategy for mixed mode operation is therefore fundamentally the same as for natural ventilation. During night cooling, the strategy set out in 5.14.3.2 is adopted; mechanical ventilation is not used during night cooling. During winter heating, the natural ventilation system is closed and mechanical ventilation is used to provide the required ventilation supply. During summer daytime ventilation, the natural system operates preferentially as described in 5.14.3.1 above. Only if the zone temperature rises more than a predetermined amount above the cooling set point and all natural inlets and outlets are driven fully open is the mechanical ventilation brought in to supplement natural ventilation. In situations where the external temperature is significantly above internal air temperature, increased ventilation may result in additional heating of both the space air temperature and the building fabric, so reducing the beneficial radiant cooling from exposed ceiling slabs. If any zones are fitted with mechanical cooling, it will be necessary to shut down the natural ventilation openings and revert to mechanical ventilation before enabling the cooling system.

5.15 Lighting controls

Lighting is often the largest consumer of electricity in a building. In a naturally ventilated building lighting may account for 40% or more of the total electricity cost. The overall lighting design together with the choice of lamps and fittings pays the major part in determining the energy cost of the lighting system; further reductions can be achieved by the use of an effective lighting control system(48). Part L of the *Building Regulations*(49) requires the use of efficient light sources and specifically requires the use of controls, stating:

> where it is practical, the aim of lighting controls should be to encourage the maximum use of daylight and to avoid the unnecessary use of lighting when spaces are unoccupied.

The lighting conditions produced by the controls system must at all times provide satisfactory illumination for the occupants of the building. Safety aspects are dealt with by *Lighting at Work*(50). The CIBSE *Code for Interior Lighting*(51) states 'Any control system must ensure that acceptable conditions are always provided for the occupants. Safety and visual effectiveness and comfort must take priority over energy saving'. The control system must default to a safe condition if a fault develops and return to the pre-fault condition after the fault has been rectified. The emergency lighting system must not be connected to the lighting control system. While the major benefit of lighting control is usually seen in terms of energy saving, a good control system brings other important benefits:

— *Occupant satisfaction and productivity*. A good working environment is essential for comfort and productivity. The widespread use of display screen equipment has made lighting design more critical.

— *Planning flexibility*. The use of electronic controls allows redesign or reallocation of work layouts without alteration to fixed luminaires or wiring. Increased controllability allows for shift working and unpredictable work patterns.

— *Better management information*. A centralised lighting control system can provide information on lighting use by different zones, assist with maintenance and can assist in security when linked to presence detectors.

5.15.1 Control methods

Four basic methods of lighting control may be described, which may be used separately or in combination with each other. Not all types of lamp are suitable for dimming or frequent switching; this will be considered below.

5.15.1.1 Localised manual controls

Local switches and dimmers can either be permanently hardwired as wall-mounted switches or ceiling-mounted pull switches, or may be operated by remote control. Remote control devices are similar to the handsets used to operate television sets and are based on infra red or ultrasonic signals. The receiver may be ceiling mounted near to the luminaires being controlled, so that walls or partitions may be moved without the need for rewiring. In a large space, the luminaires should be switchable in groups. Luminaires controlled by a single switch should cover an area with roughly constant daylight factor; in a large space this would mean that rows of luminaires parallel to the windows would each have their own switch. In other situations, it would be appropriate for luminaires related to a particular activity or work group to be controlled together. The manual switches should be placed as near as possible to the area being controlled. The *Building Regulations*[49] require that the distance from the light switch to the furthest fitting that it controls is no more than 8 m, or three times the height of the height of the floor level of the fitting, whichever is greater. Where multiple switches are located together, clear labelling is of benefit.

5.15.1.2 Timeswitches

For buildings with fixed occupancy times a simple timeswitch can be effective, set to switch lights on shortly before work commences and to switch the full lighting off after working hours. Sufficient lights may need to be left on for security and safety, and some form of override should be provided for people working late; this should incorporate an extension timer to prevent the lights being left on all night. Timeswitches are effective for external lighting, but require adjustment with season.

Reset control is an extension of a timeswitch, relying on the fact that people are motivated to switch lights on rather than off. A time control switches the whole lighting system off at predetermined times. Occupants are provided with local controls to reset individual luminaires or small groups of luminaires back on as required. Typical switch-off times are lunch time, shortly after the end of the working day and after the cleaning period. A variety of local reset switching means is possible, including pull cords and infra red transmitters.

5.15.1.3 Occupancy detection

Occupancy or presence sensors can be used to switch on lights as a person enters the controlled area. The lights are switched off again when no occupancy has been detected for a set interval; the interval is chosen to avoid frequent switching, which can shorten lamp life. A minimum on period of 10 minutes should be set; general practice is to use on-times of up to 30 minutes. An alternative strategy is to use a manual switch to switch the lights on; the presence sensor switches the light off when no occupancy has been detected for a set time. This is sometimes known as absence sensing. Presence sensing finds application in infrequently used areas such as store rooms or spaces where people may have their hands full when entering.

The most common detectors are passive infra red (PIR), which detect the movement of a warm body across their field of view. The range of operation of a PIR is up to about 10 m. For larger spaces a microwave detector may be used, which operates by the Doppler shift of radiation reflected from a moving body; operating ranges of 50 m are possible. Acoustic sensors are available which respond to noise. Presence detectors are normally combined with a photocell, which prevents operation of the lighting when there is sufficient ambient light.

5.15.1.4 Lighting level

For continuously occupied spaces which have a substantial proportion of daylighting, it is possible to control the artificial lighting so that it is turned off when daylight is sufficient and on again to produce the required amount of light as the daylight contribution falls. Photocells are used to monitor the light level in the space and regulate the lighting accordingly. Where simple on/off switching is used, there is a danger that switching will cause noticeable and annoying changes in light level. To avoid complaints, the switching levels need to be set fairly high; a switch-on occurs when the daylight illuminance is about twice the required task illuminance and switch-off takes place when the combined artificial and daylight illuminance is three or four times the required task illuminance. It is also necessary to incorporate a time delay into the control system to avoid rapid switching caused by rapidly varying light levels, caused, for instance, by passing clouds.

More effective control can be effected by the use of dimmers. Suitable lamps can be smoothly dimmed down to 10% of their maximum light output before becoming unstable. A ceiling-mounted photocell looking downwards responds to the combined daylight and artificial illumination and the control system is set to provide a constant level of illumination. This system not only takes account of varying daylight, but also the fall-off in light output during the maintenance cycle of the lamp and fitting. Since it is necessary to design an installation to provide the required illumination at the end of the cycle, lamps are oversized when new. The use of an automatic control system will show a saving in energy at the beginning of the maintenance cycle, when the lamps are new and the luminaires clean.

5.15.1.5 Scene set control

The use of lighting is an important part of interior design. In commercial applications such as restaurants, hotels etc. different lighting schemes may be required at different times of day. Other examples would include multi-function rooms requiring a range of lighting scenes to suit different activities. A scene-set controller can be used to store the individual luminaire settings and to switch or cross-fade between settings as desired. Such a controller is unlikely to be combined with other automatic control systems and the different scenes are manually selected as required. Display lighting is excused from some of the provisions of Part L of the *Building Regulations*.

5.15.2 Systems and components

5.15.2.1 System selection

Factors influencing the selection of a lighting control system include:

- the expected occupancy pattern
- the availability of daylight
- the desired level of control sophistication
- the capital cost of the system versus potential savings
- need to accommodate changes in the building or use patterns.

A building with little daylight and a workforce working regular hours will require little more than a simple timeswitch to turn off the lights at the end of the working day and a provision to ensure that they are not all turned on again and left on by the cleaners or security staff. Figure 5.83

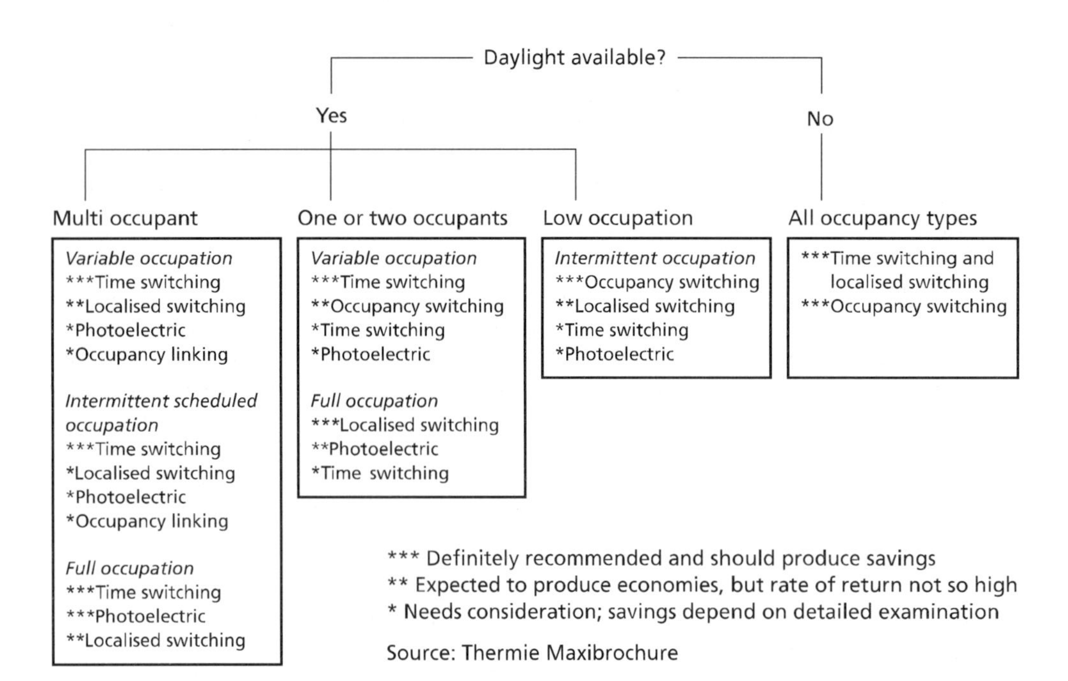

Figure 5.83 Selection of lighting control strategy

shows an initial decision tree for selecting suitable forms of lighting control for a number of applications.

An additional factor to be considered is the interaction between the lighting and HVAC systems. In some buildings constructed on the principles known as integrated environmental design (IED), heat recovered from lights played an important part in the thermal control of the building; a refurbishment in such a building which produced a lower and fluctuating heat gain from the lighting system could create problems with the HVAC control system.

5.15.2.2 Lamps and control gear

The selection of lamp type for a lighting installation will depend on may factors, such as efficiency, colour rendering, physical dimensions and appearance. The *Building Regulations*[52] require that at least 95% of the installed capacity measured as circuit watts should be restricted to the lamp types indicated in Table 5.6; display and emergency lighting are excluded. Further information on lamp types is available in the *LIF Lamp Guide*[53]. Lamp types differ greatly in their suitability for control applications. High intensity discharge (HID) lamps, while offering high luminous efficacy, offer only a limited range of dimming possibilities and are unsuitable for control by presence detection, since restart of a warm lamp may take several minutes. Conventional tungsten filament lamps (GLS) are easily dimmed and offer immediate switching with no penalty in lamp life. However, their low luminous efficacy prevents their being used as the main light source in commercial buildings. The same is true for tungsten halogen lamps, which have the additional disadvantage of requiring special control gear for the dimming of low voltage lamps.

Tubular and compact fluorescent lamps form the mainstay of most lighting schemes in commercial buildings. Fluorescent lamps require electrical control circuits for their operation, which may be classified as follows.

— *Switch start*. The simplest and cheapest option, with penalties in lamp efficiency and life where frequent switching occurs.

Table 5.6 Principal lamp types and their suitability for control applications

	Lamp	Control gear	Dimming	Frequent switching	Part L ?
GLS	Tungsten filament	None	***	***	No
	LV tungsten halogen	None	**	***	No
	Compact fluorescent	Yes			Yes
	Tubular fluorescent	Switch start	*	*	Yes
		Cold start electronic	*	*	Yes
		Warm start	*	**	Yes
		High frequency	*	**	Yes
		High frequency regulating	***	***	Yes
SOX	Low pressure sodium	Yes	*	*	No
HPS	High pressure sodium	Yes	**	*	Yes
MBI	Metal halide	Yes	**	*	Yes

***Suitable; **possible; *unsuitable.

— *Electronic start.* The electronic circuit gives instantaneous and flicker-free starting. The warm start version warms the cathode for up to 2 s before starting, giving improved lamp life.

— *High frequency.* The lamp is run at a frequency of about 30 kHz instead of mains frequency. This eliminates flicker and improves the luminous efficacy by up to 20%. In addition, the power factor of the lighting load is improved.

— *High frequency regulating.* As for high frequency, with the added advantage of being smoothly dimmable down to about 10% power. This is the control gear of choice for lighting control systems.

The life of a fluorescent lamp is adversely affected by frequent switching. Neither conventional starter switch circuits nor cold start electronic circuits are suitable for use with presence detection. Warm start electronic ballasts preheat the cathode for a period ranging from a minimum of 0.4 s to about 2 s. The Lighting Industry Federation[54] recommends that where fluorescent lamps are to be switched frequently, the longer warm-up time should be considered and that a minimum on-time of 20 minutes should be employed to ensure that both lamps and ballast reach normal operating temperature before recycling.

5.15.2.3 Lighting Control Systems

A lighting control system may be as simple as a lightswitch by the door, which may be entirely satisfactory in some situations. At the other end of the scale of sophistication, it is possible to have each luminaire individually addressable and controlled via the BMS. Such a system allows the full range of control possibilities. Luminaires may be individually controlled by permitted users over the telephone system or from a PC. The BMS can record lamp burning hours for planned maintenance or detect lamp failure. The energy of individual or groups of luminaires may be recorded, making individual billing possible in multi-occupied buildings. Since all control is via a low voltage bus system, there is no need for rewiring if the building layout is altered. Several manufacturers provide lighting control equipment which is compatible with one or more of the major field network protocols.

The four basic control methods described in 5.15.1 are brought together in a lighting control system using the so-called intelligent luminaire. The luminaire incorporates a presence detector and a downward-looking photocell, which measures the level of illumination in the area served by the luminaire. The built-in controller ensures that illumination is only provided when the space is occupied and provides a constant level of illumination in varying ambient light levels. In addition, individual luminaires may be controlled directly by the appropriate user, who has control over dimming and brightening, typically with an infra red controller or over the telephone network. In addition, the luminaires can communicate with each other over a bus system. This makes it possible to program the luminaires to act cooperatively so that, for instance, a group of luminaires is switched on if a presence is detected by any one of them. This has application in the lighting of circulation spaces or when it is desired to anticipate movement; for example, it is possible to switch the lights on in a room as a person approaches the door, avoiding the problem of walking into an unlit area. Where a person may be working after hours in a large office, the luminaires can be programmed to provide general background illumination to avoid the person working in an isolated pool of light.

The communications receiver in the intelligent luminaire may also be used to receive information from other sensors, such as a wall mounted temperature sensor. The multi-sensor can therefore serve as a communications point for the entire BMS. With the logic of lighting control strategies becoming more complex, it is important that the system design should be tested in advance, using a 'what if' approach to indicate the operation under all possible user scenarios.

5.15.2.4 Human factors

The operation of lighting controls is immediately apparent to the building occupant and for this reason it is essential that the needs and behaviour of the users is taken into account from the start. Surveys have shown that where users had little or no control of the lights in their own work space, or low awareness of the proper use of controls, low satisfaction and energy wastage were common. High satisfaction and energy efficiency were generally associated with high levels of local control plus high awareness, by both end-users and facilities managers, of how the controls operated.

Occupant surveys have suggested some general points to be borne in mind.

— *People are good at judging whether they need the lights on.* Do not switch lights on or increase brightness automatically unless this is essential for safety, or appropriate for managed areas.

— *People are not good at switching lights off.* Try to provide automatic switch-off, but...

— *People do not like being plunged into darkness.* Where possible lights should be dimmed down, or a warning given. There should be readily accessible local override switches.

— *People dislike automatic systems which distract them or do not do what they want.* Automatic switching or stepped dimming is best done at fixed times rather than at seemingly random intervals. Occupancy detectors should be positioned to avoid nuisance triggering.

— *Local controls should be accessible and their operation intuitive.* Switches should be close to the point of decision and their operation clear.

— *Individual requirements differ.* Given the choice, users select a wide range of illuminance levels. Lower levels are often chosen for work with computer screens; people with less acute eyesight or more exacting work may need additional task lighting.

Six categories of space have been identified and classified in terms of their perceived ownership?, i.e. the person or group who are *de facto* in control of the lighting of that space[55]. Each of these types of space has different requirements, which are summarised in Table 5.7.

Table 5.7 Lighting controls for different categories of occupied space[55]

Space type	Characteristics	Examples	Issues and problems	Suitable approaches
Owned	Occupants regard the space as their own and want to make all their own decisions about the status of the lights. A close relationship between user, space and control makes this possible	Small offices, particularly for one or two persons. Residential homes	Occupants need to switch both on and off and greatly object to 'Big Brother' controls. Any automatic adjustments should be imperceptible, e.g. absence sensing once the room is vacated, or gradual dimming	Avoid automatic switch-on, except possibly in small windowless rooms. Lightswitches by the door can be used effectively. Consider absence sensing and photoelectric dimming to avoid waste. Always provide local override
Shared	Occupants regard their part of the space as their own, but cannot be in full control of environmental systems which have to suit others as well. Ideally there should be some lighting control, or task lighting available on request	Open plan offices. Some production, maintenance and laboratory areas	Systems default to convenient but inefficient states, typically with all the lights on. The status then only changes if the situation becomes undesirable or at the end of the day. Security staff and cleaners may need different controls	Separate control circuits of lighting for circulation, decoration and safety avoids the first arrival switching on everything. Ideally have local switching and adjustment at individual work stations. Consider 'last out, lights off' facility
Temporarily owned	Occupants are usually present for a few hours at a time. They can be in individual or consensus control while they are there, but may not be very familiar with the controls. Lights left on when the room is vacated; nobody feels responsible	Meeting rooms, interview rooms, quiet rooms, study carrels, laboratory writing up areas, classrooms, lecture rooms	Controls not easy to find or understand. In meeting rooms presence detection must include local override for slide presentation etc. Nuisance triggering of presence detectors. Manual 'on' often preferable	Highly visible easy-to-understand local controls are required in obvious places. Absence sensing is desirable, preferably with 'last out, lights off' facility at the exit. Teaching and presentation rooms may require dimming and control from lectern
Occasionally visited	Occupants visit occasionally. There may be long vacant periods in between. People may sometimes have their hands full or be manoeuvring trolleys. Safety lighting may be needed in larger spaces	Toilets, store rooms, equipment and machine rooms, garages, cleaners' cupboards, bookstacks, warehouse aisles	Lights get left on for long periods unnecessarily. Occupancy sensors may not see people in remote parts of the space or may be triggered by passers by	Consider occupancy sensors or time-delay switches if there are no safety implications, e.g. in bookstacks where the main aisles are lit. For small rooms a switch by the door and an indicator light may be enough
Unowned	Spaces though which people pass. People will expect their path to be lit adequately. They may be prepared to operate well-placed light switches, but they may not switch them off afterwards	Corridors, circulation areas, stairs, lobbies, lifts, loading bays, car parks. Lighting should be well-structured to suit the route, and be low energy	Lights are often left on unnecessarily, either in daylight or when the area is empty. Difficult to find lightswitches; these need to be in obvious places, well-marked and perhaps with locator/indicator lights. Too many lights switched on at once	Presence-detecting or time-delay switches allow lights to be off when people are absent. Avoid sudden darkness by dim-out, sequenced switch-off of background safety lighting. In daylit areas, photoelectric controls can give excellent energy savings
Managed	The larger public spaces in which occupants would not normally expect to operate lights. Ambient light levels can often be allowed to vary within quite a wide range	Restaurants, libraries, concourse etc. Often have well-determined use hours, so systems can be programmed	Lights are often left on unnecessarily, either when there is sufficient daylight or when there are no people about. Management are often too busy with other things to make fine adjustment to lighting	Arrange circuits according to usage and daylight. Consider photoelectric daylight linking and low energy safety or utility lighting out of hours. Where added sparkle is required, try to use low energy decorative lighting

5.16 Summary

Control strategies for entire systems can be broken down into strategies for component subsystems. A fundamental requirement of a building control system is that it ensures safe operation of the plant under its control. All strategies must provide for alarms which give warning of a variable out of limits or plant requiring attention interlocks, which automatically prevent or shut down operation of plant when necessary. In addition, the strategy must include the action to be taken on receipt of a fire alarm.

Unsuitable control strategies for boilers can lead to inefficient and problematic operation. A primary water circuit using a constant-flow circulation pump and common header is recommended for effective and reliable control. Multiple boilers are used with sequencing control to match a variable load. For modern low loss boilers, it is not necessary to isolate off-line boilers and sequence control is provided on the return temperature from a common header. For high water content boilers it may be necessary to provide individual circulation pumps and to isolate off-line boilers. Boiler protection circuits are necessary where the return flow to the boiler is of variable flow and temperature.

Chiller controls are normally provided as part of a packaged unit. Where a separate cooling tower is employed, efficiency may be improved by controlling the condenser water temperature. The technique of near optimal control operates by modulating cooling tower fan speed as a function of chiller load and leads to operating efficiencies. In cool ambient conditions it may be possible to dispense with chiller operation and operate in free cooling mode, where the cooling tower alone is able to meet the building cooling load.

In secondary water circuits the use of variable-speed pumps and two-port valves offers savings in pumping energy and initial cost but care is required if the system is to remain in balance and under control under varying load conditions. Where throttling circuits are used to vary output, speed control of the circulation pump is recommended, coupled with the use of high rangeability two-port valves. The use of differential pressure control valves can be valuable in maintaining system balance.

HVAC systems using a central air handling unit normally employ some air recirculation. Return air and outside air are mixed to produce the required supply air temperature if possible; otherwise the outside air proportion is set at a minimum or maximum value as determined by the outside conditions. Where supply air humidity is controlled, it is better to use enthalpy control of the recirculation dampers for greatest efficiency. Heat recovery may be used instead of or in addition to recirculation. Where a variable rate heat recovery device is used, it is modulated to control the supply air temperature as far as possible.

Provision of the necessary amount of outside air for ventilation purposes must always be ensured. Where a variable air volume system is used in association with recirculation, it is not sufficient to simply provide a minimum proportion of outside air, since this will be inadequate at low volumes. Several techniques are available. The outdoor air damper position may be controlled as a function of fan supply speed, so that it opens further to compensate for reduced supply volume. The outdoor air supply may be measured directly and the value used to modulate the intake damper. Where both extract and supply flow rates are available, the extract fan can be modulated, so that the extract flow exceeds the supply flow by the required amount; the balance is the outdoor air intake.

VAV systems consist of a central air handling unit which supplies conditioned air to a number of VAV room terminal units. The RTUs control room temperature by modulating dampers to vary the air supply to the room. Variable-speed control of the main supply fan is required for energy efficiency. The fan speed may be controlled as a function of the static pressure in the main supply duct. Positioning of the pressure sensor two thirds of the distance down the duct will normally be found suitable. Where intelligent VAV boxes are used, which are capable of signalling their damper position, it is possible to match the air supply to aggregate demand. Further efficiency may be achieved by resetting the air supply temperature set point as a function of total load.

Fan coil units provide room temperature control under local control, but can be linked to a BMS for functions such as time scheduling, group reset of set points or optimum start heating and cooling.

Variable refrigerant flow systems consist of a number of room units, connected to an outdoor unit containing a compressor, connected by refrigerant lines. Simultaneous heating and cooling among units powered by a single outside unit is possible, giving efficient operation. VRF systems are supplied with packaged control systems, which have limited interconnection possibilities with a BMS; control of time scheduling and set points is normally possible, together with alarm transmission.

Chilled ceilings are to some extent self-regulating and two-position control of chilled water flow to panels or beam is adequate. Prevention of condensation on the cold surface is important. Where central air handling plant is used, the most effective method is to control the room temperature dewpoint to at least 2 K below the chilled water temperature; otherwise the chilled water flow temperature should be maintained 2 K above room air dewpoint. In both cases, an additional condensation detector should be used as a precaution.

Lighting is often the largest consumer of electricity in a building. There are four basic methods of lighting control.

— *Localised manual controls*. Switches must be readily accessible; the use of pull cords or remote control handsets can be advantageous. Groups of luminaires switched together should be grouped by activity or daylight factor.

— *Timeswitches*. These can be effective for set work schedules. Systems where lights are switched off by timeswitch and switched on manually are effective.

— *Occupancy detection*. This is useful in zones which are used occasionally. The lights come on when an occupant is detected and go off when no occupancy has been detected for an interval of at least 10 minutes.

— *Lighting level*. For continuously occupied spaces with some daylight, the artificial lighting can be modulated to provide a constant level of illumination. Dimming preferable to on/off switching and rapid fluctuations in level should be avoided. Intelligent luminaires are available, incorporating their own presence detector and photocell; luminaires communicate via a bus and are individually addressable, allowing the lighting system to be integrated with the BMS.

References

1 HSE *Out of control — why control systems go wrong and how to prevent failure* (Sudbury: HSE Books) (1995)

2 *Fire engineering* CIBSE Guide E (London: Chartered Institution of Building Services Engineers) (1997)

3 CIBSE *Fire precautions: sources of information on legal and other requirements* Technical Memorandum TM16 (London: Chartered Institution of Building Services Engineers) (1990)

4 *Heating systems and their control* GIR 40 (Garston: Building Research Energy Conservation Support Unit) (1996)

5 *Boiler (Efficiency) Regulations 1993*. Statutory Instrument 3083 (London: The Stationery Office) (1993)

6 HSE *Automatically controlled steam and hot water* PM5 (Sudbury, Suffolk: HSE Books) (1989)

7 Petitjean R *Total hydronic balancing* 2nd ed (Ljung, Sweden: Tour and Andersson) (1997)

8 *Minimising the risk of Legionnaires' disease* CIBSE TM13 (London: Chartered Institution of Building Services Engineers) (1991)

9 HSE *The control of legionellosis including legionnaires' disease* HSG 70 (London: HSE Books) (1993)

10 Braun J E and Didderich G T Near optimal control of cooling towers for chilled water systems *ASHRAE Transactions* **96**(2) 806–813 (1990)

11 De Saulles T and Pearson C Set-point and near optimal control for cooling towers *Building Services Journal* **20**(6) 47–48 (1998)

12 *Ice storage systems* CIBSE TM18 (London: Chartered Institution of Building Services Engineers) (1994)

13 Trend *Ice storage. Monitoring and control for optimum performance* (Horsham: Caradon Trend) (1995)

14 Levermore G J *BEMS II* (working title) 2nd ed (London: E & F N Spon) (1999)

15 *Flow of fluids in pipes and ducts* CIBSE Guide C4 (London: Chartered Institution of Building Services Engineers) (1977)

16 Seyfert H *The control device in water circulation* 450/en/95-06 (Switzerland: Staefa Control System) (1995)

17 Petitjean R The case for hydronic balancing *Journal of the Chartered Institute of Building Services* **15**(5) 43–44 (1993)

18 *Variable flow control* GIR 41 (Watford: Building Research Energy Conservation Support Unit) (1996)

19 ETSU *Guidance notes for reducing energy consumption costs of electric motor and drive systems* Good Practice Guide No 2, 2nd ed (Harwell: ETSU) (1990)

20 Bower J R *The economics of operating centrifugal pumps with variable speed drives* C108/81 (London: I Mech E) (1981)

21 Centrifugal pumps *ASHRAE Handbook: HVAC systems and equipment* Chapter 38 (Atlanta, GA: American Society of Heating, Refrigeration and Air Conditioning Engineers) (1996)

22 Parsloe C J *The design of variable speed pumping systems* AG 14/99 (Bracknell: Building Services Research and Information Association) (1999)

23 Rishel J B Use of balance valves in chilled water systems *ASHRAE Journal* **39**(6) 45–51 (1997)

24 Pickering C A C and Jones W P *Health and hygienic humidification*. BSRIA TN 13/86 (Bracknell: Building Services Research and Information Association) (1998)

25 *Air to air energy recovery* ASHRAE Handbook: HVAC Systems and Equipment Chapter 42 (Atlanta, GA: American Society of Heating, Refrigeration and Air Conditioning Engineers) (1996)

26 *Environmental Design* CIBSE Guide A (London: Chartered Institution of Building Services Engineers) (1999)

27 Olesen B W International development of standards for ventilation of buildings *ASHRAE Journal* (4) 31–39 (1997)

28 Liddament M *A review of ventilation efficiency* Technical Note 39 (Coventry: Air Infiltration and Ventilation Centre) (1993)

29 *Building Regulations 1991 (1995 ed) Approved Documents F1, F2* (London: Department of the Environment and the Welsh Office; Stationery Office) (1994)

30 *Energy efficient mechanical ventilation systems*. Good Practice Guide GPG 257 (Watford: Building Research Energy Conservation Support Unit) (1998)

31 Potter I N and Booth W B *CO_2 controlled mechanical ventilation systems* TN 12/94.1 (Bracknell: Building Services Research and Information Association) (1994)

32 IEA *Demand controlled ventilating systems: a summary of IEA Annex 18* (Coventry: IEA) (1998)

33 Fletcher J *Ventilation control and traffic pollution* Technical Note TN 5/98 (Bracknell: Building Services Research and Information Association) (1998)

34 Kettler J P Minimum ventilation control for VAV systems: fan tracking vs workable solutions *ASHRAE Transactions* **101**(2) 625–630 (1995)

35 Kettler J P Controlling minimum ventilation volume in VAV systems *ASHRAE Journal* **40**(5) 45–50 (1998)

36 Seem J E, House J M and Klaassen C J Leave the outdoor air damper wide open *ASHRAE Journal* **40**(2) 58–60 (1998)

37 Martin A J and Banyard C P *Library of system control strategies* Applications Guide AG 7/98 1st ed (Bracknell: Building Services Research and Information Association) (1998)

38 Martin A Condensation control for chilled ceilings *Building Services and Environmental Engineer* 21 (9 May) 9(1998)

39 *Natural ventilation in non-domestic buildings* CIBSE Applications Manual AM10 (London: Chartered Institution of Building Services Engineers) (1997)

40 BRECSU *Natural ventilation in non-domestic buildings — a guide for designers, developers and owners* Good Practice Guide GPG 237 (Watford: Building Research Energy Conservation Support Unit) (1998)

41 Kolokotroni M *Night ventilation for cooling office buildings* IP 4/98 (Watford: Building Research Establishment) (1998)

42 Limb M J *Passive cooling technology for office buildings — an annotated bibliography* (Coventry: Air Infiltration and Ventilation Centre) (1998)

43 Allard F *Natural ventilation in buildings: a design handbook* (London: James and James) (1998)

44 *Mixed mode ventilation systems* CIBSE Applications Manual AM13 (London: Chartered Institution of Building Services Engineers) (2000)

45 Martin A J *Control of natural ventilation* TN11/95 (Bracknell: Building Services Research and Information Association) (1995)

46 Fletcher J *Night cooling control strategies* TA 14/96 (Bracknell: Building Services Research and Information Association) (1996)

47 Bordass W T and Jaunzens J Mixed mode: the ultimate option *Building Services Journal* **18**(11) 27–29 (1996)

48 BRECSU *Electric lighting controls — a guide for designers, installers and users* Good Practice Guide 160 (Garston: Building Research Energy Conservation Support Unit) (1997)

49 *Building Regulations Part L* (London: Stationery Office) (1995)

50 HSE *Lighting at work* (London: HSE) (1998)

51 *CIBSE code for interior lighting* (London: Chartered Institution of Building Services Engineers) (1994)

52 *Lighting requirements of Building Regulations Part L* Guidance Note GN4 (London: Chartered Institution of Building Services Engineers) (1996)

53 *LIF lamp guide* 2nd ed (London: Lighting Industry Federation) (1994)

54 *Use of PIR detectors with electronic ballasts and fluorescent lamps* LIF Technical Statement No 17 (London: Lighting Industry Federation) (1995)

55 Slater A I, Bordass W T and Heasman T A *People and lighting controls* BRE Information Paper IP6/96 (Garston: Building Research Establishment) (1996)

6 Control strategies for buildings

6.0 General

6.1 Operating modes

6.2 Design techniques

6.3 Whole-building HVAC systems

6.4 Case studies

6.5 Summary

This section deals with control strategies for whole buildings. The strategy for a complete HVAC system must take into account the interaction between subsystems. The system may operate in several different modes and the condition of each subsystem in every possible mode must be specified. Practical and theoretical design techniques are described which may be used to predict control performance under a range of conditions. The major HVAC systems are summarised with advice on avoiding conflict between subsystems and the successful incorporation of user control. The section ends with some case studies to illustrate successful applications.

6.0 General

A complete HVAC system consists of an assembly of plant modules, many of which interact with each other. Similarly, the control strategy for a whole building includes the strategies of individual plant modules and must take into account all the possible interactions between subsystems. Failure to consider these interactions may result in:

- system instabilities
- high energy consumption
- compromise of safe operation.

6.1 Operating modes

The descriptions of subsystem control strategies in the previous section indicated many of the interactions which should be included in control strategies. In the design of the HVAC and control system for a new building, there are inevitably combinations of subsystems that have not previously been documented and it is essential that the design is theoretically challenged with a range of 'what if' scenarios. A variety of possible, likely and unlikely situations should be drawn up and predicted responses of the control system followed through, to ensure that there are no unexpected effects. The BSRIA *Library of Control Strategies*[1] uses a number of operating modes in which a HVAC plant could operate at different times. Possible modes are listed in Table 6.1. The number of modes will depend on the nature of the HVAC system. A mixed mode system involving heating and cooling, plus mechanical and natural ventilation may have several distinct operating modes during occupancy. The operation of all control strategies in these modes should always be specified unambiguously.

Table 6.1 Operating modes for an HVAC system

Low outside air temperature interlock (for plant protection)
Low return water temperature interlock (for plant protection)
Low zone temperature in the conditioned zone (for building fabric and contents protection)
Plant shut-down
Fan overrun for air systems, pump overrun for water systems
Optimum start heating (OSH), boost heating
Optimum start cooling (OSC)
Night cooling
Fire
Normal operation during the occupied period, subdivided into:
— heating
— cooling
— natural or mechanical ventilation (mixed mode)
— thermal storage operation

Documentation of a complete control strategy consists of the following:

- schematic diagram of plant
- description of plant
- control strategy
- specification clauses
- BMS points list
- summary of plant operation
- control flowchart.

For reasons of space it is not possible to present a range of full control strategies in this manual. Control strategies may be obtained from a number of sources. Large consultancies maintain their own libraries of strategies which have been found reliable in practice and which may be modified to meet the needs of new design projects. Some of the major manufacturers publish control strategies, which are often of general application[2,3]. The major published source of strategies in the UK is the BSRIA *Library of System Control Strategies*[1]. This is a comprehensive document, presenting a large number of control solutions with descriptions of plant operation and control strategy. A standard specification clause is provided with each strategy, together with a full points list. The expectation is that the *Library* will provide a set of standardised solutions that will enable control systems to be designed, configured and tested in less time than would be required designing from scratch. This Guide has drawn heavily on the *Library* for the control strategies summarised in the previous section and BSRIA references have been given where applicable.

It is important to choose an HVAC system and control system that are appropriate for the requirements of the building and the operations it supports. There is no advantage in having a sophisticated BMS nor in accumulating information that is not needed and will not be used. Where simple controls, perhaps packaged controllers supplied by plant manufacturers, will perform adequately, they should be used unless there is good reason to install a customised control system. Large modern buildings benefit from a full BMS; the various options for system architecture are dealt with in section 4. The power and flexibility of a modern BMS allows virtually any control strategy to be implemented, and later modified. Changes in user requirements, developments in HVAC technology and evolving policies on energy conservation ensure that innovation and change in building design is a continuing process. Few buildings are identical in their needs to others. This has the consequence that the controls designer is faced with buildings that cannot be satisfied by off-the-shelf solutions. The control strategies for subsystems set out in section 5 have been chosen to be representative of best current practice, but must always be evaluated for suitability for any given building and environmental control system.

The requirements of energy conservation and the desire to reduce or eliminate the need for air conditioning have led to building designs which are intended to operate at the limit of their capabilities during conditions of design weather conditions; it is no longer acceptable to insure against problems by oversizing the plant. This has led to a wide range of buildings, ranging from fully air conditioned sealed buildings to full natural ventilation. In many cases, however, it has been found necessary to introduce ancillary equipment to deal with particular weather or operational conditions. This gives rise to potential control problems in operating the various systems and avoiding conflicts. Practical examples are given in the case studies.

6.2 Design techniques

Various techniques are available which will assist the designer in assessing the performance of possible HVAC plant designs and associated control strategies. The various techniques differ in their methods and application and are described below. The information that may be expected from their use includes:

- performance of building under different operating conditions, especially hot weather
- information on practical problems of installation and operation
- energy consumption
- evaluation of control strategies
- prediction of control performance.

6.2.1 Full scale mock-up

A full scale mock-up of a representative section of the building is constructed. As far as possible it should use the actual materials and items of plant that will be employed in the actual building. The mock-up is placed in a large environmental chamber that can simulate external conditions, including the effects of solar radiation; several such chambers exist in the UK. The effects of indoor activities are reproduced by placing suitable heat sources in the mock-up room. Suitable instrumentation has to be provided to monitor the internal conditions produced. The mock-up is then subjected to a range of external conditions, representing extreme design conditions and a number of intermediate conditions. The HVAC installation is used to control the indoor conditions, which are recorded and evaluated. The procedure allows the designer to:

- show the visual appearance of the finished internal space
- investigate the performance of the HVAC system
- evaluate the control system and its strategy
- ensure that all components can be installed as planned and so obviate problems on site.

Where the cost of a mock-up can be justified, it has proved beneficial to all parties involved in the building project. The use of mock-ups is usually restricted in practice to room-height sections. Where it is required to predict the performance of tall atrium type structures, recourse must be made to analogue or mathematical modelling.

6.2.2 Analogue modelling

Air movement in convective flow may be modelled in a water tank, using salt solutions of varying concentration to represent air at different temperatures[4,5]. The model is operated upside down; dense salt solution moves down through water in a way analogous to the upward movement of warm air. A scale model of a section of the building is constructed out of Perspex and mounted upside down in the test tank. Coloured salt solution is injected to represented heat sources and its movement under convective flow is recorded. The technique has been used successfully to evaluate the performance of natural ventilation systems and their control by opening and closing vents[6] but has not received widespread application.

6.2.3 Mathematical modelling

Computer-based mathematical modelling may be used to simulate the performance of a building in considerable detail. Several software applications are available and may be purchased or used via a bureau. General advice on the application of models is given in CIBSE AM11[7]. The first step is to input a geometrical model of the building. This includes the building dimensions, thermal properties of the constructional elements, and the orientation and location of the building. The HVAC plant is added; separate modules exist which model the performance of common plant and control systems. Internal heat production and occupancy schedules over the year are set up. The model may then be run to produce predictions of temperatures and energy flows, using detailed weather data for the location and season of interest. Air movement is modelled using the technique of computer fluid dynamics (CFD). This has found considerable application in the investigation of naturally ventilated buildings. The models usually operate in time steps of 1 h and so cannot be used to investigate the finer points of control dynamics. The models are most useful in predicting building performance in extreme weather conditions and establishing whether internal conditions will be found acceptable in hot weather. The ability of the models to allow for the effects of thermal mass on peak temperature is of particular value in this context.

Control strategies may be evaluated using mathematical models. It is sometimes found that control criteria may be relaxed without causing any appreciable discomfort, due to the stabilising effects of the building thermal mass. This can lead to worthwhile savings in energy consumption. Models have also been found of real value in establishing effective night cooling strategies and the operation of mixed mode systems.

6.2.4 Emulation

An emulator for a building energy management system consists of a simulation of a building and its HVAC system which may be connected to a real BMS. The real BMS controls the simulated building as if it were real, transmitting control signals and receiving simulated information back as the simulated building responds to its actions. Since an actual BMS controller is used, with its own time characteristics, it is necessary that the simulated building responds at the same rate as the real building. An emulation run therefore operates in real time, e.g. it will take a week to emulate a week's building operation. An emulator can be used for:

— evaluating the performance of a BMS

— training of BMS operators

— assisting in the development of new control algorithms

— fine tuning the control parameters.

An advantage of using an emulator is that a BMS may be tested with any type of building and HVAC system for which a simulation model is available, and tests can be run on different BMS under identical conditions. Since a real BMS is used, it is not necessary to know the algorithms employed, so that products from different manufacturers may be compared without compromising any proprietary information about the control strategies. The building simulation model is of fundamental importance for the emulation technique. As well as modelling the thermal response of the building, the dynamic behaviour of the controls and actuators must be modelled realistically; this is a more stringent requirement than is found with the mathematical models described in 6.2.3 above. Operation of plant is described in great detail and the emulation exercise may be used to predict reversals and travel of actuators as an indicator of potential maintenance costs. Six emulators were developed for an IEA exercise and are described by Lebrun and Wang[8]. They worked well but have not yet found widespread application.

6.3 Whole-building HVAC systems

Table 6.2 lists some major types of HVAC system used in modern buildings with some of their characteristics. Many variations on the basic systems are possible. Control strategies for the subsystems involved have been dealt with in section 5. The designer must consider how the different components of the HVAC system may interact under different operating conditions and ensure that the control system will maintain economic and effective operation under all conditions.

6.3.1 VAV and perimeter heating

This is a widely used combination. Conditioned air is supplied to all parts of the space via the VAV boxes. The proportion of recirculated to outdoor air in the air supply is controlled to ensure that the correct amount of ventilation air is maintained against variations in supply air volume. Temperature control in the zone supplied by a group of terminal units is provided by varying the air flow though each box. Heating may be provided via the terminal unit by fitting the box with a reheat coil, which is supplied with LTHW. The relevant control strategies are set out in Section 5. It is common to provide heating by conventional LTHW radiators or other room heat emitters. This has advantages:

— There is no water supply in the ceiling.

— The heat is supplied where required to counteract perimeter heat loss, especially under windows.

— There is better heat distribution from radiators than from the ceiling, especially at low air flow rates.

The control strategies for the combined system present no great problems. The perimeter heating LTHW flow should be weather compensated. It may be necessary to fit separate controls on different facades where there are large differences in heat loss or insolation. Thermostatic radiator valves will give extra control if required; they may be fitted in addition to, but not instead of, compensated flow temperature. Given effective control of the perimeter heating, it is not necessary to interlock the VAV cooling and the heating to prevent simultaneous operation. With a deep building, perimeter heating may be necessary to maintain comfort in the perimeter zone while the central zones

Table 6.2 Major HVAC systems

Feature	VAV and perimeter heating	Fan coil units, four-pipe	Chilled ceiling displacement vent	Natural ventilation	VRF
Component					
Heating	LTHW radiators	FCU	Perimeter radiators	Radiators	Indoor unit
Cooling	VAV box	FCU	Ceiling or beam	None	Indoor unit
Ventilation	Central AHU, mixed air	Central AHU, full fresh air	Central AHU, full fresh air	Natural	Mechanical ventilation
Spatial impact					
Plant space	High	Average	Average	High	Good
Riser shafts	High	Average	Average	High	Good
Floor space	High	Average	Low	High	High
Ceiling depth	High	Average	High	High	Average
Performance					
Temperature	Good	Good	Satisfactory	Poor	Satisfactory
Air distribution	Adequate	Adequate	Good	Adequate	Poor
Noise	Good	Adequate	Excellent	Good	Adequate
Cost					
Capital	High	Average	High	Average	Low
Operating	Low	Average	Average	Low	Average
Maintenance	Average	High	Average	Low	High
Flexibility					
Rearrangement of partitions	Good	Good	Good	Poor	Poor

require cooling. Boost heating during optimum start is provided by setting the perimeter heating flow temperature to maximum, with the VAV system off or on full recirculation.

6.3.2 Fan coil units

Fan coil units are available in a wide range of configurations, including underfloor units, console units designed to go on the wall or under windows, to the common type installed in a ceiling void. Units are connected to a supply of hot and chilled water. The distribution of water round a building is simpler and takes less space than the distribution of conditioned air. This makes the installation of fan coil units more flexible than that of a VAV system. FCUs are to be found in all types of building and control systems range from a local thermostat controlling fan operation to a fully integrated whole-building system. Control of individual FCUs was dealt with in 5.10; here, we consider the implications of integration.

The FCUs themselves recirculate room air, heating or cooling as required to maintain the room air set point. Ventilation air is provided by a central air handling unit which supplies 100% outdoor air at controlled temperature and usually controlled humidity. Incorporating the temperature of the primary supply air in the control strategy allows energy savings to be made by controlling the operation of the heat exchanger to exploit free cooling. Fan coil units have no provision for dehumidification other than moisture condensing on the cooling coil. Fan coil units must therefore always be provided with a condensate tray and a means to remove any condensate. In general, it is best to avoid running an FCU with a wet coil, especially when the unit is mounted above a false ceiling where the results of any failure of the condensate removal will be a problem. The preferred method is to deal with the latent load by dehumidifying the primary outdoor air to control room humidity at the desired level; this will normally be sufficient to prevent condensation on the FCU cooling coil. The usual practice is to duct the primary supply air to each FCU, where it is mixed with the recirculated room air by the unit fan. It is also possible to distribute ventilation air through independent terminals. This may have advantages where it is required to maintain maximum flexibility for possible repositioning of the FCUs at a later date, or where the FCUs are only used during the cooling season.

In an integrated system, each intelligent fan coil unit incorporates its own controller which communicates with the BMS. Room temperature set points and time schedules may be set remotely for each unit or group of units acting together. Once of the disadvantages of fan coil units is the potential degree of maintenance required and the disruption involved in identifying and servicing faulty units which are distributed throughout the occupied space. The BMS is used to give warning of any failures or routine maintenance needed, such as the need for a filter change. A fan coil system incorporates many small water control valves, which are susceptible to sticking or blockage. It is possible to incorporate checking routines in the BMS. Each valve is driven fully open and closed in turn, and the appropriate response of the discharge air temperature is checked.

Figure 6.1 shows an air handling unit for a fan coil system, which supplies 100% outdoor air to all room units. A heat exchanger is incorporated to transfer heat between exhaust and outdoor intake air. The air is supplied at constant flow rate during the occupancy period; the extract fan is interlocked to operate at the same time. The supply air humidity is controlled to maintain the zone humidity at a level which will provide comfort and prevent, or at least minimise, condensation on the fan coils. This is typically controlled using an RH sensor in the extract. Supply air temperature may be controlled to be near the desired zone temperature, or scheduled to the outside air temperature; this will allow full advantage to be taken of any free cooling. A suitable schedule is that the supply air temperature should have a maximum temperature of 22°C when the outside air is at 12°C or below, and a minimum temperature of 14°C when the outside air is above 21°C, with a linear relation in between these values. The heat recovery is operated according to the strategy in Table 5.5 for maximum efficiency.

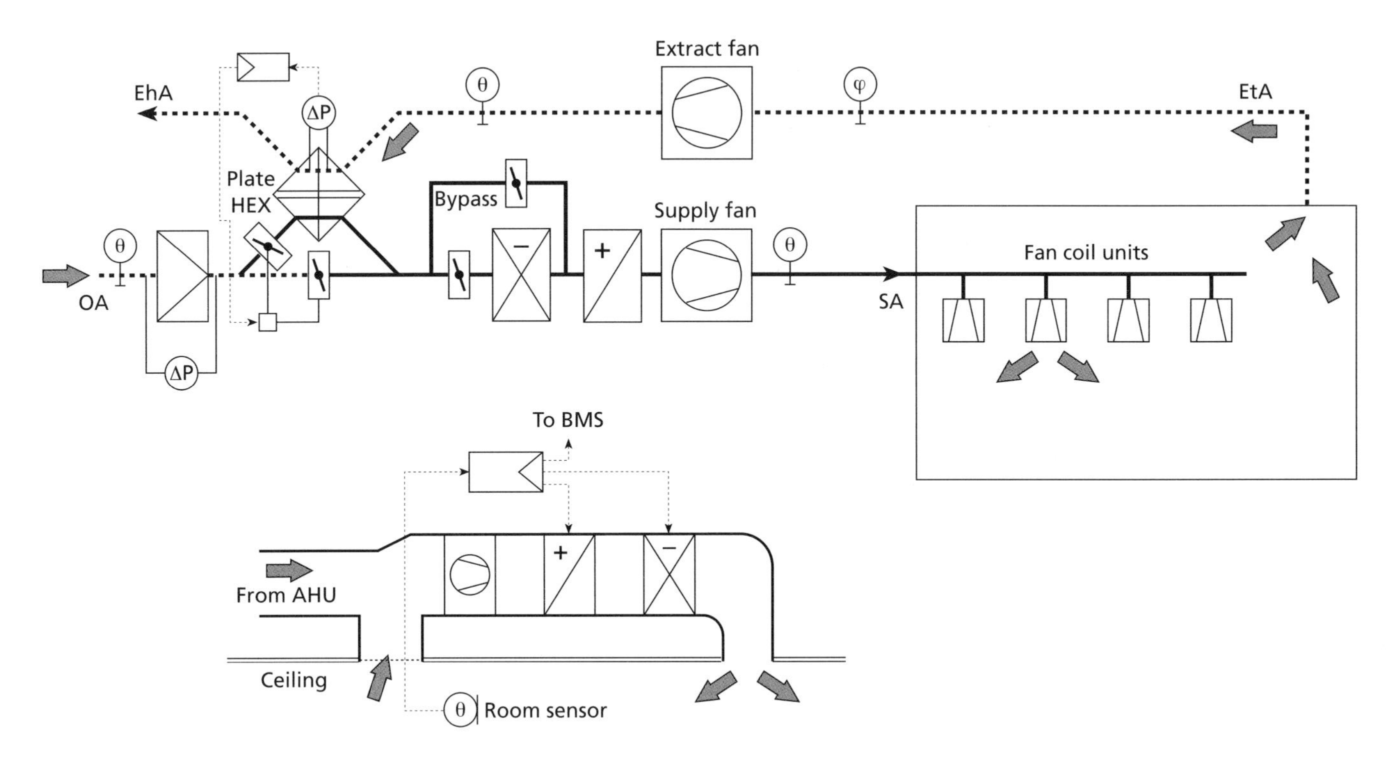

Figure 6.1 Integrated fan coil unit system. The supply air temperature is scheduled to outside temperature. Control connections not shown

Further economies may be made by resetting the flow temperatures of the LTHW and CW to the coils. This strategy requires information to be made available to the BMS on the position of the heating and cooling valves on each FCU. The temperature of the LTHW is reset so that at least one heating valve is fully open while maintaining the required room temperature in all zones. Similarly, the chilled water flow temperature is reset so that at least one cooling valve is fully open. This strategy ensures that the water being circulated to the coils is neither hotter nor cooler than required to maintain the desired conditions. Some care must be exercised when employing this type of strategy, which depends on a large number of logical criteria all being satisfied. A single sensor failure or incorrectly adjusted set point will prevent satisfactory operation. A full control strategy for a fan coil unit system is given in BSRIA *Library of System Control Strategies*, 3.16.3[1].

6.3.3 VRF systems

Variable refrigerant flow systems provide zone heating and cooling by means of indoor units which are connected via refrigerant lines to the group outdoor unit. Control strategies for VRF systems were set out in 5.13.1. The VRF indoor units do not provide ventilation air and this has to be provided separately. For some applications, such as hotel rooms, it may be adequate to use simple extract ventilation. In general it is better to provide full mechanical ventilation with controlled supply and extract. Manufacturers of VRF systems provide mechanical ventilation with heat recovery (MVHR) systems which may be used in conjunction with the indoor units to provide both temperature control and ventilation. The MVHR unit contains supply and exhaust fans, a cross-flow heat exchanger, filter and a bypass damper which can allow the air to flow straight through and bypass the heat exchanger. The unit is fully ducted and the supply air may be led to conventional diffusers or connected to the indoor units, where it mixes with the recirculated room air being heated or cooled. Each MVHR unit provides sufficient air for several indoor units.

When operating as an independent ventilation system, the unit has its own controller, offering a selection of fan speeds and heat exchange or bypass operating modes. The mode may be selected manually or automatically, based on the outdoor air and extract temperatures. Where the unit is connected to the indoor VRF units, the control may be integrated with the VRF control system, using the manufacturer's proprietary network. The modular nature of VRF systems means that the system may be extended to provide temperature control and ventilation for a large building. Humidity control is not provided in the ventilation units; some dehumidification is provided by the room units when cooling in conditions of high humidity.

6.3.4 Chilled ceiling and displacement ventilation

The range of systems coming under the general description of chilled ceilings can be divided into three groups, with different operating characteristics:

— *Chilled panels*. A chilled ceiling panel presents a flat surface to the room below. The rear surface of the panel is insulated and there is no requirement for air circulation behind the panel. Cooling is over 50% by radiation.

— *Passive chilled beam*. The chilled beam contains a cooling coil designed to cool air flowing down through it. Room air circulates up behind the beam and down through it by natural convection. Higher cooling loads are possible than with chilled panels and cooling is about 80% convection. Some beams are designed to present a cool surface to the room, enhancing the radiant proportion.

— *Ventilated chilled beam*. The beam contains an air duct which supplies the room ventilation air. The supply is 100% outdoor air and is cooled and possibly dehumidified. High velocity nozzles inject air into the beam, which induces additional room air flow thorough the cooling coil. This increases the cooling capacity of the beam over that of a passive beam.

Chilled ceilings and beams provide an effective way of producing comfort cooling in buildings with modest heat loads. Ventilation air is provided independently and this gives the opportunity of introducing mixed mode ventilation, where natural ventilation can be used at appropriate times. Mechanical ventilation with conditioned outdoor air is provided in the heating and cooling seasons; natural ventilation may be used if desired when the outside conditions are suitable. The usual form of mechanical ventilation is a form of displacement ventilation, where the air is introduced at floor level; the use of swirl diffusers ensures that air velocities fall rapidly away from the diffusers and do not cause draughts. In an ideal displacement ventilation system, the air moves steadily up through the occupied zone, taking heat and pollutants with it, to be extracted at high level. By this means, a high ventilation efficiency, potentially greater than one, can be achieved. An air temperature and air quality gradient can be achieved, whereby the occupants are in the clean cool part of the gradient, with air quality worsening at high level. This is in contrast to mixing air distribution, which gives average conditions over the whole space.

In practice, true displacement ventilation is difficult to achieve with a radiant ceiling and impossible with a chilled beam. The chilled beam provides convective cooling, with cooled air falling back into the occupied zone and mixing with the incoming ventilation air. The effect is less marked with a radiant ceiling, which cools largely by radiant exchange with the warmer surfaces in the room[9,10]. Some typical chilled ceiling combinations are shown in Table 6.3, with relevant outline strategies.

Chilled ceilings are normally arranged with a separately controlled perimeter zone. Perimeter heating using conventional LTHW radiators is used to counteract perimeter heat loss and any downdraught from windows and is controlled using weather-compensated flow temperature as described in 6.3.1 for VAV systems. A large space will require perimeter heating at the same time as central zone cooling to maintain comfort near windows.

Where the radiators are compensated, it will not always be necessary to interlock the perimeter ceiling and radiator to prevent simultaneous heating and cooling. Where a ventilated beam is used incorporating heating coils, an interlock must be present.

6.3.5 Natural and mixed mode systems

The design of mixed mode and naturally ventilated buildings is closely integrated with that of the building itself. Provision of air flow paths, control of solar gain and the need to provide adequate thermal mass, all influence the building shape and layout. Mixed mode and naturally ventilated buildings tend to be one-off designs, with unique control solutions. Such buildings are likely to include a number of controlled devices, relating to solar control, lighting, heating and mechanical ventilation and perhaps cooling. These controlled devices are operated by a combination of automatic and occupant control. There is a danger, borne out by experience, that subsystems may interact in such a way to give unsatisfactory control and excessive energy consumption. Every attempt must be made to anticipate and design out potential problems.

Full occupant control is only feasible in small cellular offices, where the control choices have limited effect on the rest of the building. The use of a weather-compensated heating circuit will avoid the worst excesses of the window-open/heating-on problem, and simple timeswitches and presence detection will limit excessive use of lighting. Most naturally ventilated buildings incorporate openable windows and other forms of occupant control. Direct control is appreciated by occupants and should lead to greater satisfaction. However, not too much should be expected of occupant control as regards optimum operation of a large building. Operation of a controlled device may have an effect remote from the occupant who, by reasons of proximity, has 'ownership' of the controlled device. For instance, the opening of leeward windows is necessary for cross-ventilation in a large building, but will have little effect on comfort nearby. There is therefore little incentive for anyone near the window to open it. While it is desirable to delegate as much control as is practical to occupants, it is

Table 6.3 Chilled ceiling systems

		Chilled panel	Passive beam	Ventilated beam
Cooling	System	Chilled panel	Coil in beam	Coil in beam
	Chilled water	14°C	14°C	14°C
	Control	Water flow on/off	Water flow on/off	Modulate water flow
Heating	System	Radiators	Radiators	Coil in beam
	Control	Compensated flow	Compensated flow	Modulated flow
Ventilation	System	Floor diffusers	Floor diffusers	Duct in beam
	Air supply	18–22°C	18–22°C	14–18°C
	Control	Constant volume	Constant volume	Constant volume
Interlocks	Heat/cool	No	No	Yes
	Windows or presence	If desired	If desired	If desired
Condensation prevention	All systems	Raise chilled water temperature above dewpoint or dehumidify ventilation air. Plus condensation detection		

not reasonable to expect them to take responsibility for the efficient operation of the whole building. The ability to take an action in response to discomfort, perhaps by opening a window, is valued. The requirement to operate windows for the benefit of staff in a remote part of the zone is unlikely to be carried out efficiently.

Most mixed mode buildings are of complementary design and are designed to operate under a concurrent or changeover strategy[11]. A concurrent system uses permanent mechanical ventilation which runs continuously to provide sufficient outdoor air for ventilation purposes. Occupants are free to open windows to provide additional cooling when needed. The natural and mechanical systems complement each other and simultaneous use is possible without conflict. A changeover mixed mode strategy aims to operate the building using the most efficient combination of available systems. The BMS selects the appropriate operating mode and enables or inhibits the relevant systems. There may be several modes, depending on season, use of building or time of day. The BMS must change from one operating mode to another in such a way that continuous control is provided and without being obvious to the occupant. Control of particular devices may be passed from manual occupant control to automatic control and this must be done without antagonising the building users. Control decisions made by the BMS must be acceptable to the occupants and appear sensible; if not, the building users may expend considerable ingenuity in outwitting the BMS. Locking windows which shut when external air temperature exceeds internal temperature may not be acceptable to the occupants in the building. Situations where the BMS promptly overrides an action taken by an occupant will be disliked and lead either to the abandonment of any involvement in environmental control or an antagonism to the building and the organisation which it represents[12].

6.4 Case studies

The control solution adopted for any individual building is likely to have its own particular characteristics. Choices are made by the designer to resolve problems or take advantage of opportunities presented by the project. It is therefore impossible for this Guide to list a comprehensive selection of whole-building control strategies that would cover all, or even most, BMSs. The remainder of the chapter presents some case studies of building management systems that have been successfully applied to actual buildings. The studies cover a range of building types and HVAC systems. The description of the control systems, while brief, serve to demonstrate the wide range of BMS applied in practice and serve to bring out some useful lessons.

6.4.1 Retrofit using a modular control system

Building Greenway School, Horsham
Type Junior School

Greenway School in Horsham teaches 400 7–11-year-old children in a group of buildings, comprising a main building and six classroom huts. Heating and hot water to the main building are provided by three gas-fired boilers and independent gas-fired convectors are used to heat the classroom huts. It was decided to keep the existing heating system and upgrade the controls to provide overall control from a single position, but allow individual teachers some flexibility in setting temperatures.

A modular control system was installed, consisting of one master and eight-slave zone controllers, plus a boiler sequence controller and the necessary sensors and actuators. All components are interconnected via a simple two-wire bus, employing the LonWorks protocol. The master zone controller is located in the school secretary's office and is used to set the time schedule and temperatures for all zones. In addition, the controller provides optimum start and weather compensation. The individual zone controllers in each classroom give a temperature display and allow limited alteration of set point.

The control system provides central control of time scheduling, bringing the entire system under central control and preventing any individual heaters being left on by mistake. No specialist knowledge is required for operation.

The modular controls chosen proved easy to install using the bus system. The straightforward interface allowed operation by non-technical staff.

6.4.2 Remote control of a group of homes for the elderly

Building 10 elderly people's homes, Clwyd
Type Residential homes

Clwyd Council operates a centralised BMS bureau and energy monitoring and targeting service, which has been successful in producing an overall reduction of 25% in energy consumption in buildings serviced. It was decided that a group of elderly people's homes operated by the social services department offered scope for improved control and monitoring. The ten homes are of modern construction and heated by a conventional boiler serving radiators and hot water services.

The energy conservation unit applies a three-year payback criterion for improvement schemes, which required a low cost solution. Each home was retrofitted with a stand-alone energy management system. This provided time and temperature scheduling, optimum start and weather compensation, including room reset control. Boiler operation includes sequencing control and improvement of efficiency by variable minimum off-time control, which increases the off-time during cycling as the flow and return temperatures get closer together, i.e. at light loads. Control of hot water services is also provided. Water heating periods are matched to kitchen operating times and the boiler flow temperature is adjusted as the duty moves between space and water heating. The controllers contain their own data-logging features and a modem, which allows communication between the controller and the BMS bureau over the public service telephone network. The controller can send routine data files automatically to the central PC and initiate alarms messages if required. If desired, alarms can be sent automatically to a fax machine. Staff at the central bureau may remotely view plant status or take direct control of operation.

The previous heating systems were poorly controlled and installation of the improved controllers produced reductions in gas consumption of between 20 and 30%. In addition to the savings produced by improved controller performance, the information provided by the monitoring and targeting service allowed other energy savings to be identified.

Lesson. The use of stand-alone controllers with built-in communication facilities allowed professional energy management of a number of dispersed buildings to be installed at low cost.

6.4.3 Low energy university building with fabric storage

Building: Elizabeth Fry Building, University of East Anglia

Type: Two floors of cellular offices and two floors of lecture and seminar rooms, plus kitchen and dining room

The building is highly insulated and well sealed. No mechanical cooling is provided. Supply ventilation air is tempered by passing though hollow core floor and ceiling slabs. A highly simplified diagram of the system as applied to the office area is shown in Figure 6.2. Gas-fired boilers provide heat to the AHU when required and a high efficiency regenerative heat exchanger is employed for heat recovery. Full fresh air ventilation is used during occupied hours. All windows are openable without restriction on their use. Ventilation to the lecture theatres is by variable speed fan controlled by the CO_2 concentration of the extract air. This compensates for the variable occupancy and gives substantial savings in fan running costs.

The building is thermally very stable and control of zone temperature is achieved by controlling the core temperature (Table 6.4), measured near the air outlets.

The Elizabeth Fry Building has successfully achieved a combination of very low energy consumption and high occupant satisfaction. The hollow core slab system achieves very steady temperatures and comfortable summer conditions without the use of air conditioning.

There were initial teething problems caused by inadequate controls and the lack of a proper BMS, which did not allow the maintenance staff to understand the operation of the various systems. Subsequently, new controls were fitted and integrated into the campus-wide BMS system. The availability of performance data was vital in allowing the control strategy and settings to be fine tuned and simplified. The commissioning and handover period extended over the first two years of occupancy. Cooperation between the building manager, controls specialist and the design team resulted in a well-configured system with a simple control strategy[10,13].

6.4.4 Mixed mode R&D facility using chilled beams

Building: Hewlett Packard, Building 3, Bristol

Type: Three-storey open plan building housing offices and computing laboratories

This building provides a research and development facility for up to 450 staff and was completed during 1998[14]. The three storeys provide two floors of open plan flexible space for offices and laboratories (Figure 6.3). The ground floor contains a presentation area, meeting rooms and a large coffee shop, which provides a centre of social interaction. A central atrium forms a central street; on the upper floors, glazed balustrades surround the central space, giving a feeling of light and openness.

The nature of the business activity produces a relatively high small power load of about 40 W/m^2. This demands mechanical cooling, which is provided by chilled beams. There is no natural ventilation. The building is well sealed, achieving a low leakage value with the assistance of concrete floor and ceiling slabs, which also provide thermal mass. Supply air is provided as 100% outside air through floor diffusers; extract is via air handling luminaires. The supply air is conditioned in the central AHU and delivered to the zones at a constant temperature of 19°C, though this may be

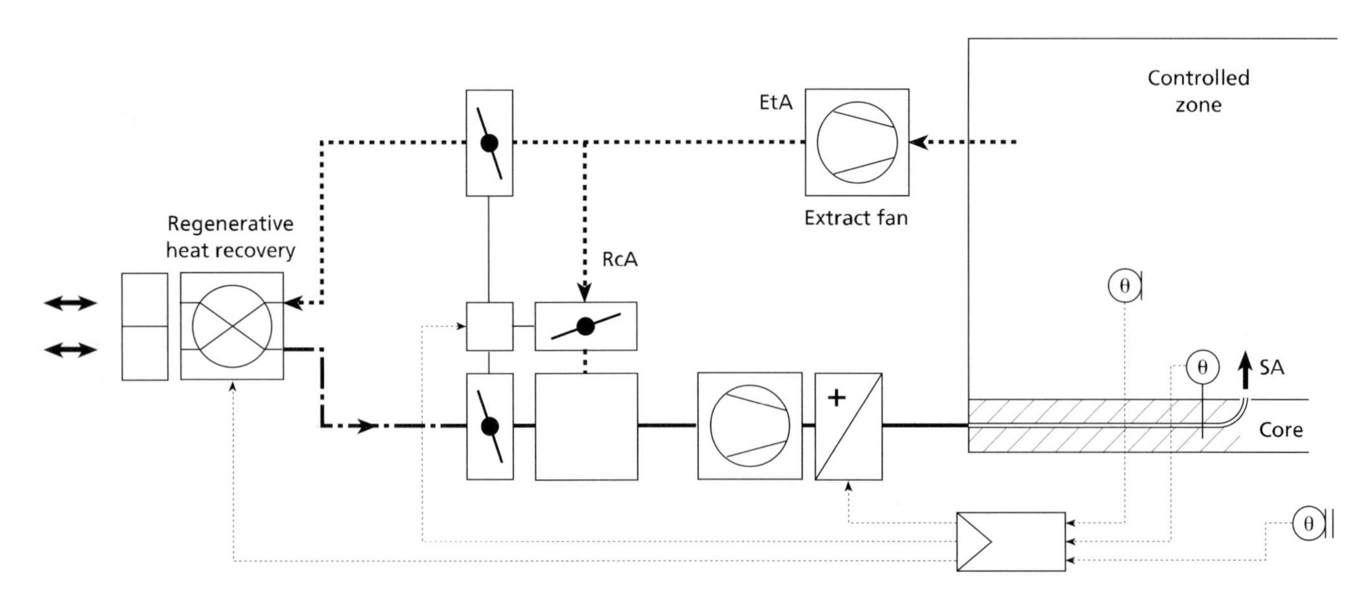

Figure 6.2 Hollow core slab system used in Elizabeth Fry Building. Full fresh air ventilation is used at all times during occupancy. Heat recovery may be deactivated

Table 6.4

Mode	Core temperature	Outside temperature	Heat recovery	Heating coil	Recirculation
Stage 2 heating	$T_c < 21.5$ for 15 min		Yes	Yes	No
Stage 1 heating	$T_c < 21.5$		Yes	No	No
Ventilation only	$21.5 < T_c < 22.5$		No	No	No
Stage 1 cooling	$T_c > 22.5$	$T_o < T_z$	No	No	No
Stage 2 cooling	$T_c > 22.5$	$T_o > T_z$	Yes	No	No
Night cooling	$T_c < 23$	$T_o < (T_c - 2)$	No	No	No
Night heating			N/A	Yes	Yes

T_c = Core temperature; T_o = Outside temperature; T_z = Zone temperature.

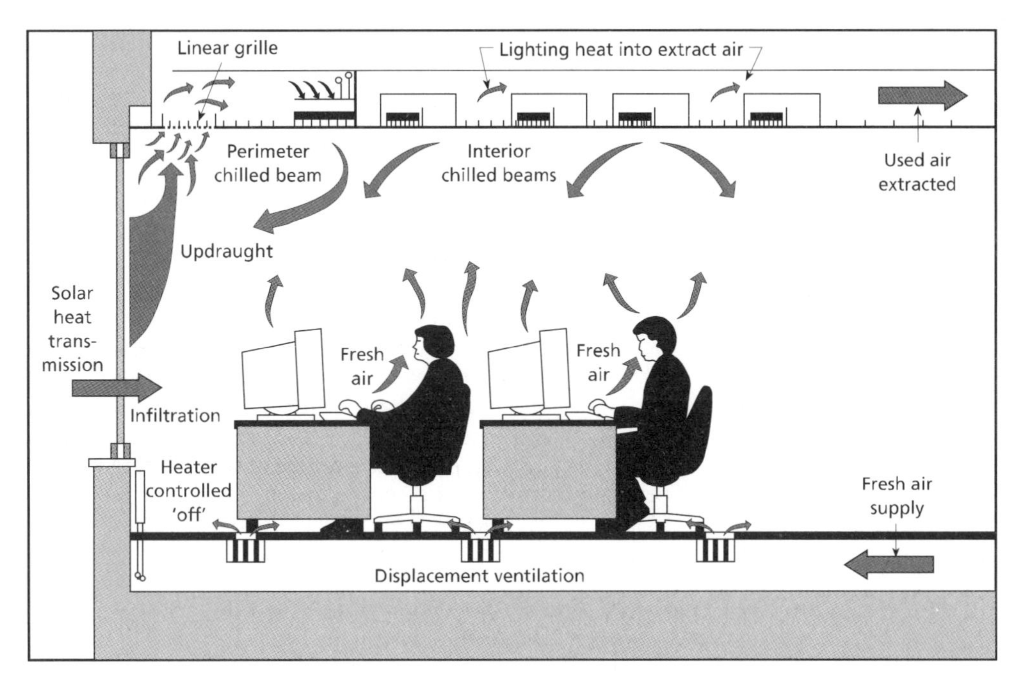

Figure 6.3 Cross-section of Hewlett Packard Building 3, showing chilled ceiling and displacement ventilation[14]

reset over the range 19–21°C if required in hot or cold weather. Full recirculation is used during warm-up. Supply air can be dehumidified to maintain the dewpoint below 12°C, to avoid condensation forming on the chilled beams; in this building no facility for humidification is provided. Perimeter heating to offset losses at the windows is provided by LTHW radiators. Additional radiators are fitted at high level in the atrium to prevent cold downdraughts in winter from the atrium roof. There is no external solar shading. Internal venetian blinds are fitted and a central blind controller sets the position and angle of the blinds. In strong sunshine, the blinds heat up and produce a convective updraught of warm air. To prevent this causing problems for the chilled beams, the perimeter is decoupled from the internal zone by drawing the air through a grille circling the perimeter. The air passes over a large cooling coil and returns to the occupied space. In hot weather, it is possible for hot air to accumulate under the atrium roof and work its way down into the second-floor offices. To prevent this, chilled water can be passed through the high level radiators.

The building has an advanced IT network. This is designed to be innovative and is subject to change. It was therefore decided to maintain the control network entirely independent of the IT system. The innovative nature of the system decided the consultants to construct a full scale mock-up of the perimeter zone to evaluate the control system under a range of operating conditions. These tests demonstrated a potential instability, where the perimeter radiators and chilled beams oscillated between heating and cooling. This was resolved by altering the control set points. Air movement in the atrium was investigated using CFD modelling, which demonstrated the need to use high level radiators, both to prevent downdraughts in winter and to limit overheating in summer. The thorough investigation of control behaviour at the design stage proved invaluable in resolving potential problems.

6.4.5 Terminal 2, Manchester Airport

Building: Terminal 2, Manchester Airport
Type: Large airport terminal

Manchester Airport's second terminal was completed in 1994. It was decided to use an integrated BMS which would provide a facilities management system which, in addition to controlling the extensive HVAC system, would interface with many other systems and provide the following control and monitoring:

- Control
  - HVAC: VAV systems in terminal, shops and offices
  - lighting
  - airport services: telephone system, baggage handling, escalators
  - runway drainage
  - access control: to plant rooms.

— Monitor
  — energy: electricity, water, gas and provide billing
  — life safety: fire fighting, smoke detection
  — apron services: ground power, battery charging.

The system is designed to provide full multi-user functionality and operate using the airport's existing fibre optic structured wiring system. It utilises a two-tier network topology. The top tier sits on the Manchester Airport fibre optic network; master controllers are located in plant rooms and workstations in various control rooms are connected directly to the network. The lower tier is then run locally to pick up controllers located in motor control centres, substations or VAV boxes, using coaxial cable or UTP as appropriate.

As far as possible, control systems were designed as generic copies of existing systems, allowing the use of existing reliable software. Commissioning was facilitated by positioning the site team's temporary buildings so that connections could be made to the structured wiring system at an early stage. This provision of a temporary location for the supervisor allowed programs to be downloaded into the controllers as soon as the network was complete. The system has proved capable of expansion. It now comprises over 8000 points and 1000 graphics panels. It can cope with multi-users. Touchscreen display panels are installed in shop and office areas, allowing adjustment of a restricted range of set points and variables.

6.5 Summary

Efficient operation of a building requires that the subsystems do not interact wastefully. All possible operating modes of the building should be listed systematically together with the operating states of the component subsystems. This will assist in identifying unsatisfactory situations, e.g. where heating and cooling may be operating simultaneously. The involvement of the building occupant in the operation of control devices can bring benefits in increased comfort and satisfaction with the environment. However, concurrent operation of automatic control systems and occupant action may produce dissatisfaction if the user is overridden by the automatic system. Careful attention is required to produce a combination of user and automatic control that is neither wasteful nor over-complex and self-defeating.

Several techniques are available to predict building performance, either analogue modelling, full scale mock-up, or various mathematical modelling techniques. Modelling can anticipate potential control problems and allow them to be corrected at the design stage. The section summarises the more common combinations of heating, cooling and ventilation and discusses the requirements for them to work together satisfactorily.

Design lessons are illustrated by a range of case studies, which range from simple heating-only classrooms to large complex integrated buildings. The choice of the appropriate system and strategy for the task in hand is emphasised.

References

1 Martin A J and Banyard C P *Library of system control strategies* Applications Guide AG 7/98 1st ed (Bracknell: Building Services Research and Information Association) (1998)

2 Honeywell *Engineering manual of automatic control* SI ed (Minneapolis: Honeywell Inc.) (1995)

3 Siemens *MSR Planungshandbuch* (CD-ROM) (Karlsruhe: Siemens AG) (1997)

4 Lane-Serff G F, Linden P F et al. Laboratory and mathematical models of natural ventilation *Roomvent'90 Proc. 2nd International Conf.* Paper A2-7 (Oslo, Norway: Norsk VVS) (1990)

5 Etheridge D and Sandberg M *Building ventilation theory and measurement* (Chichester: Wiley) (1998)

6 Brister A Cable Talk Building Services. *CIBSE Journal* **15**(11) 22–26 (1993)

7 *Building energy and environmental modelling* CIBSE Applications Manual AM11 (London: Chartered Institution of Building Services Engineers) (1998)

8 Lebrun J and Wang S *Evaluation and emulation of building energy management systems* Synthesis report AN17-SR1 (Liege, Belgium: University of Liege) (1993)

9 Alamdari F and Eagles N *Displacement ventilation and chilled ceilings* TN 2/96 (Bracknell UK: Building Services Research and Information Association) (1996)

10 Bunn R Chilling lessons *Journal of the Chartered Institute of Building Services* **17**(5) 37–38 (1995)

11 *Mixed mode ventilation systems*: CIBSE Applications Manual (London: CIBSE) (1999)

12 Probe 18: Portland Building *Building Services Journal* **21**(1) 35–40 (1999)

13 Probe 14: Elizabeth Fry Building *Building Services Journal* **20**(4) 37–43 (1998)

14 Pearson A HP sourced *Building Services Journal* **19**(12) 12–16 (1997)

7 Information technology

7.1 Energy monitoring
7.2 Fault reports and maintenance scheduling
7.3 Summary

The building management system provides an important source of information, which may be exported to other applications. Building performance and energy consumption are analysed by energy monitoring and targeting software, to assist in effective energy management. Information from the BMS is employed as part of facilities management, including the administration of planned maintenance.

7.1 Energy monitoring

A major justification for the installation of a BMS is the prevention of energy waste. Correct and efficient control of HVAC plant will itself contribute to efficient use of energy but in addition it is necessary that the energy manager receive accurate and up-to-date information on energy use in the buildings for which he is responsible. The approach generally adopted in the UK is termed energy monitoring and targeting (M&T), which is vigorously promoted by the DETR. The term monitoring and targeting does not in itself imply the use of a BMS or indeed any hardware for data collection and analysis. M&T requires that data on energy consumption be regularly collected, summarised and compared with target consumption figures. Computerised collection and analysis of data makes M&T a powerful tool for the control and reduction of energy consumption[1].

7.1.1 Monitoring and targeting

M&T has two major functions:

- the control of current energy use, by monitoring consumption and comparing it against historical data and benchmarks for similar buildings
- improvements in the efficiency of energy use by the setting of future targets.

In large buildings or complexes, effective energy management requires that areas of accountability be established, termed energy account centres (EAC). Accountability implies both responsibility for the energy consumed and the authority to control the consumption. The energy consumption of an EAC should be

- measurable
- manageable
- reconcilable against a measured activity.

An EAC could be a department, an energy intensive production process, a sublet part of a building or an individual building on a site. A standard energy performance is initially established for each EAC, which relates energy consumption to appropriate variables, such as degree-days or production output. The standard performance provides a benchmark against which energy consumption may be compared, prior to the setting of targets for future improvements. The targets must be realistic and achievable and agreed with the managers responsible for each EAC. BRECSU have published a number of Energy Consumption Guides, e.g. Guide 19[2], and consumption data for a wide range of non-domestic buildings is summarised by Jones and Cheshire[3]. Additional information is to be found in CIBSE Applications Manual AM5[4].

The monitoring part of the M&T process involves four stages:

(1) *Data collection*. Energy consumption data is collected. A comprehensive system will include electricity, gas and water meters located in each EAC, capable of sending data to a central collection point. The availability of other relevant data must be taken into account, e.g. degree-day data or production figures.

(2) *Data analysis*. The data received by the computer is checked for errors and then stored in a form suitable for further analysis. Data analysis is carried out according to the needs of the system; this is likely to include weekly and monthly totals. Any results which indicate a problem or malfunction should generate an immediate warning.

(3) *Reporting*. Management reports produced by the system which show the energy consumption of each EAC compared with targets.

(4) *Action*. It is essential that a management structure exists to make effective use of the reports generated by the M&T system.

For an M&T system to be effective, all four stages must be implemented. The collection of data, however accurate, which is not looked at and which prompts no action serves no purpose. Effective M&T requires full management support. As a rule-of-thumb, M&T may be expected to reduce energy bill by about 5%; larger savings are possible where equipment malfunction or gross control failures are identified as the result of metering information. This figure allows an appropriate level of M&T to be chosen in relation to the total energy bill. Energy management systems share many of the characteristics of a BMS. In a building the system installed for energy monitoring and targeting may share hardware with the BMS and may be operated from the same terminal by the same staff. However, the they are not the same. The BMS provides real-time control of the building services, while the M&T system is concerned with data collection and historical analysis.

7.1.2 Planning an M&T system

Energy M&T may be instituted at a simple level, using invoices or manual meter readings and simple manual or spreadsheet analysis. The major factors to be taken into account at an early stage are summarised below. The most important is the total energy bill to be monitored, since it is the potential savings on this bill that fund the M&T system. M&T systems may be classified according to their level of coverage and sophistication.

7.1.2.1 Level of coverage

— *Single site, utility based*. The site is treated as a single EAC and monitored using only utility meters and invoices. Here, a site implies a building or group of buildings served by a single utility meter. It may be possible to aggregate physically separate buildings into a single site for supply and metering purposes.

— *Single site with submeters*. Energy is monitored for each of the EACs within a single site by means of submetering.

— *Multi-site, utility based*. A number of sites are monitored using the main utility meters. Each site is treated as a single EAC.

— *Multi-site with submeters*. Several sites, each of which is subdivided into EACs and submetered.

7.1.2.2 Level of sophistication

— *Manual system*. Meters are read manually and tabulated on paper.

— *Keyboard input system*. Meters are read individually, and readings recorded by hand or by a data capture unit. Readings are then entered into a computer for analysis.

— *Automatic input system*. The meters are connected via data loggers or other system to the computer and energy consumption is collected and monitored automatically.

— *Advanced system*. Implies more sophisticated data handling, combining energy data with data from other systems or sites. May include automatic control or provide warning messages. Sophisticated bureau services are available using neural network analysis of consumption data to detect energy waste, coupled with an expert system to identify causes.

7.1.2.3 Choice of system

It is important not to choose an oversophisticated system. Not only will such a system cost more to install, but it will demand staff time to maintain it in an operational state and to take action on the results. If the requisite support is not available the system will not fulfil any purpose. The following points should be considered at an early stage in the planning:

— *Cost effectiveness*. Any proposed system will be required to meet the organisation's criteria for return on investment. The cost of a system is sensitive to the number of meters installed. Not only does this affect the installed cost, but the analysis, reporting and storage of the increased amount of data will add to the running costs. The rule-of-thumb that M&T should save 5% of energy costs may be used to provide indication of a viable level of investment.

— *Number of meters*. Each EAC requires its own meters. Electricity meters will always be required, together with gas meters where appropriate. Water meters are recommended, particularly if there are any water intensive processes. Further submetering may be desirable for particular processes or equipment. The consumption of some areas may be calculated using the concept of a virtual meter, i.e. the difference between a main and submeter.

— *Monitoring period*. This is the period between meter readings. For a fully automated system this should be half-hourly. The use of time-of-day electricity pricing is spreading from large customers to smaller or even domestic customers and half-hourly information will be necessary when considering the optimum policy for electricity tariffs. The additional information given by half-hourly metering is useful for diagnostic purposes, helping to identify reasons for anomalous consumption.

— *Reporting*. The M&T system should produce regular standardised reports, showing performance against target. Weekly and monthly periods are commonplace. There is no point in producing reports more frequently than they will be read. It is important that the information is usable by the intended audience of the report. Ideally management reports should highlight any avoidable waste and define responsibility for improvement.

— *Management organisation*. Figure 7.1[5] shows how M&T operates within the management structure. Each EAC requires a manager responsible for energy performance within the centre, and who will receive routine reports. It is essential that sufficient staff are trained in the operation of the system and a minimum of two staff should be able to operate it. If this is not practicable, the use of a bureau service should be considered. Metered data is read

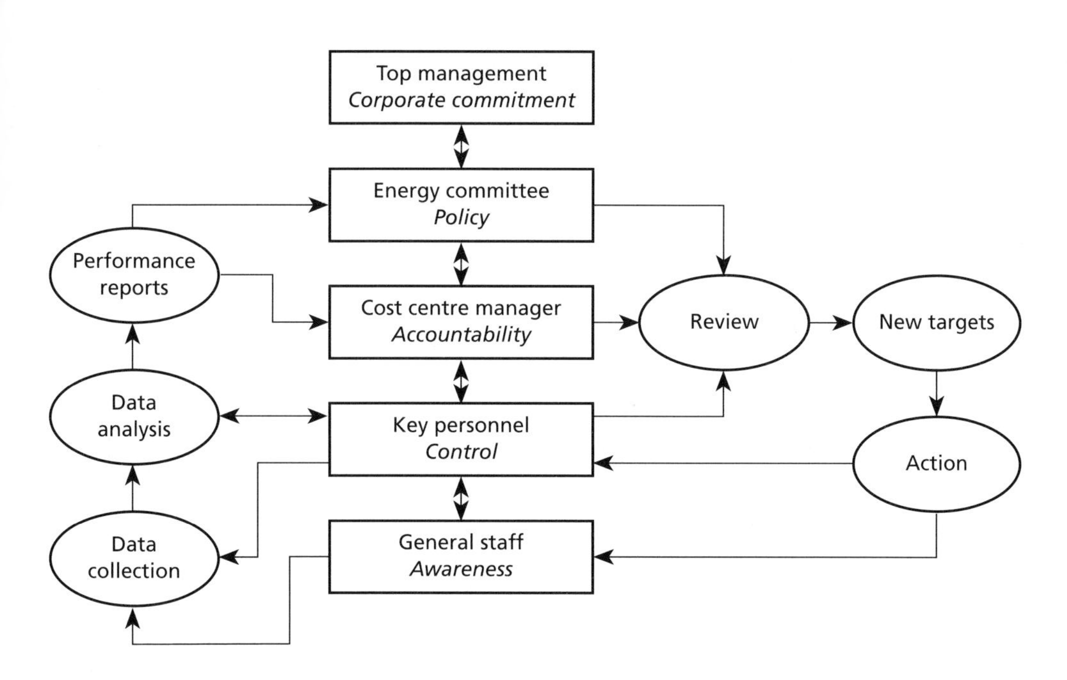

Figure 7.1 Monitoring and targeting management structure

remotely via modem or the Internet and the bureau produces the required analysis, reporting and billing as required.

7.1.2.4 Electricity Tariff

The choice of electricity tariff has its own relevance to the design of a M&T system. Since 1994, customers with a total demand of over 100 kW have been able to negotiate a supply from any electricity supplier; the freedom has since been extended to all customers. The type of tariff negotiated will have a bearing on the design of submetering and the energy management system. The major factors to be considered are:

- aggregation of sites
- choice of tariff.

It may be advantageous to group separate buildings or sites together so that they are supplied as a single unit through one main meter. The Office of Gas and Electricity Markets (OFGEM) has set out requirements for the main meter. It must comply with the Pooling Settlement Agreement document known as Code of Practice 5 and it must be maintained by an approved meter operator. The meter must record all units used each half hour and transmit the stored readings once per day to the UK Data Collection Service, which then passes the information on to the interested parties. The readings will normally be made available to the building operator. Users with a demand of less than 100 kW may continue to use a conventional meter.

The customer may install further meters downstream of the main meter; this is known as secondary metering or submetering. Secondary metering is installed for:

- tenant billing
- cost centre accounting
- energy management.

Secondary metering is not constrained by the same level of regulation as primary metering, though any meter used for tenant billing has to be OFFER approved.

With the growth in the number of competing electricity suppliers, there are several types of tariff available. The major classifications are:

- *Two-rate*. Separate rates are charged for day and night.
- *Six-rate STOD*. The unit rate varies by season and time of day.
- *Disaggregated*. This identifies the components of electricity use which contribute to the total charge, e.g. energy used, maximum demand, power factor.
- *Pool pricing*. The unit charge is based on the half-hourly spot price of electricity.
- *Tariffs requiring the measurement of reactive power*. These are mostly found in industry.

It is outside the scope of this guide to deal with the choice of tariffs. The price of electricity in the pool can vary substantially from half hour to half hour and pool pricing is only suitable for large users where the necessary active management of load can be supported. A disaggregated tariff will be suitable for sites where demand can be managed to reduce costs; this is likely to be true where there are energy intensive processes that can be controlled to contain the maximum demand. The tariff chosen will affect the level of submetering and type of energy management. Half-hourly metering is almost always worthwhile as detailed demand patterns can assist in the identification of problems or identification of excessive electricity consumption. Management of a disaggregated tariff requires information on maximum demand available in real time with some form of warning to enable action to be taken.

7.1.3 Metering equipment

Energy metering hardware in an energy management system consists of some or all of the following components:

- meter module, which measures the desired quantity and converts the value to an electrical output, typically pulses

- display module, which displays the present value of the rate of energy consumption, plus other derived quantities
- data logger, which accepts pulses from the meter, processes and stores data on energy consumption and transmits data on demand to the central computer containing the M&T software
- data transmission system, connecting one or more data loggers to the central computer
- computer, containing the analysis software.

Meter module

An electric meter module (Figure 7.2) requires simultaneous inputs of current and voltage. In a whole current meter, the entire current is wired to pass though the meter. They are suitable for single-phase applications, up to a current of about 80 A. Otherwise, a current transformer (CT) is used which is fitted around each phase conductor. Standard ring type CTs require the cable to be isolated and disconnected for installation, while split core CTs can be installed without disturbance. Clamp-on CTs are used for temporary connections with portable equipment. The secondary current output of the CT is measured by a low impedance meter; the resistance of the connection between CT and meter must be kept low or an appreciable error may be introduced. Simultaneous measurement of voltage and current is required to give an accurate measurement of power consumption, and is essential for the calculation of kW, kVA, kVAr and power factor. The harmonic content and reactive component of many modern electrical loads can give rise to substantial measurement errors if current only metering is used. However, on sites where the power factor is high and the voltage stable it may be sufficient to employ only a current connection to the meter. Where the aim of energy management is the partition of energy consumption between EACs and the detection of excess consumption, this will normally be sufficient.

Most meters provide a pulsed output in the form of voltage-free contact closures. This is true for gas, water and other meters as well as electricity meters. A defined pulse, typically about 100 ms in duration, is provided each time the meter records an equal increment of energy consumption. The scaling is chosen to give a pulse frequency of around 1 Hz at maximum power. It is possible to fit optical meter readers to existing utility meters without any disturbance to their connections. The reader employs a photoelectric device to detect the rotation of the meter disc or dial needle and provides an output pulse which may be handled in the same way as a conventional pulse output.

For most purposes it is sufficient to measure kWh alone. Where more detail is required, a dual meter module measuring kVA and kVAr may be employed. From these quantities kWh, power factor and maximum demand may be derived. It is unlikely that automatic logging of any other electrical parameters will be required unless there are concerns over the quality of supply. If necessary, a portable power analyser may be used to give detailed measurements of power quality. The measurement and display on instantaneous power (kW) can be useful for diagnostic purposes, such as identifying loads in the building and tracking down any ‘hidden loads’.

Display module

The meter may incorporate a display or be connected to a local display module. At its simplest, this shows a real-time display of the rate of energy consumption, e.g. kW. More complex displays can select kVA, kVAr and have sufficient memory to show maximum demand and cumulative energy consumption.

Data logger

The output pulses from the meter are summed in a data logger module. This will typically have several input channels for different meters. The data logger may be part of a separate energy management system. Alternatively, where the energy management system is an integral part of the main BMS, intelligent BMS outstations are able to accept pulsed inputs for energy measurement purposes (Figure 7.3). The pulses are counted, scaled and stored to give values of energy consumption per half hour for each

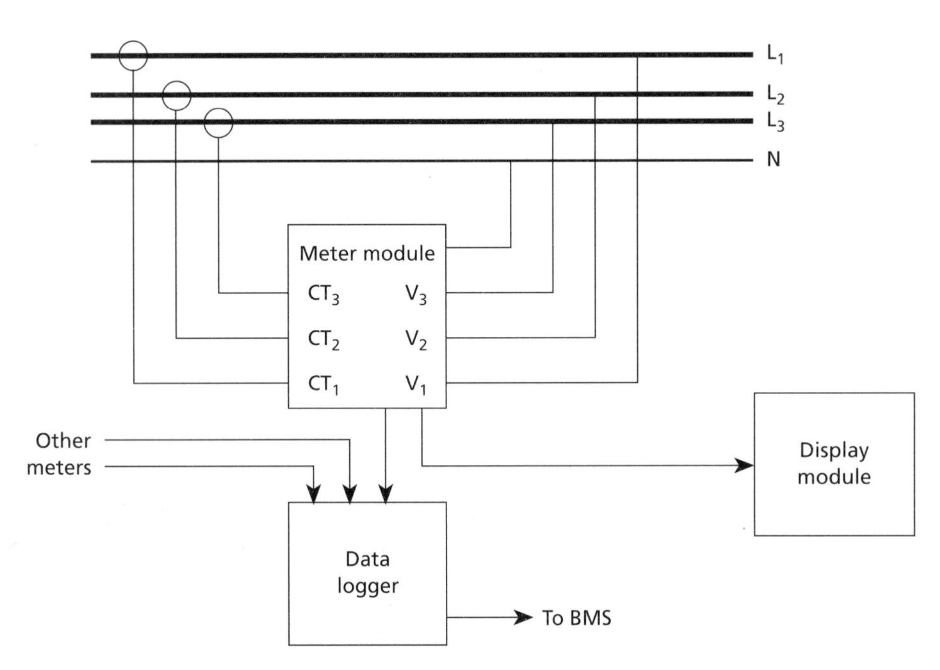

Figure 7.2 Components of an automatic electricity meter

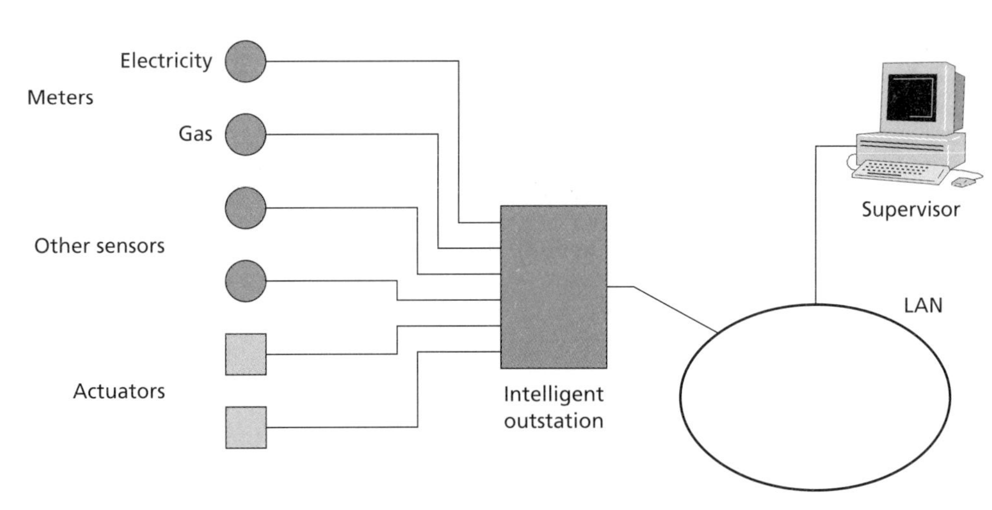

Figure 7.3 Incorporation of metering into a BMS

channel. There must be a real-time connection between the meter and the data logger. Hardwiring is the most common, being simple and reliable. Where the meters are dispersed over a large site, it is possible to use a radio link between the meters and logger. Where the energy management system is separate from the BMS, it may be advisable for the EMS data logger to accept temperature inputs. This makes possible analysis of energy consumption as a function of temperature without any additional data from sources outside the system. Meters are available with built-in communications systems such as RS 485, typically using the MODBUS communications protocol. This is preferred to pulsed output where it is required to measure variables other than simple kW h.

Computer

The central computer receives and processes the metered data. The computer may be a separate machine dedicated to M&T or the software may reside in the main BMS supervisor. Where a bureau service is used, the computer is situated off site and communicates via modem or Internet. The software can carry out some or all of the following functions:

- retrieval, checking and collation of metered data
- acceptance of data from other sources
- instant access on demand to data from individual meters
- review of data in flexible format, allowing comparison between different time periods and meter locations
- display of additional electrical data such as power factor, frequency, reactive power
- comprehensive input of tariff definitions
- automatic production of sub-billing
- automatic production of routine reports
- immediate warning of exceptions, such as parameters exceeding preset limits
- historical analysis of energy consumptions, e.g. regression or cusum.

7.1.4 Data analysis and reporting

For a simple system it may be sufficient to use a spreadsheet for data analysis. However, once there are several meters with half-hourly energy readings, it will almost certainly be better to employ a commercial software package for analysis.

7.1.4.1 Data input

The analysis software can only operate properly given a complete set of accurate data. During commissioning of the system, it is important to check that each meter channel is correctly identified and that all scaling factors have been correctly set. During routine operation, the software can be set to carry out routine checks on the incoming data:

- Is the data within preset limits?
- Is any data missing?
- Is the new reading higher than the previous reading?

Invalid readings should be flagged for the operator's attention. If the analysis proceeds with dubious input data, the subsequent reports should be so annotated. Subsequent energy analysis may involve data from other sources, such as degree-days or production figures. This data has to be provided reliably and in a form which is compatible with the metered data, e.g. the periods over which production output and energy consumption are measured must coincide. At some stage the measured energy consumptions should be compared with the fuel invoices. Where an M&T system is installed in an existing building, it is desirable to input historical consumption data to provide a basis for comparison. It may be difficult to provide data in a suitable format to match the new EAC organisation.

7.1.4.2 Routine analysis

The analysis package will be programmed to produce routine analysis every reporting period. The type of analysis will be tailored to individual needs:

- attribution of energy costs to each EAC, including tenant billing
- comparison of energy consumption with target figures.

In addition, further analysis can be undertaken which will be used in reviewing performance and setting future targets. Energy consumption often shows a simple linear relationship with an independent variable such as degree-days or production. A linear regression analysis establishes the slope of the line and intercept on the consumption axis. This can be used to estimate base consumption such as hot water loads, which do not vary with external temperature (see Figure 7.4). Once a historical regression has been established, any new point which falls off the regression line demands attention. A degree-day regression is subject to a scatter of around 10%; the analysis may be improved by refining the choice of base temperature used in compiling the degree-day figures.

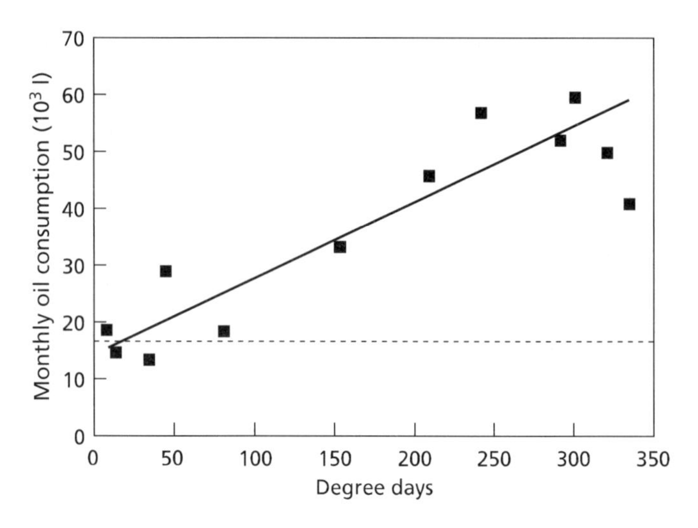

Figure 7.4 Linear regression of energy consumption against monthly degree days, showing the summer base load[4]

CUSUM

The CUSUM technique[6] compares actual performance with target performance for each measurement period and displays cumulative savings. It is necessary to be able to calculate a target consumption for each analysis period; this may be a function of other variables, such as outside temperature. By calculating the cumulative sum of the differences from the target, a trend line can be plotted that gives a clear indication of performance and changes in performance. The value of the cusum gives total saving to date and the slope gives the performance trend. A change in performance of the building system will be shown by a change in slope (see Figure 7.5).

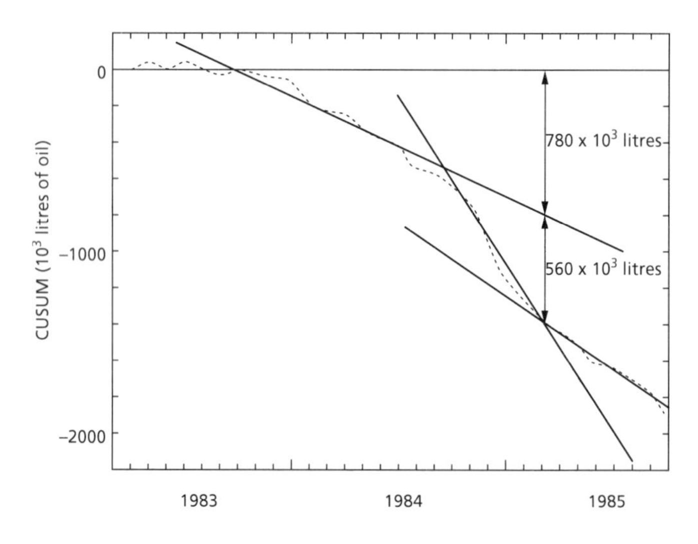

Figure 7.5 Example of a CUSUM plot showing cumulative energy savings[7]

7.2 Fault reports and maintenance scheduling

The BMS can be used to provide information on the total run time and condition of HVAC components, which can be integrated into maintenance operations. Ideally, a maintenance policy should be decided for each component of the HVAC system, taking into account risk, technical and cost issues. Five maintenance categories may be classified:

- run to fail
- install redundant units
- preventive maintenance program
- condition-based maintenance
- redesign to reduce maintenance.

The appropriate maintenance policy for each component can be chosen on the basis of the decision tree shown in Figure 7.6.

Preventive maintenance tasks are carried out either at regular calendar intervals or at intervals based on equipment run time. Preventive maintenance ranges from simple lubrication to complete tear down and rebuild of equipment. Selection of the maintenance period is of great economic significance and is usually based on manufacturers' recommendations. Planned maintenance software is available which is used to:

- catalogue maintenance tasks for various plant items
- generate and manage work orders
- manage inventory for spare parts
- store maintenance history
- generate management reports.

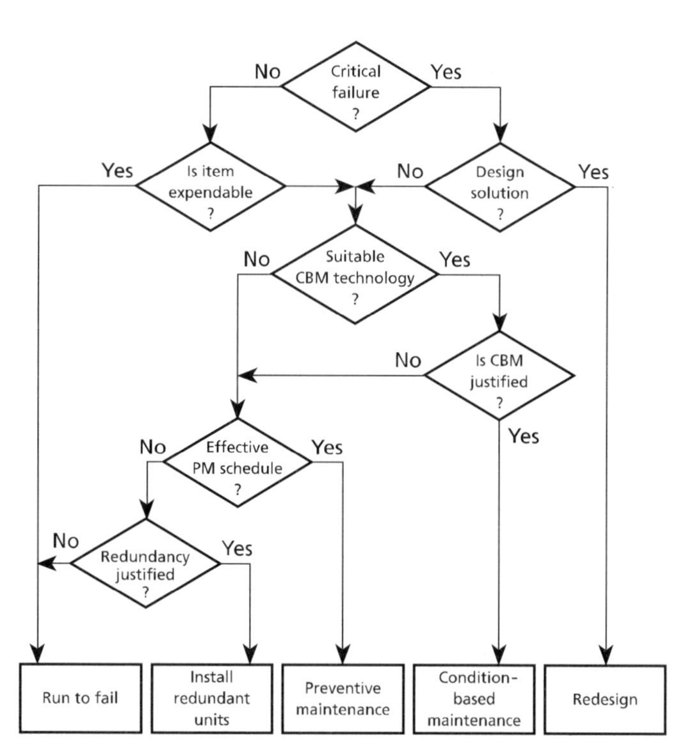

Figure 7.6 Selection of maintenance strategy

The BMS is used to provide information on equipment runtime, which is fed directly to the planned maintenance program.

Condition-based maintenance

In some situations it is possible to monitor the condition of an item of plant. This provides valuable information on system performance and reliability and can detect early signs of trouble in an item of plant, allowing maintenance to be carried out before the plant progresses to failure or when it begins to operate with reduced efficiency. This technique will normally provide greater intervals between maintenance than preventive maintenance carried out at fixed intervals. The economic justification is based on the relation between the increased cost of CBM balanced against the saving from increased maintenance periods.

CBM encompasses a set of techniques used to monitor plant condition. Some may be carried out automatically via the BMS; others require a physical inspection or measurement. Examples of direct monitoring include:

— *Pressure drop across filters*. This indicates when they need to be changed. The analysis algorithm will depend on whether the system is constant or variable flow.

— *Efficiency of heat exchangers*. This may be monitored by measuring temperatures and flows, and so detect degradation of performance caused by fouling.

— *Operating variables*. A selection of plant variables, such as temperature and pressure, is logged. A long term change in operating conditions may indicate degradation in the equipment.

Other monitoring techniques are not yet suitable for integration into the BMS, though this may become possible in special situations:

— *Vibration analysis*. Used to detect wear in rotating machinery.

— *Infra-red thermography*. Can detect high resistance connections in electrical equipment, high friction in rotating machinery and failing thermal insulation.

— *Analysis of lubricants*. Used to detect wear of gears and other mechanical parts.

— *Motor current analysis*. Used in large electrical motors (> 20 kW).

7.3 Summary

The building management system can be used to support energy management and targeting by measuring components of energy consumption and exporting the data to a EM&T applications package. The use of EM&T can save 5% of energy costs; this yardstick may be used to support decisions on the level of investment in metering. Secondary meters may be used for tenant billing as well as energy management. Monitoring via the BMS may be used to support maintenance management. Information on plant run times and condition monitoring is fed into a planned maintenance program which produces maintenance schedules.

References

1 ETSU *Computer-aided monitoring and targeting for industry* Good Practice Guide 31 (Harwell: ETSU) (1991)

2 BRECSU *Energy use in offices* Energy Consumption Guide 19. (Watford: Building Research Energy Conservation Support Unit) (1998)

3 Jones P G and Cheshire D Bulk data for benchmarking non-domestic building energy consumption *CIBSE/ASHRAE Joint National Conference*, Harrogate: 29 Sep–1 Oct 1996 1. 203-213 (London: Chartered Institution of Building Services Engineers) (1996)

4 CIBSE *Energy audits and surveys* Applications manual AM5 (London: Chartered Institution of Building Services Engineers) (1991)

5 *Energy efficiency in buildings* CIBSE Guide F (London: Chartered Institution of Building Services Engineers) (1998)

6 Harris P *Energy monitoring and target setting using CUSUM* (Cheriton: Technology Publications) (1989)

7 *Degree days* Fuel Efficiency Booklet No. 7 (London: Department of the Environment, Transport and Regions) (1993)

8 Management issues

8.1 Procurement options
8.2 Design and specification of a controls system
8.3 Tendering process
8.4 Commissioning
8.5 Operation
8.6 Occupant surveys
8.7 Cost issues
8.8 Summary

Effective planning is essential to ensure that the final building is properly controlled. This section describes different variations of the building procurement process, setting out the responsibilities of the parties involved, with emphasis on the part played by the controls specialist. The importance of good planning is emphasised in ensuring that the control system is properly specified, installed and commissioned. Operation and maintenance of the control system subsequent to handover is also dealt with. The financial benefits of the investment in a BMS may be analysed using the life cycle costing methods described in the section.

8.1 Procurement options

The form of procurement chosen by the client can have an important influence on the way in which the controls are designed and installed. Irrespective of the procurement method, we may define the main parties involved in the building process as:

- *Client*: the customer for whom the building is procured and who pays the cost.
- *Contractors*: companies that construct the building. There may be firms that manufacture and install specialist subsystems and work as subcontractors to the main contractors.
- *Consultants*: firms that offer design and cost control services and are independent of any commercial interest in construction companies.

The procurement of a large building is a complex process and over the years different forms of organisation and management have developed. *Thinking About Building*[1] identifies four main procurement options. Many variations of these basic structures are possible, and are discussed at length in Turner[2]. The major systems may be classified as follows:

- Design combined with construction
  - design and build
  - design and manage.
- Design separate from construction
  - traditional
  - management methods.

The structure of the options is shown in Figure 8.1 in simplified form. The following describes their main characteristics that are relevant to control systems.

8.1.1 Design and build

With design and build, the client places a contract with a single contractor who has responsibility for both design and construction. The client may appoint an adviser to act as employer's agent, to advise on the preparation of the client's brief, evaluation of tenders and to provide independent advice throughout the project. Design and build has advantages:

- The client has single point responsibility from one organisation.
- Because the contractor has responsibility for both design and construction, economies should be possible.

A well-written client's brief is essential to the achievement of a satisfactory controls solution in the final building; an inadequately detailed initial brief may lead to the contractor providing an absolute minimum specification. A variation is known as develop and construct, where the client uses a design consultant to produce a scope design, before obtaining tenders from contractors who develop and complete the design and then construct the building.

8.1.2 Design and manage

A single firm is appointed to design, manage and deliver a project. This is similar to design and build, except that the design and management contractor does not carry out the construction, but places this with a construction contractor. The common variations of design and manage are:

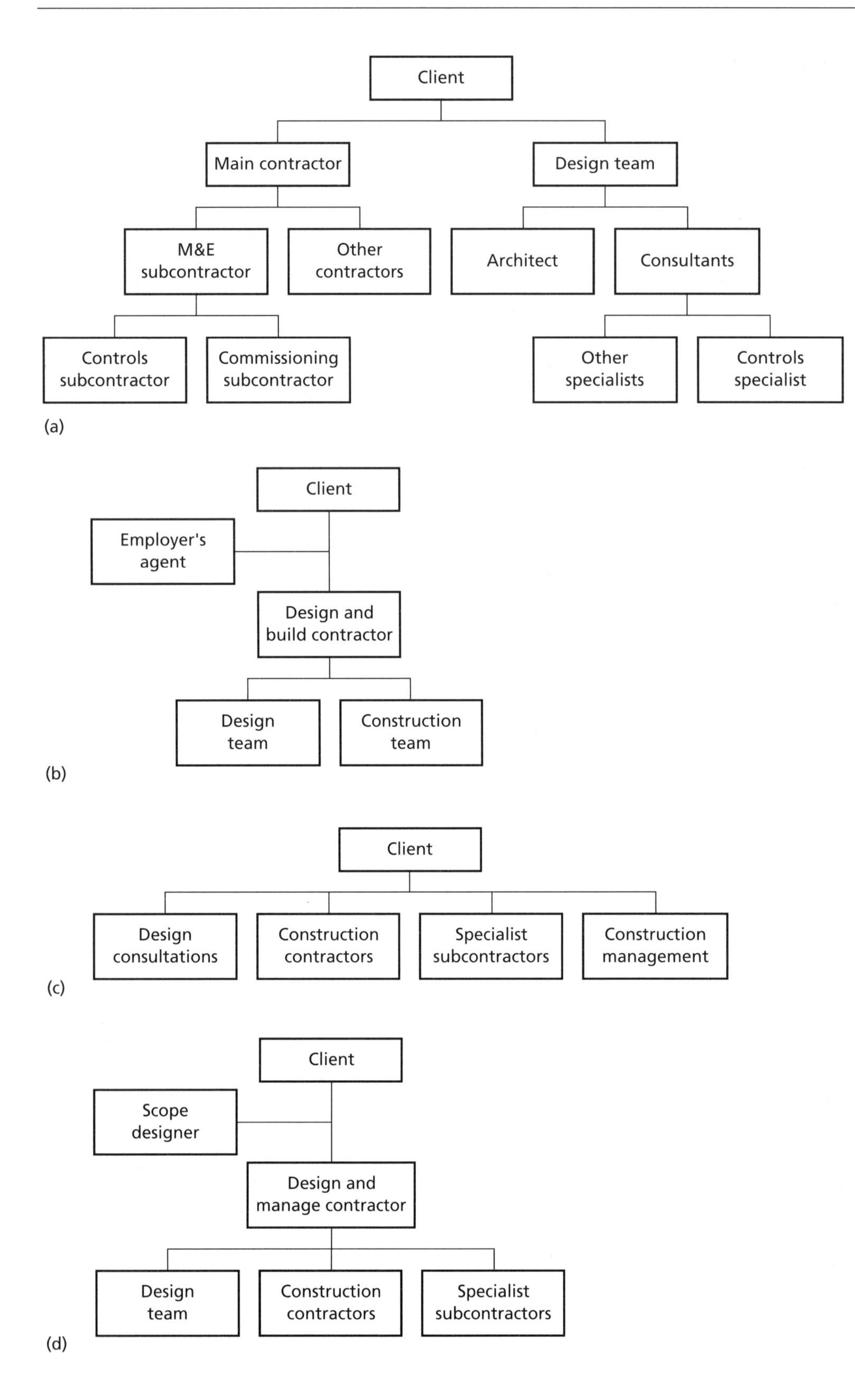

Figure 8.1 Major procurement options. (a) Traditional (lump sum) contracting relationships. (b) Design and build. (c) Construction management. (d) Design and manage

— *Contractor*. A project design and management organisation designs and manages the work, generally for a fee, and delivers the project by employing works contractors to design and or construct.

— *Consultant*. An organisation that is the client's agent designs and manages the work and obtains subcontracts from works contractors who then each enter into a direct contract with the client.

8.1.3 Traditional

The traditional procurement method separates design from construction. The client appoints design consultants who prepare a detailed design of the building. The design is put out to tender, following which a main contractor, responsible to the client, is appointed. The main contractor may appoint subcontractors. During construction, the design consultants exercise a supervisory role. This method is well understood in the UK and has the advantages of:

— providing clear contractual responsibility for each aspect of the work
— a well-understood procedure.

From the controls point of view, it has disadvantages:

— There is no involvement of specialist subcontractor at the design stage.
— Modifications to the design during construction are expensive.

8.1.4 Management methods

Management methods separate to a greater or lesser extent the function of management from those of design and construction. They have become associated with projects which are large and complex, and where early completion is required. Management methods give more flexibility to modify requirements during the course of the project and design and construction may overlap. Reduced project times are generally possible by continual re-evaluation of options and reallocation of resources. This allows variations to be introduced more easily than the traditional method but can result in increased costs. This flexibility may make it difficult to assign responsibility in case of dispute. There are two main variations of management methods:

— *Construction management*. Design and construction are still separate functions, but a construction manager is appointed to manage the whole process. The manager is appointed by the client; consultants and contractors have a direct contractual agreement with the client.
— *Management contracting*. The appointed management contractor provides the services of managing for a fee all the works contractors by employing them as subcontractors.

8.1.5 Relations between parties

There are many variations of the above, which are summarised in Turner[2]. The traditional organisation still accounts for the majority of new buildings, particularly for the smaller operations. Construction management is still relatively uncommon, except for large complex projects. Design and build is increasing. The form of organisation can affect the process of designing and installing a control system. In the traditional system, the controls specialist subcontractor is at the end of a contract chain, which may mean that available money and time are insufficient to produce the required quality of installation. The exclusion of the specialist contractor from the design process may mean that their experience is not taken into account. Fragmentation of the design consultancy can mean that controls consultants do not have an input into the design of the HVAC system itself, leading to the common complaint among controls consultants that the systems they are presented with are inherently uncontrollable.

The general awareness that procurement practices could be improved led to a review being set up under the chairmanship of Sir Michael Latham to consider:

— current procurement and contractual arrangements
— current roles, responsibilities and performance of participants, including the client.

The Latham report *Constructing the Team*[3] recognised that the efficiency of the British building industry could be increased by improving the interrelationship of the many parties involved. Many of the report's recommendations have been implemented, or are in the process of being set up. However, a subsequent report[4] found that the UK M&E contracting industry lagged well behind that of other countries. Among a number of factors, the report identified the following which are relevant to controls subcontracting:

— lack of project planning and organisation
— poor quality design of services
— lack of approved installation drawings
— inadequate use of prefabrication and pre-assembly.

Further recommendations for improvement in both quality and efficiency were contained in the Egan report *Rethinking Construction*[5], which proposed a target of a cost reduction of 10% per year plus a reduction in defects. The report strongly favoured the use of partnering methods, which are described further below.

8.1.6 Partnering

Partnering is not a different form of procurement organisation, but rather a cooperative approach which has been defined as *an arrangement whereby people are encouraged to work more efficiently together, including shared problem resolution, continuous improvement, continuity of work, fast construction, completion on time and improved profits.* Two types of partnering are in common use — project partnering, where one-off development is undertaken using a team that agrees to work in partnership, and strategic partnership, where the same team is involved over a series of construction projects. A study by Reading University[6] calculates that savings of up to 10% may be achieved with one-off partnerships, and up to 30% for strategic partnerships. Partnering aims to replace the confrontational system of watertight contracts with a more open approach to resolving problems as they arise. This often involves partnering workshops, where all parties meet on neutral territory to set goals and resolve problems. The workshops may be conducted by a facilitator, a third party skilled in team building and group dynamics.

Since partnering requires a shared approach to problems it is often linked to some form of profit sharing. While partnering may be applied to almost any form of procurement, it is best suited to a situation where the partners learn to work together over a series of projects in a stable market. The client has also an important part to play in creating the conditions where partnering can flourish. Table 8.1 indicates the situations where the partnering approach may be most useful.

The controls specialist will be expected to contribute to the design at an early stage and to take part in partnering workshops to increase the effectiveness of all members of the construction team. This will help contribute to the design of an effective, controllable HVAC system. In return the controls specialist is entitled to expect the main contractor to[7]:

Table 8.1 Situations where partnering can be used[7]

	Partnering may provide benefits	Good for partnering	Ideal for partnering
Form of contract	Traditional	Management contracting	Construction management
Initiator		Consultant, contractor	Client
Programme of work	One-off	Regular	Programme of similar projects
Project type			High value, high risk
Market conditions	Boom/bust		Stable

— offer greater predictability of work
— use standard forms of subcontract
— refrain from rebidding to help reduce subcontractor's bid costs
— draw up a preferred list of subcontractors
— honour tender commitments
— reimburse the costs of design.

Partnering does not normally involve a legally binding agreement, though it is common for everyone to sign up to a partnering charter setting out the declared aims. Care needs to be taken to ensure that a partnering agreement does not fall foul of the laws on restrictive practices or fair competition, which may prevent exclusion of potential bidders or demand that contracts must be awarded on lowest price. Partnering is by no means a panacea and it cannot compensate for other inefficiencies.

8.2 Design and specification of a controls system

8.2.1 Sequence of events

Table 8.2 illustrates the steps involved from design to completion of a building control system. The procedure is based on the traditional building procurement method, but it applies to all methods, with some variations in responsibility. The first step is for the client to set out the requirements for the building as a design concept. The more clearly the client's requirements are set out in the project brief, the better controlled will be the whole process. The outline design process should include consideration of the end user's operational requirements and the general level of control and IT systems. The client may appoint a principal adviser at this stage to supply independent advice now and later in the project. The use of a principal adviser may be crucial to the success of a project where the client has little in-house experience.

Once the brief has been agreed, the design work is passed over to consultants, whose job it is to produce the final design which will form the basis of the building contract. For a complex building, the controls design is likely to be done by a specialist consultant, either an independent firm or a specialist group within an M&E consultancy. The level of detail varies from project to project; it may be a complete specification, or a simple performance specification, which leaves the controls specialist to do the detailed design work and select the supplier. It is possible to place the controls design with a manufacturer, who will provide design, supply and installation to the project. Large building projects will be constructed by several contractors and subcontractors and it is important that their responsibilities be clearly allocated. Some level of design responsibility is normally allocated to the installing contractor, for instance by an instruction that they produce drawings. The division of responsibilities between contractors must also be made clear in the specification. An example relevant to controls is the provision of cabling. Power and signal cabling may be provided by different contractors and BMS signals may be partially carried by a shared IT network. The specification must detail responsibilities, including the routing of cabling to satisfy EMC requirements.

Potential contractors bid for the contract to construct the building. The successful bidder, the main contractor, is unlikely to provide all necessary specialists in-house, and may appoint specialist subcontractors to supply and install the control system. This subcontractor may need to provide detail design work, depending on the completeness of the

Table 8.2 Sequence of events in building procurement

Activity	Responsible
Concept and design brief of building, including control and associated requirements	Client, architect, principal adviser
Place contract for design with consultants	Client
Produce detailed design of building, HVAC and control systems, which becomes basis for tender document	Architect, consultants, controls specialist
Estimate costs and prepare bid	Contractors
Place contract with main contractor	Client
Main contractor may subcontract controls installation	Main contractor
Prepare and agree description of operation	Controls specialist
Prepare detailed installation drawings	
Obtain items from suppliers and install controls system Check operation Prepare documentation	
Test and commissioning	Commissioning sub-contractor, with consultant supervision
Staff training	Controls specialist
Acceptance and handover	Client
Fine tuning and maintenance of controls	Separate contract

controls specification. When the specialist subcontract has been appointed, they may prepare a description of operation. This document sets out how the subcontractor intends to interpret the controls specification and realise it as a practical operating control system. The description of operation is referred back through the contract chain to be agreed by the client and the consultant responsible for the controls specification. This ensures that the specification has been interpreted properly and that the client is clear as to the controls system that will be provided; the procedure also gives some protection to the controls specialist against later disputes arising from any misunderstanding. Commissioning of the HVAC and controls systems may be carried out by an independent specialist subcontractor, witnessed by the controls specialist and the control consultant. Provision of full documentation and training for the staff who will operate the systems is an essential part of the process and is completed before handover of the system to the client.

8.2.2 Design brief

The first stage is for the client to set out the requirements in a way that can be interpreted by the design team. The control system of a complex building is essential to the proper functioning of the building. It represents a substantial cost and the design of a BMS should be considered from the start. The design brief produced by the client should include any particular requirements and restrictions on the cost and scope of the controls. The client should consider carefully the intended function of the building and its occupants together with future control and IT requirements. This should include discussions between controls companies and the end user. Experience shows that the best building management systems result from situations where controls specialists and the end user have had direct contact with each other. It is never too early to include the end users, including operators of the BMS and the building occupants. The client, in association with a design consultant, produces a design brief, which forms the basis for the specification. The Building Control Group has published a pre-specification checklist, which is reproduced in Figure 8.2.

Where installation of a BMS is being considered for an existing building, planning should start with a thorough survey of the building and its operation. An energy audit will reveal sources of avoidable energy waste, which should be rectified before starting work on the control system. The need for centralised control should be considered critically; for smaller installations, the use of stand-alone controls may be a cheaper alternative compared with BMS.

8.2.3 Specification

The specification forms the basis of bids by contractors and is the basis of the eventual contractual agreement. It is therefore essential that the specification provides a complete and unambiguous definition of the building control system. Changes made to the specification after work has started are likely to be expensive. It is also important that the specification should set out clearly the allocation of design responsibilities to avoid subsequent disagreements[8]. A comprehensive treatment of the issues which should be considered when specifying a BMS is given by Pennycook and Hamilton[9]. To simplify the process of drawing up a specification and to reduce the chance of important items being overlooked, various bodies have drawn up outline specifications which may be used with the minimum of modification. Teekeram and Gray[10] reviewed the situation over several countries. The most comprehensive and widely used is the *National Engineering Specification* (*NES*), which is structured in accordance with the CPI Common Arrangement of Works Sections for Building Works[11]. The Controls and BMS sections of the *National Engineering Specification* refer to the UK standard specification for BMS which was drawn up by the BEMS Centre[12]. It is in turn based on the *Standard Specification (M&E)*[13].

The aim is to set out the specification in functional terms, rather than prescribing the hardware and software to be used. In a time of rapidly changing technology, functional objectives may be achieved by different hardware and software strategies for different systems. The functional approach should encourage suppliers to offer the most suitable systems on a competitive basis.

A project specification compiled using the *NES* has three main sections:

- Preliminaries, referring to the whole project.
- System specifications, one for each relevant work section.
- Reference specifications, relevant specifications drawn from Group Y, where Group Y is a collection of specifications of component parts and materials. These may be relevant to several sections in the system specification, so it is common practice to bind them in a separate volume for easy reference.

The Preliminaries section is drawn from Group A of the *NES* and contains an overall description of the project, together with information on organisation and contractual matters. Specification for the controls system must take its place and be compatible with the specification for the whole project. The preliminaries for the controls system specification must not conflict with the main contract preliminaries, which is the governing document for the whole project.

Each of the systems specifications is subdivided into three parts:

- *Part 1. System objectives*. The system objectives are descriptive clauses to be written by the specifier, using the following headings:
  - performance objectives
  - design parameters
  - system description
  - control requirements
  - system schematics
  - system drawings.

  The system objectives should provide a complete reference package so that the ideas behind the design are passed on in an intelligible and logical format. The first three parts should start as soon as the client's brief has been received, and kept up to date as the design proceeds. In the absence of a

BUILDING CONTROLS GROUP PRE SPECIFICATION CHECK LIST

1 GENERAL DETAILS

1.1 Have the detailed operational needs of the End User been fully considered?
1.2 Will the envisaged Form of Contract allow for the needs of the Client/User to be met?
1.3 Is there to be scope for integrating other functions/elements into the BMS?
If YES which:
a) Access/Security b) CCTV c) Lighting control
d) Chillers e) Boilers f) Variable speed drives
g) Others, please specify
1.4 Will any elements of the system need to be Acceptance Tested at manufacturers works?
1.5 Will the system be required to be expanded and/or upgraded over the next 5 to 10 years?
1.6 Will the BMS supplier have a minimum 4 week tendering?

Section 2 OUTSTATION DETAILS

2.1 Will the Motor Control Units (MCC s) contain the outstations?
If YES outstations should be contained in a separate control section
If NO please state the distance between MCC and outstation
2.2 What form of MCC will you be specifying?
2.3 Are the MCC s to be a special colour? If YES please specify
2.4 What IP ratings are the MCCs?
2.5 If OUTSIDE the MCCs what IP rating is required for outstations?
2.6 Have you specified the number of points (inputs/outputs) to be covered by the system
If YES please indicate:)
a) Analogue (resistance voltage or current) b) Digital (open or closed contacts) c) Pulse (metering)
d) Hardware e) Software f) Other (please specify)
2.7 Are the outstations to have any special features? If YES please indicate:
Memory battery back-up I/O battery back-up Special cards
Individual I/O override UPS Other features
2.8 Are the outstations to have displays/keypads?
If NO please state which are not required. Is a fully functional laptop PC required?
If YES please indicate:
a) Door mounted b) Within MCC c) Portable device d) Specially protected e) on all outstations
If NO please specify how many, which and what functionality

Section 3 CENTRAL SUPERVISOR SOFTWARE & HARDWARE

3.1 Is there to be a Central Supervisory Computer on the project?
3.2 Is the Central Supervisor dedicated to BMS or also running other programmes?
3.3 Have you specified the hardware and can it cope with the project?
3.4 What size monitor have you specified?
3.5 Have you specified ALL the software that you require for this project?
3.6 Is the Central Supervisor to be networked?
If YES please indicate the type of network.
3.7 Do you require extra computer terminals? If YES how many
3.8 Have you specified how these terminals are to be configured?
3.9 Have you indicated where these terminals are to be located?
3.10 Will remote telephone modem communication be required?

4 INSTALLATION

4.1 Will the BMS Contractor be required to supply
a) Computer hardware b) sensors/actuators c) MCC s
d) Install all the above e) Cable all the above f) Engineer BMS software
g) Supervisor software h) Power wiring i) Commission the above
4.2 Will Tendering Information accord with the Building Controls Standard Method of Measurement?
4.3 Will you be applying special conditions to the contract installation? If YES please specify

5 TRAINING

What Training will the BMS Contractor be required to provide?
Off site On site Other

6 WARRANTY AND MAINTENANCE

6.1 Have you Included for a system service & maintenance contract in your tender documentation?
If YES have you specified the type of cover you require
a) 24 hour callout b) Guaranteed response time c) 6 monthly maintenance check-up d) Annual maintenance check up e) Dedicated spares holding e) Other requirements
6.2 Have you specified the period during which you require maintenance to be provided ?

IS THERE ANYTHING ELSE YOU OUGHT TO HAVE THOUGHT ABOUT?

Figure 8.2 Pre-specification checklist (courtesy Building Controls Group)

central control system, the control requirements are listed in the system specification for the relevant work section. Where a central control system is to be used, the control requirements may either be listed in the relevant system specification, or else all details of the control system can be listed in the Central Controls Work Section, with appropriate cross-references from the systems being controlled.

— *Part 2. Selection schedules*. This part contains a list of references to the relevant Y Group Reference Specifications that are relevant to this work section. As well as invoking the relevant specification, the schedule invokes required clauses from Group Y, deleting irrelevant options; additional text may be inserted as required.

— *Part 3. Clauses specific to the system*. This contains information specific to the work section. The *NES* contains a large number of clauses and Part 3 is constructed by deleting clauses and options that are not required.

The *NES* is designed to aid production of specifications by providing a standard logical structure, which is amenable to incorporation in computer programs. As far as BMSs are concerned, it is complementary to the material presented in the BSRIA documents[12] and indeed Work Section W62 of the *NES* makes frequent reference to the BSRIA publication.

8.2.4 Conditions and tolerances

The specification of the required controlled conditions and permissible tolerances must be considered carefully. Setting unrealistically tight tolerances may result in either an unnecessarily expensive design solution to meet them, or a system which fails to meet the specification, with the attendant risk of legislation.

General comfort criteria are covered in Section A1 of the CIBSE Guide[14] and by ASHRAE-ANSI Standard 55[15]. The most important variable is the temperature in the occupied space. This may be specified in terms of air temperature, environmental temperature or one of a number of comfort indices. The adaptive model of thermal comfort[16] states that it is permissible to allow space temperatures to rise in summer, since the occupants learn to adapt and minimise discomfort. This has importance for the design of HVAC systems, since it may be possible to reduce or even eliminate the need for air conditioning. Where this principle is adopted, the acceptable limits of building performance must be understood and accepted by all parties.

Where full air conditioning is employed, the required range of air temperature and humidity is specified. The interaction of air temperature, relative humidity and moisture content must be remembered. In general, the moisture content of the air in a building is less subject to variation than the air temperature. Changes in air temperature at constant moisture content may cause the relative humidity to go outside the specified range. People are not very sensitive to the level of ambient humidity and a range of 40 to 70% RH is normally found acceptable. If closer control is required, then the tolerances must be specified in such a manner that they can be realistically met by the HVAC system. Where central control of humidity is used, it may be better to specify humidity in terms of moisture content rather than relative humidity.

Similar considerations apply to plant operation. Control limits must be within the capability of the plant and its measurement and control system. When specifying the tolerances for system variables there is no point in specifying unnecessarily narrow limits, even if they can be met. The heat output from a heat emitter may be little affected by a 20% variation in water flow; however, a narrower specification is required for chilled water systems[17].

8.3 Tendering process

8.3.1 Pre-tendering

In many cases it is advantageous to draw up a short list of potential suppliers who will be asked to tender for the project. The use of a pre-tender brief allows selection of suitable tenderers to be produced, avoiding wasted effort. A pre-tender brief is given to possible suppliers, who are asked to return a questionnaire and are then interviewed to discuss the project in outline. The pre-tender brief contains enough information to indicate the size and complexity of the project, the available budget and the proposed time scale. The brief should contain the following:

— objective of the brief

— project description

— an indication of cost

— project management and form of contract

— building and plant schedules

— questionnaire.

8.3.2 Tendering

The usual documents required for the tendering process are shown in Table 8.3. More detail, together with examples of tender documentation, is given in BSRIA[12]. If the pre-tendering process has produced a suitable short list and the specification has been properly drawn up, assessment of the tenders should be straightforward. To protect

Table 8.3 Tender documents

Document	Purpose
Instructions to tenderers	How to complete the tender documents
Form of tender	On which the tender bid is returned
Tender details and summary	Cost breakdown of the tender price. Will provide the basis for any negotiation of variations
Form of contract and special conditions	
Full specification	Full set of standard and particular specifications, together with relevant plans and drawings
Information required from tenderer	Additional requirements needed to confirm details of the tender

subcontractors against possible bad practices, a *Code of Practice for the Selection of Subcontractors*[18] has been published by the Construction Industry Board.

8.4 Commissioning

Commissioning is defined by CIBSE as 'the advancement of an installation from static completion to working order to specified requirements (as envisaged by the designer)'. Correct commissioning is vital to the satisfactory operation of the HVAC system and it is essential that sufficient time and resources be allocated to the task. Since commissioning is the last major operation in the building process and the control system is the last system to be commissioned, there is every danger that commissioning the control system will take place under great time pressure or even continue after the building is occupied. Experience has shown only too well that this can create many problems, with an unacceptably high proportion of buildings failing to operate properly, with consequences of high energy consumption and occupant dissatisfaction. This section emphasises the importance of ensuring that commissioning takes its proper place in the procurement process from the start of planning the project.

Commissioning of the control systems is dependent on satisfactory operation of the electrical and mechanical services. The control system may also interact with specialist services such as security, access control and the IT system. The greater the degree of integration, the more planning is necessary and the more care required to define areas of responsibility[19]. Sequential records of all stages of pre-commissioning and commissioning of all other aspects of the building must be issued prior to commencement of the commissioning of the automatic control system.

Table 8.4 summarises the stages of the procurement process and the activities which are relevant to the commissioning of the controls system.

Table 8.4 Commissioning activities during the procurement process

Stage	Commissioning activity
Brief	Commissioning objectives
	Appoint commissioning manager
Design	Minimise need for commissioning
	Design for commissionability
Installation	Pre-commissioning
	Commissioning
Handover	Witnessing
	O&M documentation
	User training
Operation	Monitoring and feedback
	System proving
	Fine tuning
	Recommissioning

8.4.1 Commissioning management

For the traditional UK procurement organisation, the responsibilities for commissioning the BMS are shown in Table 8.5. This arrangement has the advantage of clearly defined responsibilities. However, it does not necessarily produce the most effective commissioning. The controls specialist, who is responsible for installation and commissioning, has little or no input into the design of the BMS, nor the design and commissioning of the plant which will be controlled by the BMS.

Table 8.5 Commissioning responsibilities for traditional procurement

Party	Responsibility
Client	Write design brief
	Appoint design consultant
Design consultant	Write BMS functional specification
	Review tenders from BMS contractors
	Appoint BMS contractor
	Witness commissioning
	Approve completion
BMS contractor	Prepare tender
	Design BMS to meet specification
	Install BMS
	Commission BMS

8.4.1.1 Project management

The organisation and management of commissioning is described in detail by Pike and Pennycook[20], which gives examples of matrices of responsibilities for commissioning. Good project management is essential for large and complex projects and the subject is covered in the *Guide to the BEMS Centre Standard Specification*[12].

- A detailed commissioning programme should be written and agreed with the main contractor.
- There must be a means of monitoring progress.
- Checklists should be used to monitor progress.
- The controls specialist must have a documentation system in place for dealing with variations to contract.

Project management guidance[21] and BSRIA's Commissioning Guides define responsibilities for the personnel involved in the commissioning process. Duties will vary between projects depending on assigned responsibility. Example roles are detailed below.

The project manager has a coordinating, monitoring and controlling role for the project. The responsibilities include the following:

- aiding the client selecting the design team and other appropriate consultants and negotiating their terms and conditions of employment
- setting up the management and administrative structure for the project.

The client's commissioning manager provides the control point for commissioning for the client. The manager will be responsible for:

— arranging the appointment of the commissioning team
— establishing the commissioning objectives
— developing a comprehensive commissioning programme and preparing the roles and job descriptions for each member of the team
— arranging the sessions for user training at the end of the project.

The professional consultants must identify the services to be commissioned and define the responsibilities split between the contractor, manufacturer and client. They are also responsible for inspecting the work for which they have design responsibility (including inspecting the work at the end of the contract defects liability period) and defining the performance testing criteria to be adopted.

The site commissioning manager may be a member of the main contractor's team and provides the focus for the management of all the commissioning activities. The tasks may include the following:

— coordinating the professional team members and the client's involvement in commissioning
— ensuring that the contractors' programmes include commissioning activities and that these coordinate with the main construction activity
— acquiring appropriate information from the relevant parties to ensure that the systems can be commissioned in accordance with relevant codes of practice
— witnessing works and site testing of plant, cleaning and commissioning
— demonstrating safety systems to the local authority, fire officer, district surveyor and the building insurer
— providing a focus to collect all handover documentation.

8.4.2 The brief

The original brief will lay down the foundation for the organisation of commissioning. This is the stage at which the organisation of the whole building procurement process is decided. For large projects, the appointment of a commissioning manager should be considered, with the responsibility of ensuring that the requirements of commissioning are considered at all stages. The manager will be responsible for appointing the commissioning team and ensuring that adequate resources for commissioning are built into the specification. Decisions should be taken at this stage whether commissioning should extend beyond the handover of the building and whether provision is to be made for system proving and post occupancy feedback when the building is in operation.

8.4.3 The design stage

Commissioning should be borne in mind at the initial design stage; this applies to both the HVAC system and the control system itself. Two complementary approaches are relevant:

— designing to minimise commissioning
— designing to make commissioning easier.

The amount of commissioning needed can be reduced by appropriate design:

— Design for self-balancing wherever possible.
— Balance pressure drops across sub-branches and terminal units.
— Avoid using different terminal units on same branch.
— Use reverse return pipework layouts.
— Use automatic balancing valves.
— Use variable-speed drives for fan and pump regulation.
— Use computer analysis to give settings for preset valves.

Attention to the needs of commissioning at the design stage will make it easier. Conversely, failing to provide the necessary facilities for commissioning may make it impossible. The following items will assist commissioning:

— Include regulating valves where appropriate.
— Include isolation valves and test points.
— Include flow measurement devices at main branches and major heat exchangers.
— Ensure that the system may be cleaned and vented.
— Ensure access for commissioning and maintenance[22].
— Provide items of plant with a 'manual–off–auto' switch. The switch can be set to 'manual' for commissioning the plant itself, prior to commissioning the BMS.
— Agree at an early stage a common procedure for allocating mnemonics to identify points uniquely.
— Obtain mid and full load design values from consultant.

The design consultants are responsible for writing the system specification, which includes the commissioning specification. At this stage a commissioning specialist should review the specification.

8.4.4 Installation

Satisfactory commissioning of the BMS is of course dependent on the correct installation and commissioning of the HVAC plant. The commissioning manager will ensure that plant is in a suitable condition before commissioning of the BMS commences and is certified as such. Above all, all water systems must be flushed clean; the presence of debris in water systems is a common cause of problems in the commissioning of valves.

8.4.4.1 Pre-commissioning the BMS

Pre-commissioning is the checking of all the components of the controls system, both on the bench and on site, before putting the system into operation for commissioning proper. As much checking should be done off site before installation; this will normally be faster and allow rapid rectification of any problems that are found. Checking of sensors for accuracy is best done off site under controlled conditions. The on-site pre-commissioning consists in checking that all items have been installed in the right place, the right way round and are wired correctly. Pre-commissioning tasks include:

— maximisation of off-site testing and pre-commissioning of control panels, application software etc.

— the checking of all wiring for short circuits, continuity, identification and termination

— the checking of all sensors for correct installation

— proving of all actuators

— proving the presence of outputs and devices on the network bus system.

8.4.4.2 Commissioning the BMS

Before commissioning can start, all control hardware must be installed and pre-commissioned. Any unitary controls on plant which is to be connected to the BMS should have been commissioned, or be commissioned in parallel with the BMS. The HVAC plant itself must have been commissioned; all plant items under control of the BMS must have been commissioned and the system flushed clean and vented. All electro-mechanical safety interlocks and fail safe conditions should be implemented and operational.

Commissioning the BMS consists of two major activities:

— checking that the control system works

— setting all parameters and switches to appropriate values, including tuning control loops.

Checking the system

Both the head end supervisor and the controls of particular building services plants have to be commissioned. The supervisor requires commissioning after installation. It may be commissioned in parallel with the outstations which control specific items of plant, when it can itself be used as an aid to commissioning. Full commissioning of the central supervisor requires previous completion of plant commissioning. For specific plant commissioning, the *Code of Practice*[20] sets out five basic items, to which has been added the relation with the supervisor:

— panel controls

— hardware points

— interlocks

— control loops

— alarms

— supervisor.

This procedure is illustrated by the example of an air handling unit shown in Figure 8.3. Table 8.6 summarises the tasks of the commissioning process; the example is summarised from Pike and Pennycook[20].

Setting the values

An important function of a BMS is its operation as a timeswitch, controlling the start and stop operation of many systems throughout the building, including optimisers. All BMS supervisors allow time schedules to be readily altered at the head end. However, experience shows that the initial settings may remain unaltered. Accordingly, the client should supply as much information as possible on required time and temperature schedules, to allow the correct values to be input during commissioning. It is recommended that the controls specialist sends a questionnaire to the client asking for relevant information. A sample questionnaire is given in the commissioning *Code of Practice*[20].

All control loops require to be tuned. While the goal is minimum response time consistent with stability, it must be remembered that the operating gain of a control loop is likely to vary with operating conditions, so that the gain will vary with season. Where possible, the loop should be tuned under conditions of maximum gain; this will ensure stability under all conditions. Various ways of tuning loops including the use of automatic tuning are discussed in 2.4.1.

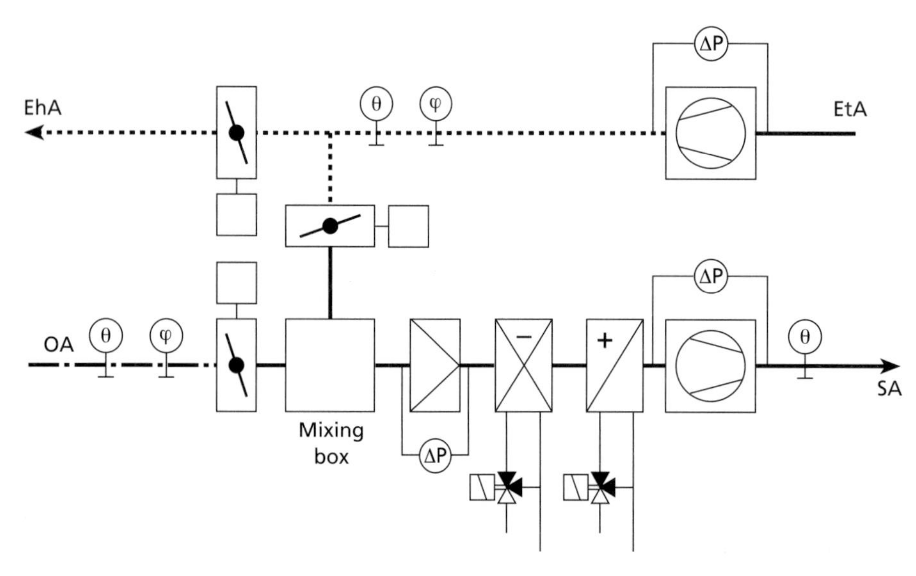

Figure 8.3 Air handling system to be commissioned

Table 8.6 Commissioning an air handling unit

Step	Function	Check
Panel controls	Power	Phases healthy, panel live, BMS power on
	Fire alarm	Reset button, lamp test
	Supply fan	Run, trip, test/off/auto switch, overload range and set
	Extract fan	Run, trip, test/off/auto switch, overload range and set
	Fuses	Correct size, spares in panel
Hardware points	Sensors	Outputs, correct connections
(at pre-commissioning)	Actuators	Full stroke travel
	Connections	Continuity and polarity
Interlocks	Fans	Supply and extract fan interlock
	Dampers	Supply, extract and recirculation dampers synchronised
	Coils	Heater and cooler batteries move to default position on shut-down
	Fire	Appropriate action on fire signal, e.g. close supply and recirculation dampers, open exhaust damper, stop fans
Control loops	Optimiser	Set parameters
	Frost	Set and test frost protection and morning boost
	Control loop	Set proportional and integral constants
		Operation of valves and dampers
		Enthalpy control if fitted
		Run-time totalisation
	Software	Check application software
	Timeswitches	Run through time schedules
Alarms	General	Operation, labelling
	Filter block	Set pressure switch and check operation
	Supply fan	Set pressure switch and inhibit time. Check operation
	Extract fan	Set pressure switch and inhibit time. Check operation
Supervisor	Graphics	Points labelling, plant depiction
		Control functions, correct values

8.4.5 Handover

Once commissioning of the control system has been completed, it is demonstrated in front of witnesses and a completion certificate issued; the client's representative may attend. The witnessing officer will require a completion method statement which details the arrangements, methods and a list of items to be demonstrated in front of witnesses. Witnessing is just one part of completion. Documentation and training arrangements are also required. Table 8.7 summarises the items to be covered during completion. When all items have been completed to the satisfaction of the testers and witnesses, a certificate to that effect is signed off by both parties and handover takes place. Handover represents change of ownership from contractor to client.

8.4.5.1 Witnessing

The organisation of witnessing will depend on the nature of the system and is decided by the witnessing officer. For large projects, there are advantages in a phased approach, so that witnessing of completed subsystems can begin before final completion of the whole system. It is not essential that every point be checked. A sufficient number should be checked to give the witnesses confidence in the overall reliability, using a quality control approach. The method suggested in the BEMS commissioning *Code of Practice*[20] is:

Table 8.7 Completion checklist[20]

Item	Description
1	Audit of cabling and hardware installation
2	Demonstration that sensors and actuators are correctly connected and addressed
3	Demonstration of the physical and logical integrity of the system
4	Demonstration of the sensor calibrations
5	Demonstration of all control actions
6	Demonstration of successful system software commissioning. (This should include loading and subsequent operation)
7	Verification of specified graphics
8	Verification of specified training arrangements
9	Verification of handover of all specified manuals, documentation and drawings
10	Verification of handover of back-up copies of software
11	Verification of handover of consumable spares

- The BMS controlling the main plant (e.g. boilers and chillers) and other important points should be witnessed completely.
- The proportion of other points to be witnessed depends on the size of the system, e.g. if fewer than 300 points, witness all the points, between 300 and 1000 points, witness 50%, and if more than 1000 points, witness 20%.
- If the failure rate is greater than 15%, the supervising officer should have the right to witness 100% of the points.
- Where there are several similar plants, one can be witnessed in detail and the others on a sampling basis.

8.4.5.2 Documentation

Good operations and maintenance (O&M) documentation is necessary for successful:

- day-to-day operation of the system
- effective maintenance
- system refurbishment.

There is a statutory requirement under the *CDM Regulations*(23,24) to provide proper documentation. The client is required to appoint a planning supervisor, whose function is to ensure that a health and safety plan is prepared, to advise the client on the necessary resources required to implement the plan, and to prepare the health and safety file. The health and safety file is a record of information for the end user, which tells those who might be responsible for the building in the future of the risks that have to be managed during maintenance, repair or renovation. Operation, maintenance and commissioning documentation make up part of the health and safety file. The file must include details of hazards and risks, and the safety certificates collected during construction. The documentation must be presented in the format required by the regulations(25).

While all building services require proper O&M documentation, the complex nature of a BMS, which may have control strategies and other information stored in computer memories, makes it essential to provide written details of what function the BMS performs and how it achieves them. Advice on the preparation of O&M manuals is given in Armstrong(26). The documentation should include:

- written description of plant operation
- control strategy or logic diagrams recording the final version of installed software
- details of system application software configuration
- points list, including hard and soft points
- copies of certificate of compliance with relevant standards
- data sheets for all control components and equipment
- instructions for switching on, operation and switching off, isolation, fault finding and for dealing with emergency conditions
- instructions for any necessary precautionary measures
- instructions for servicing
- instructions in the use of software routines for creating procedures, graphics reports etc., where applicable
- description of user adjustable points
- provision for updates and modifications.

The file may be in two parts: operating manuals for day-to-day use and sets of drawings for use during alterations. Copies of manufacturers' data may be incorporated, but are not a substitute for proper instructions. Details of spares and sources of supply should be included. In addition, there should be a complete set of record documents supplied at handover, containing records of all the commissioning values and the checklists completed during commissioning and witnessing.

8.4.5.3 Training

The successful operation of a BMS depends on the skill and knowledge of the operator. The installation of a modern BMS represents a substantial investment and proper operator training is necessary to realise the full value of this investment. The client has to decide who will operate the BMS. In many cases this will be staff, but the contracting out of facilities management services is becoming more common. Training is often part of the contract and so contract completion is not possible until the training has been carried out. This may result in financial pressure to perform the training too early, before the end user is ready or staff have been appointed. It is recommended that training costs should be determined and held as a PC sum outside the main contract, to be paid when the training is completed.

It is recommended that:

- at least two BMS operators attend an 'in-house' training course run by the controls specialist before completion of the BMS
- the BMS operators are invited to attend the commissioning of the BMS
- the BMS operators should be present during the handover period to learn about the system
- all new operators who may subsequently be appointed should also receive proper training.

8.4.6 Costing

The importance of commissioning to the satisfactory operation of the plant and control system has been emphasised. It therefore requires proper provision of time and cost to enable the process to be carried out satisfactorily; allowing commissioning to absorb time and cost overruns is not satisfactory. Effective commissioning has long lasting benefits, which need to be given full value when making out a business case for expenditure; inadequate commissioning is sure to prove expensive over the long run. The benefits of effective commissioning may be summarised:

- increased building value

— improved occupant satisfaction and productivity
— reduced energy costs
— reduced maintenance requirements and longer plant life
— sound database for future refurbishment.

When costing proposals, all activities should be taken into account, including provision for post-occupancy evaluation and fine tuning.

8.5 Operation

8.5.1 Fine tuning

After handover, the building operates under control of the appropriate staff, whether in-house staff, a contract facilities management team or other form of organisation. It is rare that a new building works perfectly without further attention; it would be unreasonable to expect this, since the manner of building use by the occupants cannot always be predicted and it will not have been possible during commissioning to experience the full range of weather conditions. It is therefore to be expected that the BMS will require attention and tuning over a period of at least a year after occupation. Proper provision should be made for this, in terms of effort and cost. This should include an allowance for a return visit at least twice during the first year. In most cases, it will be of great benefit for all concerned if the client installs a dedicated telephone line to allow remote monitoring of the BMS by the controls specialist. As well as assisting in fault diagnosis, the link can be used for training and other purposes.

8.5.2 Maintenance

While the microelectronics components in a BMS are in general very reliable, a controls systems contains moving parts which are subject to wear and devices such as sensors which may be subject to physical damage or degradation from their environment. Maintenance is therefore essential to ensure that a BMS stays in efficient working order throughout its life. Maintenance includes updating and maintenance of software and documentation.

Maintenance may be obtained from different types of provider:

— *In-house staff*. Only large organisations with a sophisticated building are likely to carry the necessary technical staff to provide BMS maintenance. In-house staff may be used to provide the first line of investigation into failures and carry out simple repairs, calling on specialist contractors for more complex work.

— *Maintenance divisions of M&E companies*. Many M&E contractors operate specialist maintenance divisions that will undertake maintenance of BMS systems, whether installed by the parent company or not. The maintenance companies are generally autonomous companies, so it should not be assumed that there is a continuity of responsibility between installer and maintenance company.

— *Systems integrators*. Many of the larger systems integrators provide comprehensive maintenance and 24 hour bureau services. This may include remote monitoring of plant condition.

— *Specialised maintenance companies*. There is a large number of small specialised maintenance companies, often specialising in a particular range of equipment. Although not having the resources of the large M&E companies, they can supply a service tailored to the customer's needs.

— *Consultants*. An increasing number of engineering consultants offer maintenance services, including training or documentation production, and the management of maintenance on behalf of the building operator.

— *Controls manufacturers*. Several major manufacturers offer maintenance contracts of systems using their equipment.

Whichever method is chosen, it is important that the maintenance is specified carefully(9). As a minimum, a maintenance specification should include requirements for:

— software upgrades
— data back-up and archiving
— checking of sensors and actuators
— arrangements for emergency callout
— performance standard for the building control system, including delivered conditions and operating efficiency.

As with commissioning, it is advantageous if the needs of maintenance are considered at an early stage; the several approaches to the organisation of maintenance are reviewed in Smith(27) and CIBSE TM17(28). Plant and equipment should be designed and installed in a manner to assist future maintenance; recommendations are given by MoD(22) and Parsloe(29). Advice on specification and contracts is to be found in Smith(30) and *Standard Maintenance Specification*(31). Experience shows that the best results are obtained in situations where the BMS is 'owned' by someone in the building operator's organisation. There should be an individual who feels responsible for the proper operation of the BMS and will take whatever initiative is necessary to ensure its successful working.

8.5.3 Recommissioning

Existing buildings may operate below optimum performance and recommissioning the building services may represent the most effective way to bring operation up to the required standard. Recommissioning can often be justified by:

— evidence of occupant dissatisfaction
— high operating costs or energy consumption
— high maintenance costs
— a change in building use
— major plant overhaul or replacement
— repeated system failures.

8.6 Occupant surveys

The experience of the building occupants plays an important part in the evaluation of the performance of a building control system. Building users consistently place good environmental control at the top of the list of desirable features in a workplace. A well-controlled environment not only produces comfort, but contributes to productivity and health as well as influencing the occupants' general opinion of other facilities. The series of PROBE post-occupancy studies of buildings has emphasised the importance of a well-controlled environment and emphasised that the occupants themselves are part of the system. Disregarding the responses of the users will itself create problems, even if the mechanical system is performing adequately. The published summary of the first PROBE series produced a set of key design lessons about the requirements of the occupant[32].

— *Rapid response to changes*. There should be a rapid response to the need for a change, whether this is provided automatically by the BMS or by management intervention.

— *Offer choices and trade-offs*. If ideal conditions cannot be provided, the occupants should be offered a choice of trade-offs, e.g. in hot conditions, openable windows allow a choice between being hot or suffering from external noise.

— *Good feedback mechanisms*. These are essential so that action can be taken quickly. Complaints or problems need to receive rapid attention. The trend towards contracting out services such as facilities management can work against rapid response.

— *Management resources*. These are vital; complicated buildings can outstrip the limited capabilities of the occupiers to run them effectively.

— *Ownership of some problems by the occupants*. This is so that they may participate in the solution without suffering a sense of alienation.

— *Perception of good control*. This is important to occupants, who require simple effective controls or else rapid and effective management response to requests.

The above design lessons emphasise that the users must perceive that they are part of the control system. However good the control system may be in technical terms, if the manner in which it is operated ignores the feelings and responses of the building occupants, then it will not be found satisfactory. The application of the above lessons will result in working methods whereby early indications of unsatisfactory operation of the building control system come to the attention of the building operator. Other indications may come from the energy monitoring and targeting procedures. In any case, it may be advantageous to initiate a regular review of system performance, say one year after installation and at longer intervals thereafter. When it is desired to initiate a proactive review of the operation of the building services operation, BRECSU propose a three-level strategy to review building performance, which can be initiated at any stage of the building's life.

— *Level 1*. Review performance, using a help desk or focus groups to obtain the views of occupants. Assess energy performance against targets or benchmarks and ensure that health and safety requirements are met. If there is cause for concern, move on to Level 2.

— *Level 2*. Use a more formal procedure to identify any problems. Use occupation satisfaction surveys and energy surveys. If performance problems are established, go to Level 3.

— *Level 3*. Employ plant-focused troubleshooting procedures to find the cause of troubles. Systems can then be fine tuned, recommissioned or refurbished as required.

The relation between building controls and complaints of poor indoor environmental quality is set out by Fletcher[33]. Reports of poor conditions are classified into 16 divisions and an action checklist for each is given. The suggested actions only apply if the problem is controls related or can be mitigated by modifying the control set-up.

A detailed occupant satisfaction survey demands a formal approach. The use of a standardised questionnaire will provide a satisfaction score for the building that will enable its performance to be compared with national benchmarks, as well as identifying areas of control and management that require attention. The PROBE standard questionnaire is copyright and may be used under licence; this ensures that it is applied according to standard procedures. Formal occupant surveys should be undertaken with caution, both to ensure results that can be compared with benchmarks and to avoid possible staff-related problems.

— The questionnaire should be short and easy to answer in at most 10 minutes.

— In large buildings, a representative sample of occupants should be surveyed, aiming at as high a response rate as possible.

— The provisions of the Data Protection Act must be borne in mind.

— The staff association or other relevant body should be kept fully informed.

— Make the goal of the survey clear to respondents; do not raise unrealistic expectations of improvement in conditions.

— Resolve any questions of confidentiality in advance.

— If the survey is to be conducted by external researchers, resolve questions of ownership of data and possible publication in advance.

One type of questionnaire provides a visual fingerprint of occupants' response to their working environment as well as an overall building score. Respondents are asked to rate up to 24 factors on a double Likert scale. This asks for two responses for each factor. One is a seven-point scale running from 'like' to 'dislike'; the other is a seven-point scale describing how important that factor is. The factors cover physical aspects of the indoor environment such as temperature and humidity, but also include some organisational factors. The responses from all respondents may be combined to give an overall liking score (OLS) for the building. The use of the importance scale means that appropriate weighting is given to each factor when computing the OLS. The overall liking score is expressed as a percentage, running from +100% to –100%. The questionnaire has been evaluated in several surveys and has been found to differentiate well between buildings(34). In practice, OLS scores range from about +20% for a well-liked building, down to –20% or worse for one disliked by its occupants. Total scores for the individual factors may be displayed as a bar chart, termed the building fingerprint; an example is shown in Figure 8.4. This gives an immediate picture of the good and bad features of the building and can be a useful tool in ascertaining which aspects of a building require attention during a post-occupancy survey.

8.7 Cost issues

8.7.1 Cost benefit analysis

A comprehensive BMS represents a substantial investment, which is expected to bring a variety of benefits to the client. In order to make a rational investment decision, it is necessary to have some method of comparison between alternative systems. Cost benefit methods of analysis have been developed by economists, which aim to express all costs and benefits associated with a plan of action in a common unit, i.e. money, and to make rational comparisons between plans which may have different size and time scales. One important principle is that the benefits and costs must be expressed from a single point of view, e.g. better indoor air quality may be beneficial to the office worker in terms of health and comfort and it may produce a benefit for the office tenant in terms of increased productivity. However, neither is a direct benefit for the building owner, who may be able to achieve the benefit as an increased rent. Cost benefit analysis is most straightforward where the building owner, operator and employer are the same. Where they are not, it must be borne in mind that the organisation receiving the benefit, e.g. of a reduction in energy consumption, may not be the organisation that paid for the BMS.

The IEA Annex 16 task, *Cost Benefit Assessment Methods for BEMS*(35) set out to produce a coherent analysis procedure which could be used to calculate savings from BMS and assist in making investment decisions. The report concluded that no single model would become accepted; the more common methods of analysis are given in the report and summarised below. More important than the method chosen is the ability to provide good quality input data, particularly where it is necessary to compare different types of costs and savings, e.g. investment cost versus productivity gain. It is recommended that a detailed listing is prepared of the costs incurred and the benefits expected from the BMS. Table 8.8 summarises the suggested headings, which may be expanded into considerably more detail.

In a modern office building, staffing costs are much greater than energy costs. The benefits of improved staff utilisation may therefore be as much or more than the savings in energy consumption. The IEA report quotes estimates which state that the benefits of an intelligent building in terms of energy savings, increased productivity and reduction in building management costs are roughly equal; put another way, this means that the overall benefits of a good BMS may be three times the direct saving in energy consumption. Another rule-of-thumb quoted is that an overall productivity improvement of 2.5% would be enough to pay for the entire BMS.

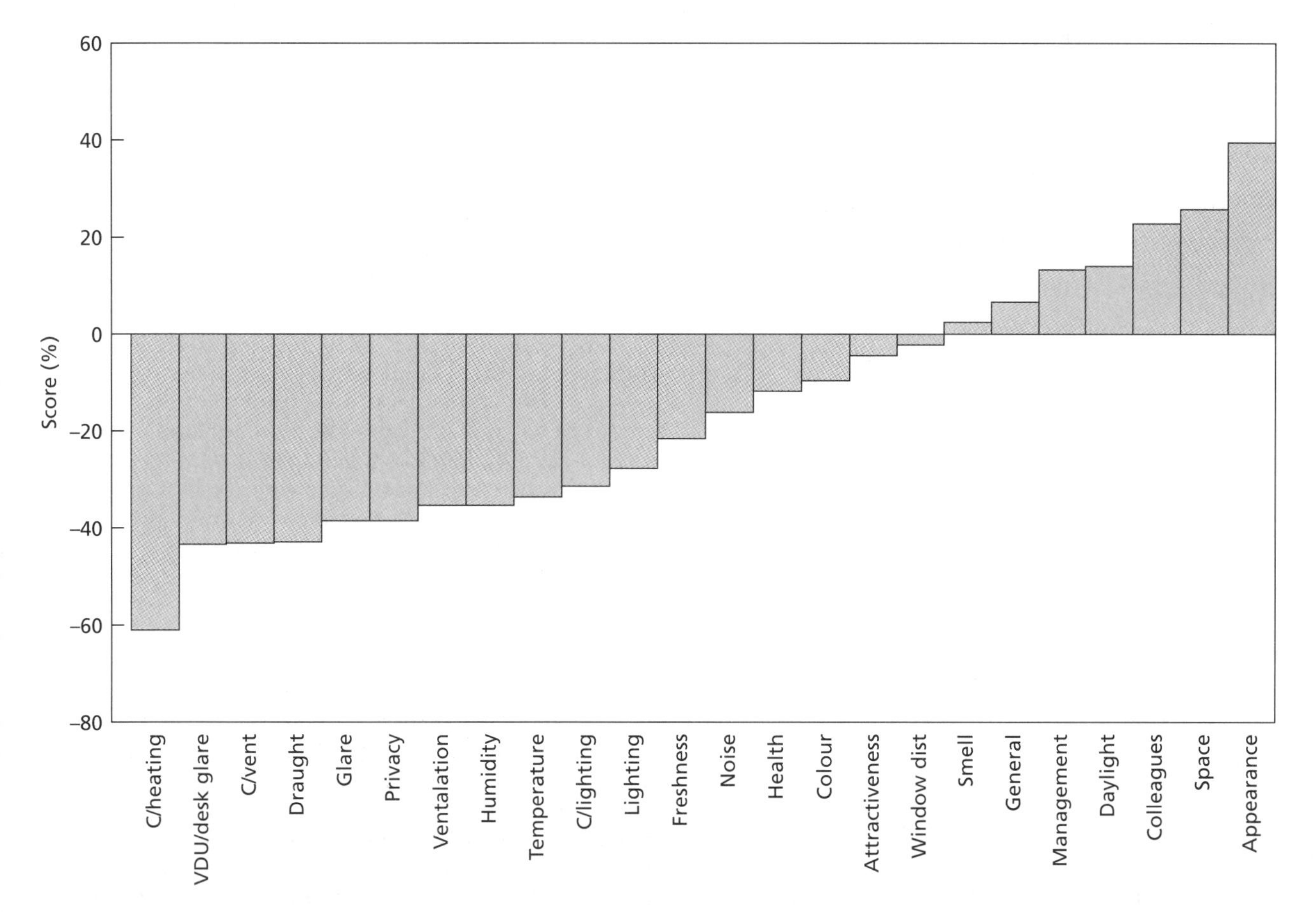

Figure 8.4 Fingerprint of a deep plan, modern, naturally ventilated building. (OLS –15%; C/ = control of)

Table 8.8 Listing of costs and benefits of a BMS

First cost	Specification and design
	Hardware
	Software
	Installation, commissioning
	Training
On-going costs	Maintenance
	Communication costs
	Staffing
Benefits	Reduction in avoidable waste
	Savings in staffing costs
	Better maintenance and fault detection
	Improved productivity
	Tax implications
Life cycle factors	Replacement cost
	Scrap value
	System economic life
	Life of related systems

8.7.2 Life cycle costing

Life cycle costing is an emerging technique that seeks to extend the cost benefit analysis over the whole life of the project, including final disposal. It has been defined as a method that collects together all tax allowances, capital and revenue costs for each system from original installation to abandonment. As with all such analyses, the quality of the data is of vital importance. For life cycle costing to take place, the tender invitation should include separate price breakdowns as follows:

- installation cost
- annual maintenance cost of components
- cost and rate of increase of maintenance contracts
- cost of replacement items
- economic life of system and components
- minimum guaranteed period that spares will be available
- system training costs.

The relation between BMS lifetime and the lifetime of related systems may be important. IT systems in general have a shorter life than BMS. If the BMS is to be closely integrated with the IT system, there may be implications for the BMS if the IT system is renewed. Similarly, replacement cycles for the BMS should be related to the likely lifetime of the HVAC system.

8.7.3 Assessment methods

There are several assessment methods available which are to be found in standard economic texts, e.g. Thuesen *et al.*[36] and ETSU[37]. It is emphasised again that good quality data is important, and that the costs and benefits should be assessed from the same standpoint. The main methods are summarised as follows.

8.7.3.1 Net cash flow

Cash flow analysis is the basis for all cost benefit analyses. For a number of periods, typically yearly, the costs and benefits of the investment opportunity are written down. The total period may extend up to the expected life of the investment, including disposal costs for full life cycle costing. Thus the cash flow table shows both the total costs and benefits and when they are expected to be achieved.

8.7.3.2 Payback method

The payback period is commonly defined as the length of time to recover the initial investment from the benefits produced by that investment; no account is taken of interest rates. Using the cash flow table described above, the net cash flow is accumulated. The year in which the total benefits equal the total cost is the simple payback period, i.e. it is the time after which the investment cost is repaid. There are several variants on the payback period and it is important to state which one is used when making comparisons. For instance, the payback period calculated by dividing the initial cost by the annual benefits does not take into account the initial period, e.g. construction phase, during which no benefits are received.

In general the payback period fails to consider the time value of money, i.e. the value of money to be received in the future is less than that of money received now, nor does it consider what happens after the payback period, e.g. the magnitude and timing of the cash flows. The payback method tends to favour shorter lived investments. Acceptable values of payback period vary with the type of organisation. Typically, payback period of three years or less are required by industry and up to 10 years by Government projects.

8.7.3.3 Discount techniques

There are several techniques based on discounted cash flow (DCF), all of which use a discount rate in the calculations, reflecting the time value of money[38,39]. The present worth method, also known as net present value, is typical. All future amounts in the cash flow table are discounted back to the start of the project. The discount rate may be thought of as the return available on money from other investments; 10% is typically used. With this method of analysis, it is possible that an investment which shows a positive net cash flow after a few years may show a negative present worth; the implication is that it would be more profitable to invest the money elsewhere. The techniques can be used for life cycle costing; the effect of the discount rate is to reduce the present value of costs and benefits which will occur more than a few years in the future.

8.8 Summary

The procurement process chosen by the client strongly influences the way in which the control system is designed and installed. The traditional procurement process separates design from construction. This provides for clear contractual responsibilities, but means that the controls specialist may be appointed late in the process and have little opportunity to contribute to the design. Design and build brings design and construction together in the same organisation. Partnering methods aim to build a stable relationship between those involved in the procurement process, enabling all parties involved to make a full contribution.

The initial project brief produced by the client is of great importance and the type of control system and the organisation of commissioning should be considered from the start. The manual includes a pre-specification check list to assist in clarifying ideas about the required building management system.

It is helpful if the specification follows a standard form, such as the *National Engineering Specification*. Once the HVAC and control system has been installed, effective commissioning is crucial. In modern complex buildings, the controls system interacts with other systems and care is necessary in planning the commissioning process and assigning areas of responsibility. The original specification should be designed both to reduce the amount of commissioning required and to make that commissioning easier. The design should be reviewed at an early stage by a commissioning specialist.

There are statutory requirements for the preparation of proper documentation, which is necessary for system operation, maintenance and refurbishment.

After the building is occupied, it must be run to provide a satisfactory working environment for its occupants. Prompt response to any user problems is important in producing satisfaction. Where there appear to be problems a survey of the building users can be helpful; standard questionnaires can be used to provide comparison with other buildings.

Decisions on how much to spend on control systems should involve the concept of life cycle costing, taking future running costs, maintenance expenditure and energy saving into account. Standard discounted cash flow techniques can be used to evaluate proposals.

References

1 *Thinking about building* (Business Round Table) (1995)

2 Turner A E *Building procurement* 2nd ed (London: MacMillan) (1997)

3 Latham M *Constructing the team. Final report of the Government/industry review of procurement and contractual arrangements in the UK construction industry* (London: Stationery Office) (1994)

4 Hawkins G *Improving M&E site productivity* TN 14/97 (Bracknell: Building Services Research and Information Association) (1997)

5 Egan J *Rethinking construction* (London: Department of Environment, Transport and Regions) (1998)

6 RCF *Trusting the team — the best practice guide to partnering in construction* (University of Reading: Centre for Strategic Studies in Construction) (1995)

7 Anderson D Partnering: all together now *Journal of the Chartered Institution of Building Services* **17**(11) 31–34 (1995)

8 Parsloe CJ *Allocation of design responsibility* TN21/97 (Bracknell: Building Services Research and Information Association) (1997)

9 Pennycook K and Hamilton G *Specifying building management systems* BSRIA TN 6/98 (Bracknell: Building Services Research and Information Association) (1998)

10 Teekeram A J H and Grey R W *Specifications and standards for BEMS* (St Albans: Oscar Faber) (1991)

11 NES *National engineering specification* (London: Building Services Research and Information Association) (1997)

12 BSRIA *Guide to BEMS Centre Standard Specification AH1/90* Vol 1 (Bracknell: Building Services Research and Information Association) (1990)

13 PSA *Building management systems, Standard Specification (M&E) No 15* (London: Property Services Agency) (1986)

14 *Environmental Design* CIBSE Guide A (London: Chartered Institution of Building Services Engineers) (1999)

15 *Thermal environmental conditions for human occupancy* ANSI/ASHRAE Standard 55 (Atlanta, GA: American Society of Heating, Refrigeration and Air Conditioning Engineers) (1992)

16 Nicol F, Humphreys M A, Sykes O and Roaf S *Indoor air temperature standards for the 21st century* (London: E & F N Spon) (1995)

17 Parsloe C J and Spencer A W *Design and commissioning of pipework systems* TN 20/95 (Bracknell: Building Services Research and Information Association) (1995)

18 Sneath C *Code of practice for the selection of subcontractors* (London: Thomas Telford Publishing) (1997)

19 Wild L J *Commissioning HVAC systems. Division of responsibilities* AG 3/89 (Bracknell: Building Services Research and Information Association) (1989)

20 Pike P G and Pennycook K *Commissioning BEMS — a code of practice* AH 2/92 (Bracknell: Building Services Research and Information Association) (1992)

21 Chartered Institution of Building *Code of practice for project management for construction and development* (Chase Production Services) (1992)

22 MoD *Space requirements for plant access, operation and maintenance.* Defence Works Functional Standard: Design and Maintenenance Guide 08 (London: Stationery Office) (1996)

23 *The Construction (Design and Management) Regulations 1994* (London: Stationery Office) (1994)

24 *Managing construction for health and safety. Construction (Design and Management) Regulations 1994 Approved Code of Practice (L54)* (Sudbury, Suffolk: HSE Books) (1995)

25 Nanayakkara R *Standard specification for the CDM Regulations* Health and Safety File AG 9/97 (1997)

26 Armstrong J H *Operation and maintenance manuals for building services installations*. Application Guide 1/87.1 2nd ed (Bracknell: Building Services Research and Information Association) (1990)

27 Smith M H *Decisions in maintenance* TN 14/92 (Bracknell: Building Services Research and Information Association) (1992)

28 *Maintenance management for building services* TM 17 (London: Chartered Institution of Building Services Engineers) (1994)

29 Parsloe C J *Design for maintainability* AG 11/92 (Bracknell: Building Services Research and Information Association) (1992)

30 Smith M H *Maintenance contracts for building engineering services. A guide to management and documentation* AG 4/89.1 (Bracknell: Building Services Research and Information Association) (1991)

31 *Standard maintenance specification for mechanical services in buildings. Volume 3: Control, energy and BMS* (London: Heating and Ventilating Contractors Association) (1992)

32 Leaman A PROBE 10 occupancy survey analysis *Building Services Journal* **19**(5) 37–41 (1997)

33 Fletcher J *Building control and indoor environmental quality — a best practice guide* (Bracknell: Building Services Research and Information Association) (1998)

34 Leventis M and Levermore G J Occupant feedback — important factors for occupants in office design *Chartered Institution of Building Services Engineers/ASHRAE Joint National Conference Part Two* Harrogate UK 1996 II. 192–200 (London. Chartered Institution of Building Services Engineers) (1996)

35 Hyvärinen J *Cost benefit assessment methods for BEMS* (St Albans: Oscar Faber) (1992)

36 Thuesen H G, Fabrycky W J and Thuesen G J *Engineering economy* 5th ed (New Jersey: Prentice Hall) (1977)

37 ETSU *Economic evaluation of energy efficiency projects* (Harwell: ETSU) (1994)

38 Wright M G *Using discounted cash flow in investment appraisal* 3rd ed McGraw-Hill) (1990)

39 Dixon R *Investment appraisal. A guide for managers* Revised ed (Kogan Page) (1994)

Appendix A1: Bibliography

Relevant references are listed as they occur in the text. The following publications contain a great deal of useful information.

A1.1 Control theory

Levermore G J *Building energy management systems* 2nd ed (London: E & F N Spon) (2000)

Underwood C P *HVAC control systems: modelling, analysis and design* (Andover: E & F N Spon) (1998)

A1.2 Control strategies

Levenhagen J I *HVAC controls system design diagrams* (New York: McGraw Hill) (1998)

Martin A J and Banyard C P *Library of system control strategies.* Applications Guide AG 7/98. (Bracknell: Building Services Research and Information Association) (1998)

MSR Planungshandbuch (CD-ROM) (Karlsruhe: Siemens AG) (1997)

A1.3 General

Engineering manual of automatic control SI edition (Minneapolis: Honeywell Inc.) (1995)

A1.4 Control valves

Petitjean R *Total hydronic balancing* (Ljung, Sweden: Tour and Andersson AB) (1997)

Appendix A2: Tuning rules

A2.1 The PID control loop

The basic feedback controller is illustrated in Figure A2.1. The error signal x_e is processed by a controller which produces an output signal Y, which in turn drives the controlled device, e.g. actuator and fan coil, resulting in a change in the controlled variable x. In 3.3.3.1 a description was given of how the components of the controlled device, e.g. valve and heat emitter, may be matched to linearise the heat output, resulting in an approximately linear relation between the controller output Y and the heat output to the conditioned space. However, even if the change in the output is roughly linear with controller output, the gain of the system can change with operating conditions.

When a controlled system running in a steady state is disturbed by a change in load, a change in set point or some other disturbance, the system will react. If the system is operating in a stable manner, it will sooner or later settle down in a new stable state. If the system is unstable, it will go into indefinite oscillations of increasing amplitude; in practice the size of the oscillation will be limited by the components of the system and the system will continue oscillating or hunting.

The high thermal inertia of most buildings, coupled with the slow response of traditional heating systems, has usually meant that reasonably stable temperature conditions can be achieved with simple control systems. In some circumstances, it is possible for satisfactorily steady conditions to be achieved in the controlled space even when the HVAC system itself is hunting in an unstable manner. This is bad practice, leading to excess wear of components. The trend towards lighter buildings and the use of air handling systems has increased the responsiveness of systems and so increased the possibility of unstable operation.

The general equation for the output of a three-term controller is

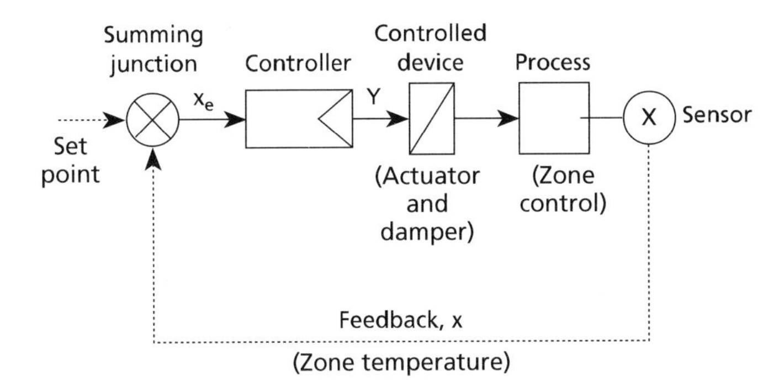

Figure A2.1 Feedback control loop

$$Y = Y_0 + K_p x_e + K_i \int_0^T x_e \mathrm{d}t + K_d \frac{\mathrm{d}x_e}{\mathrm{d}t} \quad (A2.1)$$

where:

Y	output from controller (ND, 0–1)
Y_0	constant controller output (bias) (ND)
x_e	error i.e. difference between set point and measured value (K)
K_p	proportional gain constant = (1/proportional band) (1/K)
K_i	integral gain constant (1/(Ks))
K_d	derivative gain constant (s/K).

The equation has been written in terms of the controller output as a non-dimensional (ND) quantity scaled from 0 to 1. In practice, the output would be a standard voltage or current output, or else a digital value for DDC. The controlled variable is illustrated as a temperature, but could be any other controlled variable.

The equation may be rewritten

$$Y = Y_0 + K_p \left(x_e + (1/T_i) \int_0^T x_e \mathrm{d}t + T_d \frac{\mathrm{d}x_e}{\mathrm{d}t} \right) \quad (A2.2)$$

where:

$T_i = K_p/K_i$	integral action time (s)
$T_d = K_d/K_p$	derivative action time (s)

Most controllers are set up this way, with independent adjustments for T_i and T_d, which are usually calibrated in minutes.

The stability of a simple system may be analysed by setting up the differential equations which describe its dynamic behaviour[(1)]. Standard analysis methods, based on Laplace transforms, are then used to check for stable operation. However, most systems are too complex for this analysis to be practical and empirical rules have been derived to assist in setting up a three-term controller to achieve the desired compromise between stability and speed of response. For completeness, the general case of a three-term controller is given. However, in most HVAC applications the derivative term is not required and is disabled by setting the derivative time to zero.

A2.2 Digital control

The above equations are written in terms of continuous analogue quantities. Digital control systems, which constitute the majority of modern systems, operate with discrete sample values of the measured quantities. The general equation for a three-term PID controller becomes

$$Y_{k+1} = Y_0 + K_p \left\{ x_{k+1} + \frac{1}{T_i} \sum_{j=1}^{k} x_j \Delta t + T_d \frac{x_{k+1} - x_k}{\Delta t} \right\} \quad \text{(A2.3)}$$

which is similar to Equation A2.2, but with the following changes in symbols:

Y_k output from controller at time step k (ND, 0–1)

Y_0 controller offset (ND)

x_k deviation from set point at time step k (K)

Δt sampling interval (s).

The discrete sampling of the controlled variable produces some effects not found with analogue controllers. Noise or interference can cause sudden changes in the controller output. Noise affecting a single sampled value will produce a sudden change in controller output due to the phenomenon known as derivative kick; this is caused by the change in value of the derivative term $(x_{k+1} - x_k)/\Delta t$. A change in set point will also produce a derivative kick, together with the associated proportional kick; the kick lasts for only one time step but can cause unwanted movement in the controlled device. While in general a short sampling time increases the accuracy of control, it increases the effect of derivative kick.

Software techniques may be used to reduce the effect of derivative kick. A moving average of the last four sensor values may be used to reduce the effects of noise; logical filters can also be beneficial. It is possible to set an error deadband around the set point. Control action is only taken when the controlled variable moves outside the deadband; this prevents operation of the controlled device due to minor noise signals in a system that is controlling satisfactorily near the set point. Another technique is to apply the derivative calculation to the change in value of the controlled variable, rather than the change in deviation of the controlled variable from the set point as shown above; this avoids any derivative kick when the set point changes. In practice, derivative control is not implemented in most BMS controllers, since the benefits are not sufficient to justify the extra effort of setting up the controller and the risk of introducing instabilities.

Equation A2.3 is termed the position algorithm. The position of the controlled device is related directly to the output signal Y from the controller; a constant position offset Y_0 can be used if required. There may be situations where the error x is large for an extended period of time; this can easily arise during a long warm-up period. The integral term may then become very large. This will cause the controller to overshoot the set point and proper control is not established until the integral term has become dissipated. This phenomenon is known as integral wind-up and may be prevented in software by limiting the integral term to a maximum value, typically that which will produce between 50 and 100% output from the controller.

The problem of wind-up may be avoided by using the incremental control algorithm. This is based on the change between successive values of the controlled variable.

$$\Delta Y_{k+1} = Y_{k+1} - Y_k = K_p \left\{ x_{k+1} - x_k + \frac{\Delta t}{T_i} x_k \right\} \quad \text{(A2.4)}$$

Equation A2.4 has been written without a derivative term, which is not normally used in incremental control. The change in output ΔY_{k+1} is used to reposition the controlled device, e.g. by holding on an actuator motor for a time interval proportional to ΔY_{k+1} or by controlling a stepper motor. There is no positional feedback from the actuator. The algorithm has the advantage of avoiding integral wind-up.

A2.3 Tuning

The aim of tuning a controller is to restore control following a disturbance as quickly as possible and to achieve stable control with a small or zero offset from the set point. These criteria interact; increasing the speed of response risks introducing instability. The controller parameters must be set to values appropriate to the system. A full analysis of the dynamic response of a controlled system is complex[2] and not often attempted in practice. Several methods have been developed to estimate suitable values of the control constants.

The ultimate cycling method, also known as the ultimate frequency method, can be carried out on the intact control system. To use this method, the integral and derivative actions of the controller to be tuned are first disabled. The proportional gain setting is then increased (i.e. the band is reduced) in small steps from a low value; at each value of the band, the system is given a small disturbance by adjusting the set point. At low values of the proportional gain the controlled variable will settle down to a stable value when the set point is changed; as the gain is increased, a value will be reached when the system starts hunting. The oscillations should be steady, with neither increasing nor decreasing amplitude. The value of the proportional gain K_p^* at which this happens is noted, together with the period of oscillation T^*. The settings of the controller may then be obtained using the Ziegler–Nichols method set out in Table A2.1. This method produces a response to a step change in the set point similar to that shown in Figure A2.2. This is also known as the quarter wave method, because the amplitude of the first overshooting wave is four times that of the second.

Reaction curve methods depend on measuring the open loop response. This requires the controller to be taken out of the control loop and the response of the controlled variable to an artificial step change ΔY in controller output to be measured. Most systems will respond with a combination of time delay and first-order response. Figure A2.3 shows how the time delay T_t, time constant T_g and final change in controlled variable x_∞ may be estimated

Table A2.1 Ultimate frequency method controller settings to produce a quarter-wave response (Ziegler and Nichols). K_p^* is the gain which just produces hunting with period T^*

Controller mode	K_p	T_i	T_d
P	$0.5K_p^*$		
PI	$0.45K_p^*$	$0.8T^*$	
PID	$0.6K_p^*$	$0.5T^*$	$0.125T^*$

Underdamped quarter-wave response

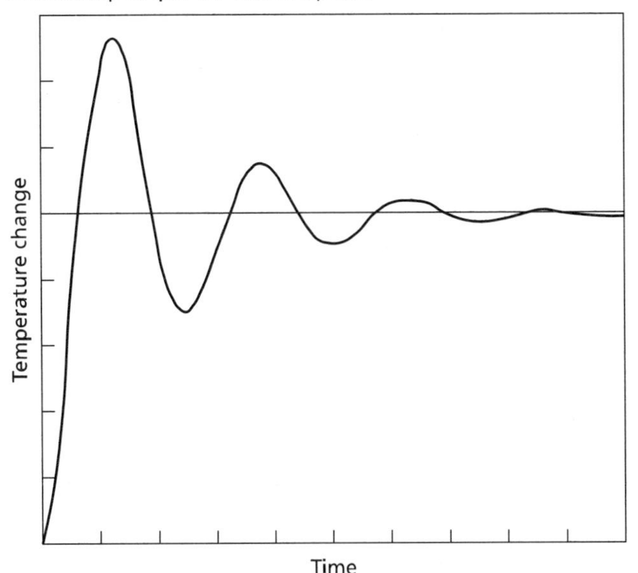

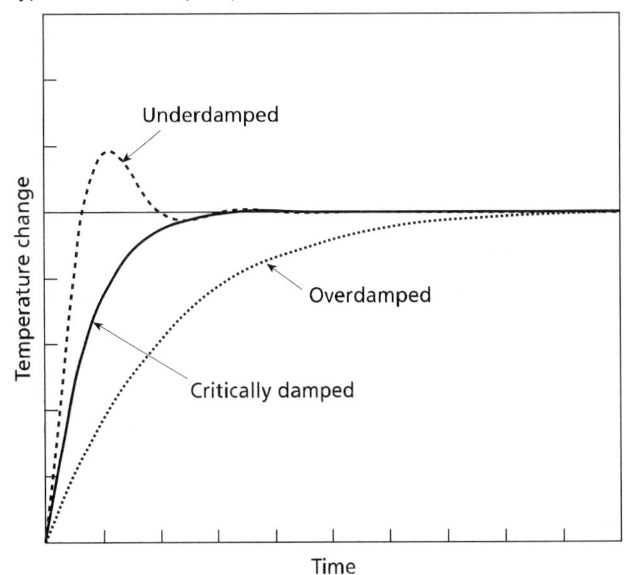

Figure A2.2 Response to a change in set point, showing underdamped and critically damped responses

from the reaction curve. These values may then be used to calculate the optimum controller settings. Table A2.2 shows values calculated according to a method developed by Cohen and Coon[3] quoted by Kreider and Rabl[4], which produces a quarter-wave response.

The quarter-wave response produced by the Ziegler–Nichols or Cohen–Coon methods may be considered too oscillatory and a more damped response is often more appropriate in practice. Control settings to produce a critically damped response may be estimated from the open loop response by a method due to Bekker[5] and are set out in Table A2.2. It was pointed out at the beginning of the section that the gain of a controlled system is likely to vary with operating conditions. Control settings should be established with the system in a high gain condition. Setting the controller in conditions of low gain could result in system instability when the conditions move to high gain.

Digital controllers add a fourth setting, the computation time step, which is the interval between digital controller updates. The sensitivity of the integral term increases and the sensitivity of the derivative term decreases as the sampling time interval increases. A long sampling interval means that the controller does not receive information at a sufficient rate to operate properly and may result in unstable control; on the other hand, a very short sampling interval may waste execution time resources.

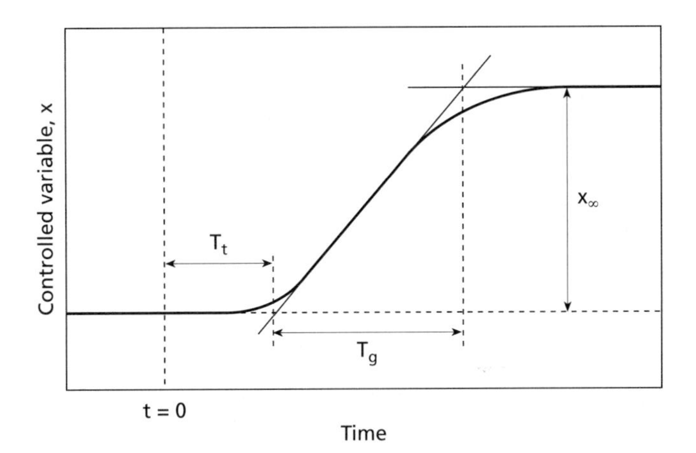

Figure A2.3 Reaction curve. Curve shows open loop response of a system to a step change in controller output ΔY applied at time $t = 0$. The values of time delay T_t, time constant T_g and final change in output x_∞ are estimated as shown

A2.4 Step-by-step tuning procedure

The procedure given below, to be followed for tuning a controller by either of the two methods analysed above, has been adapted from a controller manual. Settings have been quantified in terms of the proportional band, rather than gain, since this corresponds to the settings on most controllers. This description should be read in conjunction with the advice given in section 2. An alternative simplified procedure is given by Borresen[6].

A2.4.1 Closed loop ultimate cycling

This tuning method is carried out on the intact control loop.

Table A2.2 Controller settings: open loop reaction curve method. The measured variables are defined in Figure A2.3

Controller mode	K_p	T_i	T_d
Quarter-wave response[4]			
P	$AB(1 + R/3)$		
PI	$AB(1.1 + R/12)$	$T_t\dfrac{30 + 3R}{9 + 20R}$	
PID	$AB(1.33 + R/4)$	$T_t\dfrac{32 + 6R}{13 + 8R}$	$T_t\dfrac{4}{11 + 2R}$
Critically damped response[5]			
PI	$0.37AB$	T_g	

$A = \Delta Y/x_\infty$, $B = T_g/T_t$, $R = 1/B$.

(1) Connect a chart recorder to measure the controlled variable.

(2) Set the controller to give proportional action only, by the selector switch if fitted, or by setting the integral time to maximum and the derivative time to zero.

(3) Start with a wide proportional band, i.e. low gain setting.

(4) Reduce the proportional band in steps. After each reduction, make a small adjustment to the set point and observe the response. To start with the loop will stabilise with the controlled variable at a steady value.

(5) Note the proportional band setting $X_p^\star$ when the response to a set point change is for the loop to go into oscillation with a steady amplitude. Measure the period of the oscillation $T^\star$.

(6) Calculate the control parameters from Table A2.3.

(7) Set the parameters on the controller and check how the control loop behaves in response to a change in set point. Adjust the parameters if necessary.

A2.4.2 Open loop response

(1) Connect a chart recorder to measure the controlled variable.

(2) Open the control loop by disconnecting the controller from the controlled device.

(3) Make a sudden change ΔY in the position of the controlled device, either by hand or by applying a constant voltage to the device input to simulate the controller output. ΔY is measured as a fraction of full output.

(4) Record the change in controlled variable with time, which will have the form shown in Figure A2.3. Calculate the time lag T_t, the time constant T_g and the final change in controlled variable.

(5) Calculate the parameter settings from Table A2.4.

(6) Reconnect the controller and observe the reaction of the control loop to changes in set point. Adjust parameters if necessary.

Table A2.3 Control parameters, ultimate cycling method

Controller mode	X_p	T_i	T_d
P	$2X_p^\star$		
PI	$2.2X_p^\star$	$0.9T^\star$	
PID	$1.7X_p^\star$	$0.75T^\star$	$0.15T^\star$

Table 10.4 Control parameters, reaction curve method

Controller mode	X_p	T_i	T_d
P	$\frac{X_\infty}{\Delta Y}\frac{T_t}{T_g}$		
PI	$1.1\frac{X_\infty}{\Delta Y}\frac{T_t}{T_g}$	$3.5T_t$	
PID	$0.9\frac{X_\infty}{\Delta Y}\frac{T_t}{T_g}$	$2.5T_t$	$0.5T_t$

References

1 Underwood C P *HVAC control systems: modelling, analysis and design* (Andover: E & F N Spon) (1998)

2 Levermore G J *Building energy management systems* 1st ed (London: E & F N Spon) (1992)

3 Cohen G H and Coon G A Theoretical considerations for retarded control *Trans ASME* **75** 827–834 (1953)

4 Kreider J F and Rabl A *Heating and cooling of buildings. Design for efficiency* International ed (New York: McGraw Hill) (1994)

5 Bekker J, Meckl P and Hittle D C A tuning method for first order processes with PI controllers *ASHRAE Transactions* **97**(2) 19–23 (1991)

6 Borresen B A *Controllability — back to basics* ASHRAE Transactions **96**(2) 817–819 (1990)

Appendix A3: Glossary

Adaptive control system	A system in which automatic means are used to adjust system parameters to achieve optimum performance
ANSI	American National Standards Institute
ASHRAE	American Society of Heating, Refrigeration and Air Conditioning Engineers
Authority	The authority of a valve or damper describes the degree of control it exerts when installed in a circuit
BACnet	The Building Automation and Control networking communication protocol: ANSI/ASHRAE Standard 135-1995. Provides a means by which building automation devices from various manufacturers can share data and work together
Baud rate	A measure of rate of transmission of data in bits per second
BMS	Building management system. A building-wide network which allows communication with and control of items of HVAC plant. May also include other systems such as lighting
Boost period	The period when plant is operated at full capacity, normally in the preheat period before occupancy
Bridge	A device that connects two or more LAN segments at the physical and data link layers. This device may also perform message filtering if programmed to do so
Bus	A common communications link which carries messages between devices (nodes) which are connected to the bus. The term field bus refers to low level communications
Cascade control	A control system consisting of a slave controller controlling a system variable and a master controller controlling the set point of the slave. Also known as reset control or master–submaster (US)
CDM	Construction (Design and Management) Regulations, 1994
CEN	European Committee for Standardisation
Clean contact	A contact not connected to any voltage source. Also called a voltage free contact
Client	The person who commissions the design and construction of a building. May not be the end user
Compensator	Normally implies weather compensation: a control device whose function is to vary flow temperature as a function of external temperature
Control function	A term used to describe type of control, e.g. compensation, night set back etc.
Control parameter	A preset variable used in a control algorithm, e.g. width of the proportional band
Controlled variable	The physical quantity being controlled by the system, e.g. air temperature
CPU	Central processing unit
Deadband	(1) The range of values of the controlled variable within which a floating controller produces no change in control action, also called neutral zone (2) The range of values of the controlled variable in which HVAC plant is off, e.g. the interval between heating and cooling operation (3) The range between upper and lower switching values for a two-position controller. Better termed the differential
Dead time	The time required for a response in the controlled variable to be observed following a step change in input to the controller
Dead zone	See deadband, sense 2
Default value	Stored value of a control parameter or other system variable to which the system reverts in the absence of an instruction
Derivative action time	Time taken for the proportional term to match the derivative term of a PID controller when the error changes linearly with time
Derivative control	A control method in which the control output signal is proportional to the rate of change of the controlled variable. Normally used in conjunction with other modes, e.g. PID
Desired value	The desired value of the controlled variable. Not necessarily the same as the set point

Deviation — The difference between the set point and the actual value of the controlled variable

Differential — The difference between the higher and lower operating levels of an on/off controller

Direct digital control (DDC) — Implies that the control algorithms are in the form of software. Analogue inputs or outputs may coexist with digital control

Distance–velocity lag — The time taken for a heating or cooling medium to travel from its controlled source to the point of delivery

Distributed intelligence — A system where data processing is carried out at local controllers or outstations as well as at the central station

Distributed processing unit (DPU) — See universal controller

Double regulating valve — A valve for the regulation of flow rate having an established valve characteristic and provided with indicated positions of the valve opening and an adjustable stop such that the valve can be closed for isolation purposes and reopened to the previously determined set position

Duty cycling — A control method which alternates or cycles the sequence of plant use to ensure standby plant is brought into use. See also sequencer

EAC — Energy account centre. An area of energy accountability, used in energy monitoring

Economiser control — Controlled mixing of recirculated and outdoor air to give economy in heating or cooling costs. See free cooling

EMC — Electromagnetic compatibility

Emulation — Representation of a building system by a computer simulation, in such a way that a real BMS may be connected to the emulator as if it were a real system

Enthalpy control — Controlled mixing of recirculated and outside air based on their enthalpies. Only used in association with humidity control of supply air

Events (BACnet) — Changes in the state of an object, typically between normal, off-normal and fault condition. Events may be reported to other BACnet devices

Exhaust air — Air which is discharged to outside

Extract air — Air which is extracted from an occupied zone

Feedback control — Also called closed loop control, where a measure of the controlled variable is fed back to the controller

Feedforward control — See open loop control

Field processing unit — See universal controller

Floating action — A control mode where the output signal changes at a steady rate when the input signal falls outside the neutral zone

Flow coefficient (valve) — K_v: The flow of water through a flow measurement device or double regulating valve at a temperature between 5 and 40°C and measured in cubic metres per hour, that will induce a pressure loss of 1 bar
K_{vs}: The flow of water through a flow measurement device, of either fixed or variable orifice types, at a temperature between 5 and 40°C and measured in cubic metres per hour, that will induce a differential pressure, or signal, of 1 bar across the pressure tapping points

FND — Firmenneutrales Datenuebertragungssystem (FND non-proprietary data communications system). German DIN standard designed for BMS communication. Low uptake outside Germany

Free cooling — (1) The use of outside air in an economiser control system to minimise or avoid the use of chiller plant
(2) The provision of chilled water from a cooling tower or dry cooler, without the use of chiller plant

Gateway — A device that connects two or more dissimilar networks by message translation and signal conditioning, permitting information exchange between them. A gateway would be necessary to translate LonWorks messages to the BACnet format

Head end — Another term for the supervisor of a BMS

Health and safety file — Documentation prepared under the CDM regulations and handed to the client on completion of construction. Includes operating manuals of building services

HTHW — High temperature hot water (used in heating systems; water temperature over 120°C)

Hunting — A control state where the system does not settle to a steady value but oscillates about the set point

HWS — Hot water services, i.e. hot water to taps

Integral action time — The time it takes for the integral term of the control output equation to match the offset caused by the proportional term on a step change in error

Integral control — A control mode where the output signal changes at a rate which is

	proportional to the deviation of the input from the set point. Usually found combined with proportional control
Internetwork	Two or more networks connected by routers
Interoperability	The coordinated operation of various pieces of equipment from various manufacturers on the basis of standardised interfaces and services
Jacketing software	Software used in conjunction with adaptive controllers to prevent out-of-range operation
JCT	Joint Contracts Tribunal
Let-by	Residual flow through a valve when closed
Load error	The difference between the set point and the steady state value of the controlled variable of a proportional or other controller
Local area network (LAN)	An arrangement that enables computer or processor based equipment to be linked so that they may communicate with each other. Implies both hardware, e.g. cabling, and software
LON	Local operating network using the LonWorks protocol
LonTalk	Protocol using a neuron chip for communications. It covers a variety of physical media including twisted pair, power line carrier, and wireless transfers
LonWorks	LonWorks is a registered trade mark and is the generic term for bus systems based on the LonTalk protocol
LPHW	Low pressure hot water (used in heating systems; water at approximately atmospheric pressure)
LTHW	Low temperature hot water (used in heating systems; water temperature 70–100°C)
M&T	Monitoring and targeting. Used to refer to energy management information systems
Modem	A device which converts data into a form suitable form transmission over a telephone or other network
MS/TP	Master–Slave/Token-Passing EIA-485 LAN developed for BACnet by ASHRAE.
MTHW	Medium temperature hot water (used in heating systems; water temperature 100–120°C)
NBS	National Building Specification
NES	National Engineering Specification
Neutral zone	See deadband
Night set back	The reduction in set point of a heating system during unoccupied hours
NIST	National Institute for Standards and Technology, which is located in Washington DC. The NIST was substantially involved with the development of BACnet
Node	A computer or digital device connected to a network, with an address on that network
Object (BACnet)	BACnet 'Objects' represent in a standardised way the information required for a building automation control system. The 'analogue input' object, for example, represents sensor inputs such as temperature sensors
Open communication	A term used for the exchange of data between various pieces of equipment on the basis of open or standardised protocols
Open loop control	A control system where the controller does not receive information on the results of its actions cf. feedback control
Open protocol	A protocol available for use by anyone. It may be protected by copyright and licence fees may be charged for its use. Changes in an open protocol may be made at any time by the original author
Open system	Apart from open communications, an open system requires standardised formats, services and interfaces in order to transfer software simply to other platforms and to achieve a uniform operation philosophy. Open communication is also a prerequisite for an open system
Optimiser	A control system employing an optimum start algorithm
Optimum start	A control system or algorithm which starts plant operation at the latest time possible to achieved specified conditions at the start of the occupancy period
Optimum stop	A control system or algorithm which stops plant operation at the earliest possible time such that internal conditions will not deteriorate beyond preset limits by the end of the occupancy period
Outstation	A device connected to sensors and actuators which can perform local control and other functions. It can exchange information with the central station
PID	Control incorporating proportional, integral and derivative actions
Planning supervisor	Person appointed by the client under the CDM regulations to oversee safety of construction

PLC — Programmable logic controller

Point — Generic name for a source of data (input) or control output in a control system. May be analogue or digital. A virtual point is computed from real points

Points schedule — The list of all points in a control system

Pressure tapping point — Normally two pressure tapping points are provided on a flow measurement device either side of a fixed or variable orifice to enable the pressure differential to be measured

Principal adviser — Person or organisation that can impartially advise a client on the need to build and how to go about procurement

Proportional band — The range of input values corresponding to the full range of output variation of a proportional controller. May be expressed in actual units (e.g. temperature) or as a percentage of the full input range of the controller

Proportional control — Control action where the control output is proportional to the deviation of the deviation of the input from the set point

Proportional gain — Reciprocal of proportional band

PSDN — Public switched data network. A data communications network set up by a public telephone service

PSTN — Public switched telephone network: the normal telephone service

PTP — Point-to-point protocol. Provides serial communications between two BACnet half-router devices, typically for dial-up communications over phone modems. It may also be applied as a hard-wired interface. For the duration of the connections, messages may be passed between networks on either side of the PTP connections

PWM — Pulse width modulation, used in frequency inverters for motor speed control

Rangeability — Operating range of a valve, expressed as the ratio of maximum and minimum controllable flows

Recirculated air — Air which has been extracted from a zone and which is mixed with outdoor air to make up the supply air

Regulating valve — A valve which is set during commissioning to achieve the required design flow

Repeater — A device that connects two or more physical segments of a LAN at the physical layer and exactly reproduces the electrical signal on either side so that nodes either side of the repeater are considered to be on the same network

Reset action — Integral control action (US usage)

Reset control — Non-standard term for cascade control

Reset rate — Inverse of integral action time

Resolution — The smallest change in input signal that can be discriminated in a device

Return air — Air extracted from a zone and returned to the main air handling plant. Not necessarily the same as recirculated air. Extract air is preferred term

Router — A device which connects two or more networks, of same or different types, so that a message may be transferred across the appropriate networks to reach its destination. The router translates between protocols where required

SCADA — Supervisory control and data acquisition

Self-optimising control system — See adaptive control system

Sequencer — A controller which controls the operation of multiple boilers or chillers, both for modulation and duty cycling

Set point — The value to which a controller is set to achieve the desired value of the controlled variable

Shoe valve — Rotating shoe that slides over internal ports. Normally three-way

Slipper valve — See shoe valve

Soft point — See virtual point

Standardised protocol — A standardised protocol is put forward by a national or international standardisation committee. Member countries are obliged to accept these standards. Any changes to the standards must be approved by the standardisation committee. The certification procedure is often standardised at the same time. Examples are BACnet, FND, PROFIBUS etc.

Stroke — The range of movement of a linear actuator

Supervisor — The user interface for a BMS. Also called head end

Supply air — Air supplied by a central air handling unit to the occupied zones

TCP/IP — Transmission Control Protocol/ Internet Protocol. It encompasses media access, packet transport, session communications, etc. TCP/IP is supported by a large number of hardware and software vendors

Three term control	Another term for PID control
Throttling range	See proportional band
Time constant	The time required for a quantity to reach 63% of its new final equilibrium value following a change or disturbance.
Unitary controller	Device which controls and/or monitors a single piece of equipment. See dedicated function controller
UTP	Unshielded twisted pair
Virtual point	A variable which is computed rather than directly measured, e.g. enthalpy
Voltage	Circuits are classified as: (1) Extra-low voltage (ELV). Not exceeding 50 V between conductors and not exceeding 30 V AC or 50 V DC between any conductor and earth (2) Low voltage. Between 50 and 250 V (3) Medium voltage. Over 250 V and not exceeding 650 V
Weather compensation	The control of heating medium temperature as a function of external temperature
Wind-up	An excessive or out-of-range control action produced by an integral action controller following start up or other period when there has been a deviation from the set point
Zone	A part of a building which is controlled independently of other zones

Index

Absence sensing, 5-57
  see also Occupancy
Absorption chillers, 5-7
Accelerator heater, 2-3
Accuracy of sensors, 3-2
Active analogue sensors, 3-1–2
Actuators, 3-7, 3-9
  electrical connections, 3-7–8
  mechanical characteristics, 3-7
  pneumatic, 3-8
  valves, 3-12
Adaptive control, 2-12–13
Adaptive tuning techniques, 2-12–13
A/D conversion, 3-2
Air-cooled condensers, 5-8
Air flow tracking (volume matching), 5-39, 5-40
  variable-volume extract fans, 5-36
Air handling systems, *see* Central air handling plant
Air handling units (AHUs), 5-28–9
Air quality, 1-3
  poor outdoor, 5-35
  sensors, 3-5–6
Airside fan coil units, 5-47, 5-48
Alarms, 5-1
  user interface, 4-18
American Society of Heating, Refrigerating and Air Conditioning Engineers (ASHRAE), 4-9–10
Analogue alarms, 5-1
Analogue modelling, 6-2–3
Analogue sensors, 3-1–2
Angle of rotation, rotary actuators, 3-7
Architecture, building management systems, 4-1
  bus systems, 4-3
  centralised intelligence, 4-1–2
  conventional controls, 4-1
  modular systems, 4-3
  packaged systems, 4-3
  programmable controllers, 4-2–3
  three-level model, 4-4
  utilisation of IT network, 4-4–5
Arcnet, 4-6, 4-7
Artificial intelligence, 2-13–14
Atomised spray humidifiers, 5-31–2
Authority
  damper, 3-19
  valve, 3-13–15, 5-14
Automatic balancing valves, 3-9
Automatic flow-limiting valves, 5-25
Autotransformers, 3-21
Autotuning, 2-12
Auxiliary heat exchanger, free cooling, 5-11
Averaging, 2-7

Back end protection of boilers, 5-4, 5-18
BACnet, 4-7, 4-9–10
Baffles, 3-19
Base energy requirement, 1-3, 1-4
Batibus, 4-7, 4-12
Batibus Club International, 4-12
Bekker tuning method, A2-3
Best Practice in the Specification for Offices, 1-3
Blanking plates, 3-19
Bleeding, cooling towers, 5-11
Boiler (Efficiency) Regulations 1993, 5-3
Boilers, 5-3
  controls, 5-4
    boiler protection, 5-4–5
    burner controls, 5-4
    multiple boilers, 5-5–7
    water heating plant, 5-5
  primary hydraulic circuits, 5-18–19
  safety strategies, 5-2
  system pressurisation, 5-7
  types, 5-3–4
BRE Digest 424, 4-13, 4-14
BRESTART, 2-8
Brief
  commissioning, 8-9
  design, 8-5, 8-6
BS 5389, 4-16
BS 6651, 4-13
BS 7807, 4-16
BS CEN 60439, 3-24
BS EN60947-4-1, 3-21
BS EN7730, 1-3
Buffer vessels, *see* Decoupling, hydraulic
Building Automation and Control Network (BACnet), 4-7, 4-9–10
Building backbone cabling, 4-7, 4-8
Building management systems (BMSs)
  architecture, 4-1
    bus systems, 4-3
    centralised intelligence, 4-1–2
    conventional controls, 4-1
    modular systems, 4-3
    packaged systems, 4-3
    programmable controllers, 4-2–3
    three-level model, 4-4
    utilisation of IT network, 4-4–5
  benefits, 1-5–6
  commissioning, 8-10–11
  definition, 1-2
  pre-commissioning, 8-9–10
Building operation, *see* Operation
Building Regulations (1995)
  energy conservation, 1-4
  lighting controls, 5-56, 5-57, 5-58
  natural ventilation, 5-54
  optimum start, 2-8
  ventilation, 5-34
  weather compensation, 2-8, 5-27
Burner controls, boilers, 5-4
Bus network topology, 4-6
Bus systems
  BMS architecture, 4-3
  standardising bodies, 4-9
Butterfly valves, 3-9, 3-10
  actuators, 3-7
Bypass ports, valves, 3-10, 3-11
  authority, 3-15

Cabling
  classification, 4-7
  electromagnetic compatibility, 4-14
  sensors, 3-6
  specification, 8-4
  structured, 4-7–8
Calibration of sensors, 3-6
Calorifiers, 5-19–21
Campus backbone cabling, 4-8
Capacitative humidity sensors, 3-3, 3-4
Capacitative pressure sensors, 3-4
Capacity index, 3-12
Carbon dioxide
  concentration
    air quality sensors, 3-5, 3-6
    mechanical ventilation, 5-34
    natural ventilation, 5-53, 5-54
  emissions, 1-2
Carbon monoxide
  sensors, 3-6, 5-35
  ventilation, 5-34, 5-35
Cascade control (reset control), 2-6
  central air handling plant, 5-29
  lighting, 5-57
  tuning, 2-11
  underfloor heating, 5-27
  zone temperature, 2-9
Case studies, 6-7
  low energy university building with fabric storage, 6-8
  mixed mode R&D facility using chilled beams, 6-8–9
  remote control of elderly people's homes, 6-7–8
  retrofit using a modular control system, 6-7
  Terminal 2, Manchester Airport, 6-9–10
Cash flow analysis, 8-16
Cavitation, valves, 3-16
Ceilings, *see* Chilled ceilings
Central air handling plant, 5-28–9
  cooling coil, 5-30–1
  heating coil, 5-29–30
  humidification, 5-31–2
  preheat coil, 5-29
  safety strategies, 5-2
Central heating system, 5-18
Centralised intelligence, 4-1–2
Central processing unit (CPU), 4-2
Centrifugal compressors, 5-7
Characterised V-port valves, 3-11
Characteristics
  damper, 3-18–19
  valve, 3-11–12
Charge to full, ice storage, 5-13
Chilled ceiling panels, 6-5
Chilled ceilings, 5-49–50
  case study, 6-8–9
  and displacement ventilation, 6-5–6
Chilled mirror dewpoint sensors, 3-3, 3-4
Chilled water systems, 5-2
Chiller downstream ice storage system, 5-12
Chiller priority control, ice storage, 5-13
Chillers, 5-7
  control, 5-8–9
  cooling towers, 5-10–11
  free cooling, 5-11
  ice storage, 5-12–13
  multiple, 5-9–10
  primary hydraulic circuits, 5-18–19
Chiller upstream ice storage system, 5-12
Chlorinated fluorocarbons (CFCs), 1-2–3
Chlorine sensors, 3-6
CIBSE Code for Interior Lighting, 5-56
CIBSE Code of Professional Conduct, 1-3
CIBSE Guide A1, 1-3
CIBSE Guide to Energy Efficiency in Buildings, 1-3
CIBSE Policy Statement on Global Warming, 1-3
Closed loop systems, 2-1
  cooling towers, 5-11
Closed loop ultimate cycling method, 2-11, A2-2, A2-3–4
Close-down, *see* Shut-down period
Closing torque, dampers, 3-18
Coaxial cable
  electromagnetic compatibility, 4-14
  networks, 4-6–7
Code of Practice for the Selection of Subcontractors, 8-8
Coexistence, networks, 4-8
Cohen–Coon tuning method, A2-3
Commissioning, 8-8
  brief, 8-9
  costing, 8-12–13
  design stage, 8-9
  handover, 8-11–12
  installation, 8-9–11

management, 8-8–9
Commissioning valves, 3-9
Common ports, valves, 3-10, 3-11
authority, 3-15
Communication media, networks, 4-6–7
Compatibility, networks, 4-8
Complementary ventilation, 5-56
Components of a control system, 2-1, 2-2
Computer fluid dynamics (CFD), 6-3
Computers, *see* Information technology
Condensation prevention, 5-50
Condensing boilers, 5-3–4
multiple boiler systems, 5-5
Condition-based maintenance (CBM), 7-7
Conditions, 8-7
monitoring, 2-14
Configuration mode, intelligent outstations, 3-25
Constant volume, dual duct, single supply fan, 5-48
Constant-volume room terminal units, 5-46
Constructing the Team, 8-3
Construction management, 8-3
Contingency ventilation, 5-55
Contractual problems with systems integration, 4-16
Controllers, *see* Intelligent outstations
Control loops, interaction between
autotuning, 2-12
stability issues, 2-10
Control modes, *see* Modes of control
Control of Fuel and Electricity, 1-3
Control panels, 3-23–4
electromagnetic compatibility, 4-14–15
Control ports, valves, 301-, 3-11
authority, 3-15
design, 3-11
Conventional controls, 4-1
Cooling coils, 5-30–1
Cooling towers, 5-8, 5-10–11
Cost benefit analysis, 8-15–16
Cost effectiveness, energy monitoring and targeting, 7-2
Costing, 8-12–13
Cost issues
assessment methods, 8-16
cost benefit analysis, 8-15–16
life cycle costing, 8-16
Coverage, energy monitoring and targeting, 7-2
Critically damped response, 2-11
Current source inverters (CSIs), 3-23
Current transformers (CTs), 7-4
Cusum technique, 7-6

Damper authority, 3-19
Damper characteristic, 3-18–19
Damper ratio, 3-19
Dampers
applications, 3-20
face and bypass control, 3-20
return air mixing, 3-20
fire protection system, 4-16
modulating, 3-18–19
selection, 3-17–18
ventilation, 5-36–8, 5-40
Data analysis, energy monitoring and targeting, 7-1, 7-5–6
Data collection, energy monitoring and targeting, 7-1
Data-gathering panels, 4-2
Data logging, 7-4–5
user interface, 4-18
Deadband, *see* Differential gap/band
Decoupling, hydraulic, 5-18–19, 5-20
incompatible flows, 5-14
Dedicated function controllers, *see* Unitary controllers
Dehumidification
chilled ceilings, 5-50
cooling coils, 5-30
displacement ventilation, 5-41
with mixing damper control, 5-40
VRF systems, 6-5
see also Humidity control
Demand-controlled ventilation (DCV), 5-34–5, 5-54
air quality sensors, 3-5
Derivative action, 2-7
PID control, 2-5
Derivative action time, 2-11
Derivative kick, A2-2
Design and build, 8-1, 8-2, 8-3
Design and manage, 8-1–2
Design and specification of a controls system
commissioning, 8-9
conditions and tolerances, 8-7
design brief, 8-5, 8-6
sequence of events, 8-4–6
specification, 8-6–7
techniques, 6-2
analogue modelling, 6-2–3
emulation, 6-3
full scale mock-up, 6-2
mathematical modelling, 6-3
Design brief, 8-5, 8-6
Develop and construct, 8-1
Dewpoint sensors, 3-3, 3-4
chilled ceilings, 5-50
Differential gap/band (deadband), 2-2
fan control units, 5-48
heat pumps, 5-52
Differential pressure, 3-4
valves, 3-8–9
Differential pressure control valves (DPCVs), 3-9
mixing circuits, 5-16
thermostatic radiator valves, 5-28
variable-volume systems, 5-24–5
Digital alarms, 5-1
Digital control, 2-5–6
tuning, A2-1–2, A2-3
Digital data technology, 1-2
Digital inputs (DIs), 2-1
Digital outputs (DOs), 2-1
Dimmers, 5-56, 5-57
Direct digital control (DDC), 2-1, 2-6–7
BMS architecture, 4-1–2
origins, 1-2
Direct expansion (DX) cooling coils, 5-30–1
Direct humidification, 5-31
Direct ice storage, 5-12
Direct on-line (DOL) starting, 3-21
Discounted cash flow (DCF), 8-16
Discount techniques, 8-16
Displacement ventilation, 5-41
and chilled ceilings, 6-5–6
Distance–velocity lag, 2-6
Distributed processing units, *see* Intelligent outstations
Diverting circuits, 5-15
Diverting valves, 3-10
Documentation
control strategy, 6-1
operations and maintenance, 8-12
Drip loops, sensor cables, 3-6
Dual duct systems, 5-48–9
Dual mixing circuits, 5-16
Dumb outstations, 4-2
Dynamic balancing valves, *see* Flow limiting valves
Dynamic torque, dampers, 3-18

Earthing, 4-13–14
Economiser cycle, 5-37, 5-38
Eddy current coupling, 3-21–2
Egan report, 8-3
EHS, 4-9
EIBus, 4-6, 4-7, 4-11–12
Electrical connections, actuators, 3-7–8
Electricity tariffs, 7-3
Electro-hydraulic actuators, 3-7
Electromagnetic compatibility (EMC), 4-12
EMC Directive, 4-13
installation, 4-13
cabling, 4-14
control panels, 4-14–15
location, 4-13
power supply and earthing, 4-13–14
Electromagnetic humidity sensors, 3-3
Electronic soft start, 3-21
Electronic start, fluorescent lamps, 5-59
Elizabeth Fry Building, 6-8
Emulation, 6-3
Energy
conservation, 1-3–4, 1-5
efficiency, motors, 3-20–1
monitoring, 7-1
data analysis and reporting, 7-5–6
metering equipment, 7-3–5
neural networks, 2-14
and targeting (M&T), 7-1–3
recovery, 5-32
fixed plate heat exchangers, 5-33
run-around coils, 5-33
thermal wheel, 5-32–3
use, lighting, 5-56
Energy account centres (EACs), 7-1, 7-2
Enthalpy control
cooling coils, 5-30
mechanical ventilation, 5-38, 5-39
Environment
global, 1-2–3
indoor, 1-3
Equal percentage valves, 3-11–12
Ethernet, 4-6, 4-7
European Committee for Standardisation (CEN), 4-4, 4-9
European Home Systems (EHS), 4-9
European Installation Bus (EIBus), 4-6, 4-7, 4-11–12
European Installation Bus Association (EIBA), 4-11
European Standards, 1-3
External air supply, 5-38–40
External air temperature, 5-53
Extract fans, variable-volume, 5-35–6

Face and bypass control, dampers, 3-20
Fan-assisted VAV units, 5-43, 5-44
Fan coil units (FCUs), 5-46–7
airside control, 5-48
interaction between control loops, 2-10
waterside control, 5-47–8
whole-building HVAC systems, 6-3–5
Fans, 3-23
dual duct systems, 5-48–9
VAV systems, 5-43–5, 5-46
variable-volume extract, 5-35–6
Fan speed control, fan coil units, 5-47
Fault reports, 7-6–7
Feed and expansion (F&E) tanks, 5-7, 5-8
Feedback mechanisms, 8-14
Feedforward systems, *see* Open loop systems

Fibre optic cable, 4-6–7
Field Instrumentation Protocol (FIP), 4-9
Field processing units, *see* Intelligent outstations
Filters, 5-35
Fine tuning, 8-13
Fingerprint, building, 8-15
FIP, 4-9
Fire protection system (FPS)
  dampers, 3-18
  safety, 5-2, 5-3
  systems integration, 4-16
Firm Neutral Data Transmission (FND), 4-9
Firmware, intelligent outstations, 3-26
Fixed plate heat exchangers, 5-32, 5-33
Fixed-speed circulating pumps, 5-22–3
Floating control, 2-3
  time lags, 2-6
Flow, hydraulic, 3-8–9
  sensors, 3-5
Flow coefficient, 3-12, 3-13
Flow dilution
  multiple boilers, 5-6
  multiple chillers, 5-9
Flow limiting valves, 3-9
Fluorescent lamps, 5-58–9
Flux vector control, 3-23
FND, 4-9
Form 2 control panels, 3-24
Form 3 control panels, 3-24
Form 4 control panels, 3-24
Four-pipe water systems, 5-47, 5-48
Four-port valves, 3-10
Free cooling, 5-11
Freely programmable controllers, *see* Intelligent outstations
Frost protection
  air handling systems, 5-2
  central air handling plant, 5-29
  cooling towers, 5-11
  water systems, 5-2
Fuel Efficiency Booklet 10, 1-4
Full outside air damper system, 5-35
Full scale mock-ups, 6-2
Full systems integration, 1-5, 4-15
Fundamentals of control, 2-1, 2-14
  artificial intelligence, 2-13–14
  modes of control, 2-1–7
  optimum start, 2-7–8
  stability and tuning, 2-9–13
  weather compensation, 2-8–9
Fuzzy logic, 2-13

Gain scheduling, 2-12
Gas sensors, 3-5–6
Gauge pressure, 3-4
Global environment, 1-2–3
Global warming, 1-2, 1-3
Globe valves, *see* Plug and seat valves
Graphics, 4-18
Greenway School, Horsham, 6-7
Group remote controllers, heat pumps, 5-51–2

Handover, 8-11
  documentation, 8-12
  training, 8-12
  witnessing, 8-11–12
Hardware
  configuration, intelligent outstations, 3-25
  user interface
    plug-in keypad, 4-17
    stand-alone controllers, 4-17
    supervisor, 4-18
    touchscreen, 4-18
HBES, 4-12
HCFCs, 1-2–3
Health and safety file, 8-12
Heat flow monitoring, IEE inventory estimation, 5-13
Heating coils, 5-29–30
Heat pumps
  reverse cycle units, 5-52–3
  VRF split systems, 5-51–2
Hewlett Packard, Building 3, Bristol, 6-8–9
High frequency control, fluorescent lamps, 5-59
High frequency regulating control, fluorescent lamps, 5-59
Home and Building Electronic System (HBES), 4-12
Horizontal cabling, 4-7, 4-8
Hot water systems
  floating control, 2-3
  pressurisation, 5-7
  safety strategies, 5-2
  see also Boilers
Hot wire anemometers, 3-5
Hours run alarms, 5-1
Humidification, 5-31–2
Humidistats, 3-1, 3-3
Humidity control, 1-3
  conditions, 8-7
  sensors, 3-3–4
  VRF systems, 6-5
  see also Dehumidification
Hunting behaviour, 2-10
Hydraulic circuits, control of, 5-14
  basic circuits, 5-15–18
  calorifiers, 5-19–21
  network design, 5-14–15
  primary circuits, 5-18–19
  thermostatic radiator valves, 5-27–8
  underfloor heating, 5-26–7
  variable-volume systems, 5-21–6
Hydraulic flow, 3-8–9
  sensors, 3-5
Hysteresis
  actuators, 3-8
  logic control, 2-7
  sensors, 3-2

Ice bank, 5-12
Ice builder, 5-12
Ice storage, 5-12
  control techniques, 5-12–13
  system configuration, 5-12
Ideal damper characteristics, 3-18
IEE Wiring Regulations, 4-14
Incremental algorithm, 2-6
Index circuit, 5-23, 5-24
Indirect humidification, 5-31
Indirect ice storage, 5-12
Individual room unit controllers, heat pumps, 5-51
Indoor environment, 1-3
Induction VAV terminal units, 5-43
Inductive pressure sensors, 3-4
Information technology (IT), 1-2, 1-4–5
  energy monitoring, 7-1–6
  fault reports and maintenance scheduling, 7-6–7
  network, 4-4–5
  systems integration, 4-4–5, 4-15, 4-16–17
Infra-red thermography, 7-7
Ingress protection system, 3-24–5
Inherent characteristics of dampers, 3-18, 3-19
Injection circuits, 5-16–17
Injection fan system, 5-39
Installation, 8-9
  commissioning the BMS 8-10–11
  pre-commissioning the BMS, 8-9–10
Installed characteristics of dampers, 3-19
Integral action time
  PI control, 2-4, 2-5
  tuning, 2-11
Integral control, 2-4
  time lags, 2-6
  see also PI control; PID control
Integrated environmental design (IED), 5-58
Integration, *see* Systems integration
Intelligent building system (IBS), *see* Full systems integration
Intelligent fan coil units, 5-47
  whole-building HVAC systems, 6-4
Intelligent luminaires, 5-59
Intelligent outstations (programmable controllers; universal controllers), 2-7, 3-24–5
  BMS architecture, 4-3
  configuration, 3-25
    hardware, 3-25
    strategy, 3-26
    system, 3-25
  heat pumps, 5-52
  stand-alone controllers, 4-17
Intelligent sensors, 3-2
Intelligent VAV boxes, 5-42, 5-44–5
  interaction between control loops, 2-10
  supply air temperature control, 5-45–6
Interchangeability, networks, 4-8, 4-9
Interconnected circuits, free cooling, 5-11
Interference, A2-2
Interlocks, 5-1–2
  cooling towers, 5-11
Internal bypass, 5-16
International Organisation for Standardisation (ISO), 4-9
Internetworks, *see* Wide area networks
Interoperability, 1-6, 4-15
  networks, 4-8, 4-9
Inverter drives, 3-22
  current source inverter (CSI), 3-23
  motor control centres, 3-24
  pulse amplitude modulation (PAM), 3-23
  pulse width modulation (PWM), 3-22–3
IP protection numbers, 3-24–5
ISDN telephone line, 4–6
ISO, 4-9

Keypad, plug-in, 4-17

Lamps, 5-58–9
  see also Lighting control
Latham, Sir Michael, 8-3
Leakage rating, dampers, 3-18
Legionnaires' disease, 5-8, 5-10, 5-21
Let-by, 3-12
Library of System Control Strategies, 6-2
Life cycle costing, 8-16
Lift and lay valves, *see* Plug and seat valves
Lighting control, 5-56
  methods, 5-56–7
  systems and components, 5-57–61
Lightning protection, 4-13–14
Linear actuators, 3-7
Linear regression analysis, 7-5–6
Linear valves, 3-11
Load error (offset), 2-3, 2-4
Local area networks (LANs), 1-2, 4-5
  communication, 4-6
  intelligent outstations, 4-3
  OSI Basic Reference Model, 4-9
Localised manual lighting controls, 5-56–7

Logic control, 2-6–7
Logic operators, 2-7
LonMark Interoperability Association, 4-10
LonTalk protocol, 4-3, 4-10
LonWorks, 4-6, 4-7, 4-10–11
  case study, 6-7
  systems integration, 4-16
Look-up tables, 2-7
Loop reschedule interval, 2-6
Low-loss headers, 5-16
Low pressure header systems, 5-18, 5-19
Lubricant analysis, 7-7

Maintenance, 8-13
  remote operation, 1-5
  scheduling, 7-6–7
Managed space, and lighting controls, 5-60
Management contracting, 8-3
Management issues, 8-16–17
  commissioning, 8-8–13
  cost issues, 8-15–16
  design and specification of a controls system, 8-4–7
  occupant surveys, 8-14–15
  operation, 8-13
  procurement options, 8-1–4
  tendering process, 8-7–8
Management methods, procurement, 8-3
Manchester Airport, Terminal 2, 6-9–10
Manual differential, 2-2
Master controller, 2-6
Mathematical modelling, 6-3
Mechanical differential, 2-2
Mechanical ventilation, 5-34
  chilled ceilings, 5-50, 6-6
  dehumidification with mixing damper control, 5-40
  demand-controlled ventilation, 5-34–5
  displacement ventilation, 5-41
  full outside air damper system, 5-35
  mixing damper systems, 5-36–8
  outdoor air supply, 5-38–40
  poor outdoor air quality, 5-35
  variable-volume extract fan, 5-35–6
  see also Mixed mode ventilation
Mechanical ventilation with heat recovery (MVHR) systems, 6-5
Metering equipment, 7-2, 7-3–4
  computer, 7-4, 7-5
  data logger, 7-4–5
  data transmission system, 7-4
  display module, 7-4
  meter module, 7-3, 7-4
Mineral-insulated copper-sheathed cable (MICC), 4-16
Minimum flow rate control, 5-25–6
Mixed mode ventilation, 5-55–6
  and chilled ceilings, 6-6
  whole-building HVAC systems, 6-6–7
Mixing
  circuits, 5-15–16
  dampers, 3-18
    ventilation, 5-36–8, 5-40
  valves, 3-10
Mock-ups, full scale, 6-2
Modelling
  analogue, 6-2–3
  mathematical, 6-3
Modern control systems, 1-2
Modes of control, 2-1
  cascade control, 2-6
  choice of, 2-7
  digital control, 2-5–6
  integral control, 2-4
  logic control, 2-6–7
  PI control, 2-4–5
  PID control, 2-5
  proportional control, 2-3–4
  time lags, 2-6
  two-position (on/off) control, 2-2–3
Modular systems, 4-3
  case study, 6-7
Modulating dampers, 3-18–19
Modulating valves, 3-9
  actuators, 3-7
  selection, 3-17
Montreal protocol, 1-2
Motor control centres, 3-23–4
Motor current analysis, 7-7
Motors
  energy efficiency, 3-20–1
  speed control, 3-21–3
  starting, 3-21
  variable-speed, 3-23
Mounting of sensors, 3-2, 3-6
MPT1329, 4-7
MPT1340, 4-7
Multi-gas sensors, 3-5, 3-6
Multiple boilers, 5-4, 5-5–7
Multiple chillers, 5-8, 5-9–10
Multiple pressure sensors, VAV supply fan control, 5-44, 5-45
Multi-speed motors, 3-22
Multi-split systems, heat pumps, 5-51

National Accreditation of Measurement and Sampling (NAMAS), 3-6
National Engineering Specification (*NES*), 4-13, 8-6–7
Natural ventilation, 5-53
  chilled ceilings, 5-50, 6-6
  controlled devices, 5-53
  control strategies, 5-54
    night cooling, 5-54–5
    occupation period, 5-54
  measurement devices, 5-53–4
  whole-building HVAC systems, 6-6–7
  see also Mixed mode ventilation
Near optimal control, cooling towers, 5-11
Need for controls, 1-2, 1-6
  benefits of a BMS, 1-5–6
  building operation, 1-5
  energy conservation, 1-3–4
  global environment, 1-2–3
  indoor environment, 1-3
  information technology and systems integration, 1-4–5
Net cash flow, 8-16
Net present value, 8-16
Network analysis, 5-15
Networks, 4-5
  communication media, 4-6–7
  hydraulic circuits, 5-14–15
  IT, 4-4–5
  protocols, 4-7
  standardisation, 4-8–9
  standard systems, 4-9–12
  structured cabling, 4-7–8
  topology, 4-5–6
Neural networks, 2-13–14
Neutral head end, 4-15
Nickel resistance thermometer, 3-3
Night cooling
  chilled ceilings, 5-50
  mixed mode ventilation, 5-56
  natural ventilation, 5-54–5
Night setback control, 5-27
Nitrogen dioxide sensors, 3-Noise, A2-2
Non-oscillatory instability, 2-9

Obscuration, 3-6
Occasionally visited space, and lighting controls, 5-60
Occupancy
  calculation, 3-5
  lighting, 5-57
  mechanical ventilation, 5-34
  natural ventilation, 5-53, 5-54, 5-55
Occupant surveys, 8-14–15
Offset, 2-3, 2-4
On/off control, *see* Two-position (on/off) control
Open loop reaction curve method, 2-11, A2-3, A2-4
Open loop systems
  components, 2-1
  cooling towers, 5-11
  motor speed control, 3-22
  variable-volume extract fans, 5-35–6
  weather compensator, 2-9
Open Systems Interconnection (OSI) Basic Reference Model, 4-9
  BACnet, 4-10
Operating conditions, 2-10
Operating differential, 2-2, 2-3
Operating modes, 6-1–2
Operation, 1-5
  conditions, 8-7
  fine tuning, 8-13
  maintenance, 8-13
  recommissioning, 8-13
Opposed blade damper, 3-18, 3-19
  applications, 3-20
Optical fibre, 4-6–7
Optimisation, 2-14
Optimum charge, ice storage, 5-13
Optimum start, 2-7–8
  adaptive control, 2-12
  chilled ceilings, 5-50
  underfloor heating, 5-27
  water systems, 5-2
Optimum stop, 2-8
  underfloor heating, 5-27
Orifice plates, 3-5
Oscillation test, *see* Closed loop ultimate cycling method
Oscillatory instability, 2-9–10
Outdoor air supply, 5-38–40
Outdoor air temperature, 5-53
Overall liking scores (OLS), 8-14–15
Overview of Guide, 1-1–2
Owned space, and lighting controls, 5-60
Ozone layer, damage to, 1-2

Packaged split units, heat pumps, 5-51
Packaged systems, 4-3
Parallel blade damper, 3-18, 3-19
  applications, 3-20
Parallel ice storage system, 5-12
Parallel pumping, 5-21–2
Partnering, 8-3–4
Passive analogue sensors, 3-1
Passive chilled beams, 6-5
Patch cabling, 4-8
Pattern recognition adaptive controller (PRAC), 2-12–13
Payback method, 8-16
PC-based controllers, heat pumps, 5-52
Peltier effect, 3-3
Percentage of full stroke (spindle lift), 3-11–12
Performance reviews, 8-14
Perimeter heating
  and chilled ceilings, 6-6
  and VAV, 6-3–4

PI control, 2-4–5
choice of, 2-7
interaction between control loops, 2-10
time lags, 2-6
tuning, 2-12, A2-1
PID control, 2-5
choice of, 2-7
digital control, 2-6
tuning, A2-2
Piezoelectric transducers, 3-5
Pitot tubes, 3-5
Platinum resistance thermometer (PRT), 3-3
Plenum-pressure system, 5-39–40
Plug and seat valves (lift and lay valves), 3-9, 3-10
actuators, 3-7
Plug and shoe valves, *see* Shoe valves
Pneumatic actuators, 3-8
Point to point network protocol, 4-7
Point to point network topology, 4-5, 4-6
Polling, 4-2
Position algorithm, 2-6, A2-2
Positioners, 3-8
Position indication, actuators, 3-8
Potentiometer pressure sensors, 3-4
Power line carrier, 4-6, 4-7
Power supplies, 4-13
Practical tuning, 2-11
Pre-commissioning the BMS, 8-9–10
Predicted mean vote (PMV) index, 1-3
Preheat coils, 5-29
Presence sensing, 5-57
see also Occupancy
Present worth method, 8-16
Pressure and dampers, 3-18
Pressure dependent VAV units, 5-42
Pressure drop, 3-9
Pressure independent VAV terminals, 5-42–3
Pressure sensors, 3-4–5
Pressure switches, 3-1
Pressurisation of hot water systems, 5-7, 5-8
Pre-tendering, 8-7
Primary hydraulic circuits, 5-14
boiler and chiller connections, 5-18–19
Primary resistance starters, 3-21
PROBE post-occupancy studies, 8-14
Procurement options, 8-1
design and build, 8-1
design and manage, 8-1–2
management methods, 8-3
partnering, 8-3–4
relations between parties, 8-3
traditional, 8-2–3
Programmable controllers, *see* Intelligent outstations
Project management, 8-8–9
Project patterning, 8-3
Proportional band, 2-11
Proportional control, 2-3–4
choice of, 2-7
interaction between control loops, 2-10
time lags, 2-6
ventilation, 5-34–5
Proportional plus integral control, *see* PI control
Proportional plug integral plus derivative control, *see* PID control
Proportional-speed floating control, 2-3
Protection of boilers, 5-4–5
Pseudo-proportional control, 2-3
Pulse amplitude modulation (PAM) inverters, 3-23
Pulse width modulation (PWM) inverters, 3-22–3
Pump differential pressure control, 5-23
Pumping energy requirements, 5-21
Pump overrun function, 5-5
Pump pressurisation, 5-7, 5-8
Pumps, 3-23
heat, 5-51–3

Quality, air, *see* Air quality
Quarter wave response, 2-11
Quarter wave tuning method, A2-2
Questionnaires, occupant surveys, 8-14–15
Quick opening valves, 3-11

Radiators
and chilled ceilings, 6-6
naturally ventilated buildings, 5-54
thermostatic valves, 5-27–8, 6-3
and VAV, 6-3–4
Radio, 4-6, 4-7
Rain intensity sensors, 5-53
Rangeability, valves, 3-12
Read only mode, intelligent outstations, 3-25
Reciprocating compressors, 5-7, 5-8, 5-9
Recommissioning, 8-13
Rectangular dampers, 3-17
Reference standards, sensor calibration, 3-6
Refrigerant pressure, 5-8
Refrigerant vapour migration, 5-11
Regenerative heat exchanger, 5-32
Regulating valves, 3-9
Remote control, 1-5
case study, 6-7–8
heat pumps, 5-51–2
lighting, 5-56
Reporting
energy monitoring and targeting, 7-1, 7-2
faults, 7-6–7
user interface, 4-18
Reset control, *see* Cascade control
Reset rate, 2-4
Response speed, 8-14
sensors, 3-2–3
Rethinking Construction, 8-3
Return air mixing, dampers, 3-20
Reverse cycle heat pump units, 5-52–3
Reviews of performance, 8-14
Ring network topology, 4-5–6
Room terminal units (RTUs)
constant-volume, 5-46
dual duct systems, 5-48, 5-49
variable air volume, 5-41, 5-42–3
Rotary actuators, 3-7
Rotary compressors, 5-8
Rotary shoe valves, *see* Shoe valves
Rotation angle, rotary actuators, 3-7
Round dampers, 3-17
Run-around coils, 5-32, 5-33
Running time, actuators, 3-7

Safety, 5-1
alarms, 5-1
documentation, 8-12
earthing, 4-13
fire and smoke control, 5-3
interlocks, 5-1–2
lighting, 5-56, 5-59
selection switches, 5-2–3
strategies, 5-2
Safety shut-off valves, 3-9
Sampling time
digital control, 2-6
stability issues, 2-10
tuning, 2-11
Scene set lighting control, 5-57
Screened cables, 4-14
Screw compressors, 5-7
Scroll compressors, 5-7
Sealed expansion vessel pressurisation, 5-7, 5-8
Secondary hydraulic circuits, 5-14, 5-18
Secondary metering, 7-3
Security issues, 4-18
Security systems, 4-16
Selection switches, 5-2–3
Self-tuning controllers, 2-11–12, 2-13
adaptive techniques, 2-12–13
autotuning, 2-12
Sensing element, 3-1
Sensitivity of sensors, 3-2
Sensors
calibration, 3-6
general categories, 3-1–2
mounting, 3-6
natural ventilation, 5-53
selection requirements, 3-2
accuracy, 3-2
response speed, 3-2–3
types, 3-3
air quality sensors, 3-5–6
humidity sensors, 3-3–4
pressure sensors, 3-4–5
temperature sensors, 3-3
velocity and flow sensors, 3-5
Set point control, 5-34
Seven-layer model, *see* Open Systems Interconnection (OSI) Basic Reference Model
Shannon's sampling theorem, 2-10
Shared space, and lighting controls, 5-60
Shielded twisted pair, 4-6
Shoe valves, 3-9, 3-10
actuators, 3-7
Shut-down period
fire and smoke control, 5-3
heat pumps, 5-53
sequencing of operations, 5-2
Sick building syndrome (SBS), 1-3, 5-34
Signal conditioning, 3-1
accuracy, 3-2
Similar modified parabolic (equal percentage) valves, 3-11–12
Single pressure sensors, VAV supply fan control, 5-44
Sizing
dampers, 3-19
valves, 3-15–16
Skipped charge, ice storage, 5-13
Slab temperature, 5-53
Slip, 3-21
Smoke protection
dampers, 3-18
safety, 5-2, 5-3
Software, 4-18
Soft wiring, intelligent outstations, 3-26
Solar radiation, 5-53
Sophistication, energy monitoring and targeting, 7-2
Space types and lighting controls, 5-59–60
Specification, 8-4, 8-6–7
Speed control, motors, 3-21–3
Spindle lift, 3-11–12
Spindles, 3-9
Stability
of control systems, 2-9–10
of sensors, 3-2
Standardisation, network, 4-8–9
Standardising bodies, 4-9
Standard Specification (M&E), 8-6
Standard systems
BACnet, 4-9–10
Batibus, 4-12
EIBus, 4-11–12

HBES, 4-12
LonWorks, 4-10–11
Star–delta starters, 3-21
Star network topology, 4-5, 4-6
Start
motors, 3-21
optimum, *see* Optimum start
Start-up period
calorifiers, 5-20
sequencing of operations, 5-2
Static pressure
dampers, 3-18
valves, 3-8
variable-volume extract fans, 5-36
Status alarms, 5-1
Status sensors, 3-1
Steam injection humidifiers, 5-31
Steam valves, 3-12–13
Stem position (spindle lift), 3-11–12
Stop, *see* Optimum stop
Storage type humidifiers, 5-31
Store exit temperature, ice inventory estimation, 5-13
Store priority control, ice storage, 5-13
Strain gauge pressure sensors, 3-4
Strategic partnerships, 8-3
Strategy configuration, intelligent outstations, 3-26
Stroke, linear actuators, 3-7
Stroke time, actuators, 3-7
Structured cabling, 4-7–8
Submaster controller, 2-6, 2-11
Submetering, 7-3
Sulphur dioxide sensors, 3-6
Summer cooling, 5-56
Supervisors, 3-25
centralised intelligence, 4-2
programmable controllers, 4-2–3
user interface, 4-18
Supply air temperature control for VAV systems, 5-45–6
Supply fan control for VAV systems, 5-43–5
Surveys, occupant, 8-14–15
Swing (operating) differential, 2-2, 2-3
Switched reluctance drive, 3-22
Switch start, fluorescent lamps, 5-58
System configuration, intelligent outstations, 3-25
hardware, 3-25
strategy, 3-26
system, 3-25
Systems integration, 1-4–5, 4-15
implementation, 4-15
with IT systems, 4-4–5, 4-16–17
problems, 4-16

TCP/IP, 4-7
Technical problems with systems integration, 4-16
Telephone line, 4-6, 4-7
Temperature control, 1-3
components of control system, 2-1, 2-2
conditions, 8-7
modes
cascade control, 2-6
two-position control, 2-2–3
sensors, 3-3
ventilation, 5-34
Temporarily owned space, and lighting controls, 5-60
Tendering process, 8-7
Terminal 2, Manchester Airport, 6-9–10
Thermal wheels, 5-32–3
Thermic actuators, 3-7
Thermistors, 3-3
Thermometers
nickel resistance, 3-3
platinum resistance (PRT), 3-3
Thermostatic radiator valves (TRVs), 5-27–8, 6-3
Thermostats, 2-3, 3-1, 3-3
Thermosyphonic method, free cooling, 5-11
Three-level model, BMS architecture, 4-4, 4-9
Three-phase squirrel cage induction motor, 3-21
Three-port valves, 3-9, 3-10
authority, 3-14–15
terminology, 3-10, 3-11
Three-position control, 2-3
Three-term control, *see* PID control
Throttling circuits, 5-17–18
Throttling range, 2-3
Thrust, linear actuators, 3-7
Time constant, 3-2, 3-3
Time lags, 2-6
Time proportioning control, 2-4
Time scheduling
thermostatic radiator valves, 5-27
user interface, 4-18
Timeswitches, lighting, 5-57
Tolerances, 8-7
Topology
cabling, 4-7–8
network, 4-5–6
Torque
dampers, 3-18
rotary actuators, 3-7
Totaliser alarms, 5-1
Touchscreen, 4-18
Traditional procurement option, 8-2–3
Training, 1-5, 8-12, 8-13
Transducers, 3-1
Transfer lag, 2-6
Transient response, *see* Open loop reaction curve method
Transmitters, 3-1
Transport delay, 2-6
Tree network topology, 4-6
Tristate control, 2-3
Tuning, 2-10, 2-11, 8-13, A2-2–3
digital control, A2-1–2
PID control loop, A2-1
practical, 2-11
self-tuning controllers, 2-11–13
step-by-step procedure, A2-3–4
Turbine flow meters, 3-5
Turbulence and dampers, 3-18
Twisted-pair copper wire
electromagnetic compatibility, 4-14
networks, 4-6, 4-7
Two-pipe water systems, 5-47, 5-48
Two-port valves, 3-10
authority, 3-13–14
design, 3-12
Two-position (on/off) control, 2-2–3
stability, 2-9–10
tuning, 2-11

Ultimate cycling/frequency method, *see* Closed loop ultimate cycling method
Underfloor heating, 5-26–7
Unitary controllers, 3-24, 4-3
United Kingdom Accreditation Service, 3-6
Universal (structured) cabling, 4-7–8
Universal controllers, *see* Intelligent outstations
University of East Anglia, 6-8
Unowned space, and lighting controls, 5-60
Unshielded twisted pair (UTP), 4-6
Unstable control systems, 2-9–10
Upload/download mode, intelligent outstations, 3-25
User interface, 4-17
hardware, 4-17–18
levels of operation, 4-17
software, 4-18
users, 4-18

Valve authority, 3-13–15
hydraulic circuits, 5-14
Valve characteristic, 3-11–12
Valves
actuators, 3-7
design
characteristic, 3-11–12
flow coefficient, 3-12
steam valves, 3-12–13
hydraulic flow, 3-8–9
selection
authority, 3-13–15
cavitation, 3-16
checklist, 3-16–17
sizing, 3-15–16
thermostatic radiator, 5-27–8, 6-3
types, 3-9–11
Vapour compression chillers, 5-7
Vapour pressure of water, 3-16, 3-17
Variable air volume (VAV), 5-41–2
intelligent, *see* Intelligent VAV boxes
operating conditions, 2-10
overlapping controlled zones, 2-10
and perimeter heating, 6-3–4
room terminal units, 5-42–3
supply air temperature control, 5-45–6
supply fan control, 5-43–5
ventilation, 5-37, 5-38, 5-39, 5-40
Variable constant volume (pressure independent VAV terminals), 5-42–3
Variable refrigerant flow (VRF) systems, 5-51–2, 6-5
Variable-speed circulating pumps, 5-23–4
thermostatic radiator valves, 5-28
Variable-speed drives (VSDs), 3-21
Variable-speed motors, 3-23
Variable-voltage control, 3-22
Variable volume, dual duct, dual supply fans, 5-48–9
Variable volume, dual duct, single supply fan, 5-48
Variable-volume extract fans, 5-35–6
Variable-volume systems
automatic flow-limiting valve, 5-25
control of minimum flow rate, 5-25–6
design recommendations, 5-26
differential pressure control valves, 5-24–5
fixed-speed circulating pump, 5-22–3
parallel pumping, 5-21–2
pumping energy requirements, 5-21
variable-speed circulating pump, 5-23–4
Velocity algorithm, 2-6
Velocity ratings, dampers, 3-18
Velocity sensors, 3-5
Ventilated chilled beams, 6-5–6
case study, 6-8–9
Ventilation, *see* Mechanical ventilation; Mixed mode ventilation; Natural ventilation
Vibration analysis, 7-7
Voltage vector control (VVC), 3-23
Volume matching, *see* Air flow tracking
Volumetric devices, ice inventory estimation, 5-13

Water, vapour pressure of, 3-16, 3-17
Water heating plant, 5-5
Waterside fan coil units, 5-47–8

Weather compensation, 2-1, 2-8–9
boilers, 5-4, 5-5
thermostatic radiator valves, 5-27
underfloor heating, 5-27
Wet and dry bulb psychrometers, 3-3
Whole-building HVAC systems, 6-3–4
chilled ceiling and displacement ventilation, 6-5–6
fan coil units, 6-4–5
natural and mixed mode systems, 6-6–7
VAV and perimeter heating, 6-3–4
VRF systems, 6-5
Wide area networks (WANs), 4-5
BMS architecture, 4-4
OSI Basic Reference Model, 4-9
remote supervision, 1-5
Wind direction, 5-53
Windows, 5-53, 5-54, 6-6–7
Wind speed, 5-53, 5-54
Wind-up
digital control, 2-6, A2-2
PI control, 2-5, 2-10
Winter ventilation, 5-56
Witnessing, 8-11–12
Work area cabling, 4-8

Ziegler–Nichols tuning method, A2-2
Zoned ventilation, 5-55–6
Zone trim, 2-9